Further volumes in this series:

Dependable Computing and Fault-Tolerant Systems

Edited by
A. Avižienis, H. Kopetz, J. C. Laprie

Advisory Board
J. A. Abraham, V. K. Agarwal, T. Anderson, W. C. Carter,
A. Costes, F. Cristian, M. Dal Çin, K. E. Forward, G. C. Gilley,
J. Goldberg, A. Goyal, H. Ihara, R. K. Iyer, J. P. Kelly,
G. Le Lann, B. Littlewood, J. F. Meyer, B. Randell,
A. S. Robinson, R. D. Schlichting, L. Simoncini, B. Smith,
L. Strigini, Y. Tohma, U. Voges, Y. W. Yang

Volume 8

Springer-Verlag Wien GmbH

C. E. Landwehr, B. Randell,
L. Simoncini (eds.)

Dependable Computing for Critical Applications 3

Springer-Verlag Wien GmbH

Dr. Carl E. Landwehr
Naval Research Laboratory, Washington, DC 20375-5000, USA

Prof. Dr. Brian Randell
Dept. of Computer Sc., Univ. of Newcastle, Newcastle upon Tyne NE1 7RU, UK

Prof. Dr. Luca Simoncini
Dip. Ingegneria dell'Informazione, Università di Pisa, I-56100 Pisa, Italia

With 91 Figures

Production of this volume has been supported by

ISSN 0932-5581
ISBN 978-3-7091-4011-6   ISBN 978-3-7091-4009-3 (eBook)
DOI 10.1007/978-3-7091-4009-3

FOREWORD

This volume contains the papers presented at the Third IFIP International Working Conference on *Dependable Computing for Critical Applications*, sponsored by IFIP Working Group 10.4 and held in Mondello (Sicily), Italy on September 14-16, 1992.

System developers increasingly apply computers where they can affect the safety and security of people and equipment. The Third IFIP International Working Conference on *Dependable Computing for Critical Applications*, like its predecessors, addressed various aspects of computer system dependability, a broad term defined as the degree of trust that may justifiably be placed in a system's reliability, availability, safety, security, and performance. Because the scope of the conference was so broad, we hope the presentations and discussions will contribute to the integration of these concepts so that future computer-based systems will indeed be more dependable.

The Program Committee selected 18 papers for presentation from a total of 74 submissions at a May meeting in Newcastle upon Tyne, UK. The resulting program represented a broad spectrum of interests, with papers from universities, corporations, and government agencies in eight countries. Much diligent work by the Program Committee and the quality of reviews from more than a hundred external referees from around the world, for which we are most grateful, significantly eased the production of this technical program.

Like past DCCA organizers, we designed the conference to promote an open exchange of ideas. All paper sessions ended with a 30-minute discussion period on the topics dealt with in the session. In addition, three panel sessions focussed on dependability issues in particular applications. The first, entitled "Safe Vehicle-Highway Systems," addressed concerns in future computer-controlled highway systems intended to permit dense, high-speed traffic. The second, "Malicious and Inadvertent Human Operator Faults," recognized that designers of critical systems must anticipate both inexperienced and malevolent users, and make provisions accordingly in their system designs. The final panel session, "Security in Intelligent Networks," drew attention to the need to protect both the confidentiality and integrity of communications over future advanced telecommunication networks.

Thanks are due to to Ettore Ricciardi for his very able management of the local arrangements, and to Cinzia Bernardeschi and Andrea Bondavalli for their excellent work on the formatting of the text of this volume. Finally, we would like to acknowledge the generous assistance provided by WG10.4 Chair Jean-Claude Laprie from LAAS-CNRS and, for most of the clerical work involved in preparing the technical program, Anke Jackson of the University of Newcastle upon Tyne.

Luca Simoncini
General Chair
University of Pisa
Pisa, Italy

Carl Landwehr
Program co-Chair
Naval Research Laboratory
Washington, DC, USA

Brian Randell
Program co-Chair
Univ. of Newcastle upon Tyne
Newcastle upon Tyne, UK

Sponsors

IFIP Working Group 10.4 on *Dependable Computing and Fault Tolerance*

In cooperation with:

IFIP Technical Committee 11 on *Security and Protection in Information Processing Systems*
IEEE Computer Society Technical Committee on *Fault-Tolerant Computing*
EWICS Technical Committee 7 on *Systems Reliability, Safety and Security*
AICA, Associazione Italiana per l'Informatica ed il Calcolo Automatico
The University of Pisa
Istituto di Elaborazione dell'Informazione del CNR, Pisa

Conference Organization

General Chair:
Luca Simoncini
University of Pisa
Pisa, Italy

Program co-Chairs:
Carl E. Landwehr
Naval Research Laboratory
Washington, DC, USA

Brian Randell
Univ. of Newcastle upon Tyne
Newcastle upon Tyne, UK

Local Arrangements Chair:
Ettore Ricciardi
IEI-CNR
Pisa, Italy

Ex Officio:
Jean-Claude Laprie
LAAS-CNRS
Toulouse, France

Session Chairs

Jacob Abraham
University of Texas at Austin
Austin, Texas, USA

Wlad Turski
Warsaw University
Warsaw, Poland

Gerardo Rubino
IRISA
Rennes, France

Jack Goldberg
SRI International
Menlo Park, California, USA

Peter Bishop
Adelard
UK

Jayanaran H. Lala
C. S. Draper Lab.
Cambridge, Massachusetts, USA

Alain Costes
LAAS-CNRS
Toulouse, France

Ian Sutherland
ORA Corporation
Ithaca, New York, USA

Program Committee

J.A. Abraham
University of Texas at Austin
Austin, Texas, USA

P. Bishop
Adelard, UK

A. Costes
LAAS-CNRS
Toulouse, France

D. Craigen
Odyssey Research Associates
Ottawa, Canada

K. Dittrich,
University of Zurich,
Zurich, Switzerland

H. Ihara
Hitachi, Ltd.
Yokohama, Japan

R.K. Iyer
University of Illinois
Urbana-Champaign, Illinois, USA

J.P. Kelly
University of California,
California, USA

R. Kemmerer
University of California,
California, USA

H. Kopetz
Technische Universität
Wien, Austria

J.H. Lala,
CS Draper Lab.
Cambridge, Massachusetts, USA

K. Levitt
University of California,
California, USA

B. Littlewood
City University
London, UK

T. Lunt
SRI International
Menlo Park, California, USA

J. Meyer
University of Michigan
Ann Arbor, Michigan, USA

M. Morganti
Italtel
Settimo Milanese, Italy

S. Natkin
CEDRIC CNAM,
Paris, France

J.J. Quisquater
Philips Research
Brussels, Belgium

R.D. Schlichting
University of Arizona
Tucson, Arizona, USA

F.B. Schneider
Cornell University
Ithaca, New York, USA

D. Siewiorek
Carnegie-Mellon University
Pittsburgh, Pennsylvania, USA

L. Strigini
IEI-CNR
Pisa, Italy

I. Sutherland
ORA Corporation
Ithaca, New York, USA

W.Turski
Warsaw University
Warsaw, Poland

Referees

K. Akita	I. Greenberg	K.T. Narayana
D. Alstein	P. Hall	P. G. Neumann
P.E. Ammann	R. Hamlet	V. Nicola
T. Anderson	W. Heimerdinger	K. Okumoto
J. Arlat	C.A.R. Hoare	T. Ozsu
H. Bal	J. Hooman	D. Parnas
G. Balboni	R. Horst	P.A. Porras
P. Banerjee	M. Jackson	D. Powell
D. Barbara	M. Joseph	D. Proutfoot
P.A. Barrett	G. Kaiser	T.R.N. Rao
W. Bartlett	B. Kaliski	K. Rothermel
G. Bernot	N. Kanekawa	G. Rubino
P. Bernstein	K. Kanoun	M. Saaltink
T. Berson	K. H. Kim	W.H. Sanders
K. Birman	J. Knight	G. Saucier
B. Bloom	L. Lamport	Z. Segall
A. Bondavalli	G. Le Lann	L. Sha
J. Bowen	R. Leveugle	N. Shankar
D.K. Branstad	B. Lindsay	J. Shen
M. Broy	S. Lipner	K. G. Shin
J. Carmo	D.D.E. Long	E. Shokri
R. Chakka	N. Lynch	J.G. Silva
P.J. Crepeau	M. R. Lyu	T. Basil Smith
F. Cristian	H.S.C. Madeira	D. Stuart
J. Dick	J. Madey	P. Thévenod-Fosse
E.W. Dijkstra	G. Madhukar	Y. Tohma
J. Dobson	M. Malhotra	K. S. Trivedi
L. Donatiello	D. Mandrioli	Y. Tsai
T. Downs	F. Manola	A. Veevers
S. Eckmann	K. Marzullo	P. Verissimo
P. Eggert	M. Mazer	Yi-min Wang
P.D. Ezhilchelvan	E.J. McCluskey	E. Weyuker
P. Frankl	C.L. Meadows	K. Wilken
E. Kent Fuchs	J.F. Meyer	J. Wing
A. Gaivoronski	D.R. Miller	J. Woodward
M.C. Gaudel	A. Mok	Y. Yacobi
J. Goldberg	A. Mourad	Y. Yokote
W. Gorke	M. Preston Mullen	C. Yount
K. Goswami	J. Muppala	
Ph. Granger	T. Nanya	

Contents

FUNCTIONAL TESTING

ON FUNCTIONAL STATISTICAL TESTING DESIGNED FROM SOFTWARE BEHAVIOR MODELS

Pascale THÉVENOD-FOSSE, Hélène WAESELYNCK
Laboratoire d'Automatique et d'Analyse des Systèmes du C.N.R.S.
7, Avenue du Colonel Roche, 31077 Toulouse Cedex, France

Abstract

Statistical testing involves exercising a piece of software by supplying it with input values that are randomly selected according to a defined probability distribution over its input domain. This paper focuses on **functional statistical testing,** that is, when an input distribution and a number of random inputs are determined according to criteria relating to software functionality. The criteria based on **models of behavior** deduced from specification, i.e., finite-state machines and decision tables, are defined. The modeling approach involves a hierarchical decomposition of software functionality. It is applied to a module from the **nuclear field**. Functional statistical test sets are designed and applied to two versions of the module: the real version, and that developed by a student. **Twelve residual faults** are revealed, eleven of which affect the student's version. The other fault is quite subtle, since it resides in the driver that we have developed for the real version in our experimental test harness. Two other input distributions are experimented with: the uniform distribution over the input domain and a structural distribution determined so as to rapidly exercise all the instructions of the student's version. The results show that the functional statistical test sets have the highest fault revealing power and are the most cost-effective.

1 Introduction

Testing involves exercising the software by supplying it with input values. In practice, testing is partial as it is not possible to exercise a piece of software with each possible data item from the input domain. Hence, the problem of selecting a subset of the domain that is well-suited for revealing the actual but unknown faults; this issue is further compounded by the increasing complexity of real software systems. Many test criteria, relating either to program structure or software functionality, have been proposed as guides for determining test cases (see e.g. [1], [10], [12]).

Using these criteria, the methods for generating the test inputs proceed according to one of two principles [11]: either deterministic or probabilistic [5], [7], [16]. In the first case, which defines **deterministic testing**, test data are predetermined by selection in accordance with the criteria retained. In the second case, which defines **statistical** (or **random**) **testing**, test data are generated according to a defined probability distribution over the input domain; both distribution and number of input data items being determined in accordance with the criteria retained.

Some previous work focused on **structural statistical testing** [17], in which input distributions were determined according to structural criteria defined in the deterministic approaches [13], [15]. These distributions, called *structural distributions*, aim at ensuring that the program structure is properly scanned during a test experiment. Structural statistical testing has been shown to be an efficient way of designing test data during a unit testing phase, i.e. for programs with a known structure which remains tractable [18]. As a result, the *functional* statistical testing approach investigated in this paper is mainly concerned with but not confined to larger software components, that is, modules integrating several unit programs.

Functional statistical testing consists of determining an input distribution as well as a number of random test cases according to criteria based on software functionality. The criteria must facilitate the determination of input distributions, referred to as *functional distributions*, which will ensure that the different software functions are well probed within a reasonable testing time. This paper presents a **rigorous method** for designing functional statistical testing, based on **criteria related to behavior models**, as deduced from software specification.

Section 2 recalls the notion of *test quality with respect to a criterion*, from which the method for designing statistical test sets according to a given criterion is stated. Section 3 deals with *behavior models* that facilitate the definition of *functional criteria*. It lays the foundation of a structured method for determining functional test data. The approach is exemplified by a case study in Section 4: statistical test data are defined from behavior models deduced from the specification of a *safety critical module* from the nuclear field. Section 5 gives the *experimental results* that support the efficiency of these test data in revealing faults. *New areas of research* in functional statistical testing are described in Section 6.

2 Background

Previous work has shown that statistical testing is a practical verification tool. Indeed, the key to its effectiveness is the derivation of a probability distribution over the input domain that is appropriate to the test objective. The theoretical framework recalled below induces a rigourous method for determining these distributions.

2.1 Basic framework

Test criteria take advantage of information on the program under test in order to provide guides for selecting test cases. This information relates either to program structure or its functionality. In both cases, any criterion specifies a set of elements to be exercised during testing. Given a criterion A_i, let S_{Ai} be the corresponding set of elements. (To comply with finite test sets, S_{Ai} must contain a finite number of elements that can be exercised by at least one input item.) For example, the structural testing approach called "branch" testing requires that each program branch be executed: $A_i =$ "branches" $\Rightarrow S_{Ai} =$ {executable program edges}. The notion of *test quality with respect to a criterion*, firstly defined for random inputs only [16], has been generalized as follows to be applicable to any test set irrespective of whether or not the inputs are deterministic or random [20].

Definition 1 A set T of N input data items covers a test criterion A_i with a probability q_N if each element of SA_i has at least a q_N probability of being exercised during the N executions supplied by T. q_N is called the **test quality with respect to A_i**.

In the case of **deterministic testing**, the tester selects a priori a number N of inputs such that each element of SA_i is exercised at least once thereby providing in essence a "perfect" test quality ($q_N = 1$) with respect to A_i. It is worth noting that deterministic sets are often built so that each element is exercised *only once*, in order to minimize the test size (number of input items). In the case of **statistical testing**, a finite number of random inputs can never ensure that each element of SA_i is exercised at least once, since no data specifically aimed at exercising these elements have been intentionally included in the test set; hence, $q_N < 1$. The test quality q_N provided by a statistical test set of size N is deduced from the following theorem [16].

Theorem 1 In the case of statistical testing, the test quality q_N with respect to a criterion A_i and the number N of input data items are linked by the relation:

$$(1-P_i)^N = 1-q_N \qquad \text{with } P_i = \min \{p_k, k \in SA_i\} \qquad (1)$$

p_k being the probability that a random input exercises the element k of SA_i.

Relation (1) is easy to justify: since P_i is the probability per input case of exercising the least likely element, each element has a probability of at least $1 - (1-P_i)^N$ of being exercised by a set of N random input cases. The result of this is that on average each element of SA_i is exercised several times. More precisely, relation (1) establishes a link between the test quality and the number of times, denoted n, the least likely element is expected to be exercised: $n \cong - \ln(1-q_N)$, whatever the value of P_i. For example, $n \cong 7$ for $q_N = 0.999$, and $n \cong 9$ for $q_N = 0.9999$.

The **method for determining a statistical test set** according to a criterion A_i is based on the preceding theorem. It involves two steps, the first of which is the corner stone of the method. These steps are the following:

(i) *search for an input distribution* which is well-suited to rapidly exercise each element of SA_i to decrease the test size; or equivalently, the distribution must accommodate the highest possible P_i value;

(ii) *assessment of the test size N* required to reach a target test quality q_N with respect to A_i, given the value of P_i inferred from the previous step; relation (2) deduced from relation (1) yields the minimum test size:

$$N = \log (1\text{-}q_N) / \log (1\text{-}P_i) \qquad\qquad (2)$$

An acute question arises from the **definition of test criteria**: a real limitation is due to the imperfect connection of the criteria with the actual faults. Due to the current lack of an accurate model for software design faults, this problem is not likely to be solved soon. Nevertheless, the criterion does not influence random data generation in the same way as in the deterministic approach: it serves as a guide for defining an input distribution and a test size, but does not allow for the a priori selection of a (small) subset of input data. Hence, one can expect that the criterion adequacy with respect to faults should lead to a less acute problem in the statistical approach; and all the more so as several test cases are involved per element to be exercised and as these test cases are unbiased by human choice. Indeed, there is a meaningful link between fault exposure and random data: from relation (1), *any fault involving a failure probability $p \geq P_i$ per execution according to the chosen input distribution has a probability of at least q_N of being revealed by a set of N random inputs*. No such link is foreseeable as regards deterministic data; and this link should carry more weight than the warrant of a *perfect* test quality with respect to *questionable* criteria. Previous work on unit testing has already supported this assumption: the main conclusions recalled below justify our present investigation of functional statistical testing for larger software components.

2.2 On the fault revealing power of structural statistical testing

The first investigations focused on current structural criteria and the theoretical results were confirmed by experiments relating to the unit testing of four real programs from the nuclear field [18], [20]. Path selection criteria were used [13], [15], each of them defining a proper set of (sub)paths to be executed. Given a program and a criterion A_i, the **method for determining an input distribution** according to A_i was applied as follows. The program flow graph analysis provides the set SA_i of (sub)paths, together with their execution probability p_k according to probabilities of input values. Then, an input distribution that lets $P_i = \min\{p_k\}$ be as high as possible is inferred by solving the equation set {pk} (see examples in [16], [18]). The structural statistical test sets thus designed ensure that the program structure is soundly probed, the level of probing being an increasing function of the criterion stringency (from "instruction" to "path" level): the more stringent the criterion, the larger the test size.

In the **experiments**, the target faults were mutations [4], seeded in the source code. The four programs were subjected to several structural deterministic, structural statistical and uniform statistical test sets; the latter sets are the 'conventional' random test sets, i.e., when data are drawn from a uniform distribution over the input domain. The efficiency of the sets was assessed in terms of percentage of mutation faults revealed, called mutation score. A total of 2816 mutations was involved, and the results were in favor of structural statistical testing for two reasons:

(i) *structural statistical test data rapidly uncovered 99.8% of the seeded faults;* both structural deterministic data and uniform random data were far from exhibiting so good fault revealing powers, failing to reveal several hundreds of mutations;

(ii) *the mutation scores provided by the structural statistical test sets were repeatedly observed,* whatever the particular test sets generated according to a same structural input distribution [20]. On the contrary, in the case of deterministic testing, the test sets related to the same criterion exhibited quite disparate mutation scores and the most stringent criteria did not always supply the highest scores; similarly, the uniform random test sets provided various scores.

This work confirmed the fact that the effectiveness of deterministic testing and uniform testing depends heavily on the particular input values chosen [7], while the effectiveness of structural statistical testing does not; the two former approaches were never more efficient than the latter one. This supports the high fault exposure power of statistical testing, as expected at the end of Section 2.1. The comparison between the uniform and structural statistical sets showed that *the structural analysis of a program does provide a relevant information.* The weakness of the deterministic sets resulted more from a limited number of data that failed to compensate for the imperfect connection between structural criteria and faults, than to a strong inadequacy of the criteria.

Indeed, in the light of the experimental results, the impact of the criteria stringency was deemed not critical in the case of statistical testing: the conclusion was that the *most cost-effective approach* is to retain weak criteria facilitating the structural analysis needed to determine an input distribution, and to require a high test quality (say, 0.9999) with respect to them [20]. Finally, the efficacy of a *mixed test strategy* combining structural statistical testing and deterministic testing of special/extremal input values was fully confirmed: the six mutations not uncovered after completion of the structural statistical tests were changes affecting the bounds of an array; and these are typical cases for extremal value testing, that are poorly catered for by statistical testing within short testing time.

As structural testing is only applicable to programs that lend themselves to a tractable analysis, this work has to include larger **modules integrating several unit programs**. This is the aim of the following sections.

3 Functional statistical testing

"The goal of functional testing of a software system is to find discrepancies between the actual behavior of the implemented system's functions and the desired behavior as described in the system's functional specification." [14]. Accordingly, relevant functional test criteria must facilitate the selection of an input distribution and a test size ensuring that the software functions, and their interactions, are properly scanned.

3.1 Functional test criteria

Functional testing approaches refer to different levels of software description: external specification [12], [14], [21], internal design [9], program code [9], or a combination of levels [6], [22]. Some of them are rather informal, based on a careful review of the documents relating to the chosen level. They facilitate the derivation of deterministic test cases that are assumed to be functionally sensible. Examples of such approaches are "equivalence partitioning", "boundary-value analysis" and "cause-effect graphing" [12]. Keeping in mind that the probabilistic approach calls for the study of the influence of the input distribution on the coverage of the chosen criterion, informal approaches are not convenient for our purpose.

On the contrary, **finite-state machines (FSM)** used for describing software behavior – see e.g. [1], [3] – possess properties that are well-suited for a probabilistic analysis. An FSM can be depicted by a graph having a finite number of states and a finite number of transitions. The principle consists in:

(i)   associating one *state* with each mode of behavior;

(ii) weighting each *transition* with the input conditions that trigger it, and eventually with the action caused when the transition is made.

Different criteria A_i may be defined from an FSM depending on the stringency of the graph coverage: state coverage, transition coverage, or sequence of transitions coverage [21]. A criterion defines a finite set S_{Ai} of elements to be exercised: S_{Ai} = {graph states}, {graph transitions}, {sequences of graph transitions}, for the aforementioned examples. The influence of the input distribution on the value P_i = min {p_k, $k \in S_{Ai}$} is studied by replacing the input conditions that weight the transitions with their probabilities of occurrence in the distribution.

Although finite state machines are helpful to describe a large number of functions and their interactions, there are behaviors for which the FSM approach makes no sense [3]. A typical example is when actions depend on a combination of conditions, thereby causing an explosion of the number of states and/or transitions. Then, **decision tables (DT)** are another modeling tool that is well-suited for describing such behaviors – see e.g. [1], [3] – and for a probabilistic analysis. A DT defines a finite set of rules, each rule specifying the actions that take place when a specific combination of input conditions is met. A natural criterion A_i related to a DT is the rule coverage, that is: S_{Ai} = {rules}, and P_i = min {p_k, $k \in S_{Ai}$} where p_k is the probability that the rule k be activated. For a rule k, the replacement of the specified combination of conditions with its probability of occurrence provides the expression of p_k.

FSMs and DTs are **complementary modeling tools**: in essence, FSMs are appropriate to describe sequential behaviors while combinational functions are easier to translate into DTs . They form the basis of our modeling approach, that involves a hierarchical decomposition of software specifications.

3.2 Behavior modeling approach

In keeping with the goal of functional testing presented above, we opted for functions deduced from module **specification**, rather than implemented functions deduced from module design. Thus, the test cases should be more likely to expose design faults and can be defined early in the development process. Since a detailed specification analysis should determine a large number of functions that may not be described by any FSM or DT of reasonable size, the modeling approach is based on a **hierarchical decomposition** of the functionalities that proceeds from a **top-down** approach; thereby involving a sequence of models M_i ($i \geq 0$), each M_i being either an FSM or a DT.

First, high-level functions and their interactions are identified, providing M_0; then, the high-level functions are refined through other models M_1, M_2, etc. For example [1]: in telephony, two-level models are common, three- and four-level models are not unusual. At a given level i, the actions caused by an FSM transition or a DT rule may be the production of output results or the transfer of control to lower level models. Hence, the set of models forms a tree network, M_0 being the root. The decomposition stops when the functions are considered elementary with respect to the outputs they supply. The function granularity at level i is the result of a compromise between the complexity of M_i and the required number of further refinements, i.e., of models M_k, $k > i$. A single limitation applies to the models: the FSM graphs must be strongly connected so that no state becomes unreachable when increasing the size of the statistical test sets. Note that, although the decomposition suggested here is independent of the module implementation, design information could easily be taken into account.

The top-down approach is well-known and used in most current specification and design methods. But the definition of statistical test sets from a hierarchical decomposition of software functionality has never been investigated.

3.3 Design of statistical testing

The coverages of **DT rules** and of **FSM states in steady-state conditions**, i.e., after a number of initial executions large enough to ensure that the transients die down, are the criteria that we retained. Hence, p+1 models {M_0, ..., M_p} provide p+1 sets of elements (rules or states) that have to be exercised. Starting with the models M_i, the first step consists of replacing the input conditions that weight the FSM transitions and that enter the DT rules with their probabilities of occurrence as function of the input probabilities. The next two steps, described below, are those identified in the method for determining a statistical test set, namely (i) *search for an input distribution* and, (ii) *assessment of the test size.*

3.3.1   Search for an input distribution

A lot of DT rules and FSM states are likely to be derived from the p+1 models. Hence it would not be realistic to attempt intensive coverage of all of them at the same time. This is because of:

(i)   *the module complexity;* when all the models are encompassed, it is difficult to assess the probabilities of the elements as numerous correlated factors are involved;

(ii) *the explosion of the test size;* even if these assessments are feasible and tractable in order to derive an input distribution, a prohibitive test size will probably be required to reach a high test quality as the probability of the least likely element remains very low due to the large number of elements.

To address this issue several distinct test sets are designed each one focusing on the coverage of a subset of models. To do this, one defines a **partition of the models M_i** into s+1 (s ≤ p) disjoint subsets PS_j (j = 0, ..., s), each PS_j gathering one or several models of consecutive levels. For each PS_j, a specific input distribution can reasonably be derived, that maximises the stationary probability of the least likely element related to the models M_i grouped in PS_j. One gets s+1 input profiles, and a proper test size N_j will be assessed for each of them (§3.3.2).

Since only a subset of elements is taken into account to determine the input distribution specific to a given PS_j, some input variables may not be involved and as a result no probability is obtained for them. Hence, the information deduced from other partition subsets must be included to define a complete input profile. To accomplish this, **the inputs that are not involved at a given partition level j** are classified according to three types:

(i) *upper level inputs* (except for j = 0), conditioning the transfer of control to a model $M_i \in PS_j$ from the upper level models; their probabilities must provide the most likely transfers to M_i;

(ii) *lower level inputs* (except for j = s), taken into account in lower level models; their probabilities are set as defined from the corresponding PS_k (k > j);

(iii) *unrelated inputs*, for which a uniform distribution may be used.

In practice, the determination of the input distributions uses a **bottom-up** approach (from PS_s to PS_0) since, from (ii), the input distribution specific to a level j may be partly defined at lower levels.

3.3.2 Assessment of a test size N

A complete test set is composed of s+1 suites of test cases, involving $N = N_0 + \ldots + N_s$ test cases. It must provide a test quality q_N with respect to the selected criteria, i.e., FSM states and DT rules. Let P_j be the stationary probability of exercising the least likely state or rule related to a partition subset PS_j (P_j is inferred from the input profile derived for PS_j). Since P_j is a probability in steady-state conditions, the assessment of the test size N_j specific to PS_j involves two factors:

(i) first, equation (2) yields the test size in *steady-state conditions*;

(ii) and second, this test size must be augmented with the number N_{j1} of initial state transitions needed to *ensure that the transients die down*, that is:
$$N_j = N_{j1} + \log (1-q_N) / \log (1-P_j)$$

4 Case study: a safety critical application

The real case study reported in the succeeding sections illustrates the feasibility of the proposed approach. The models M_i are derived from the high level specifications of the target module, whose main requirements are summarized below.

4.1 High level requirements of the target module

The module is extracted from a **nuclear reactor safety shutdown system**. It belongs to that part of the system which periodically scans the position of the reactor's control rods. At each operating cycle, 19 rod positions are processed. The information is read through five 32-bit interface cards. Cards 1 to 4 each deliver data about four rod positions; these cards are all created in the same way and are hereafter referred to as *generic* cards. The 5th card delivers data about the three remaining rod positions as well as monitoring data; this card which is therefore processed differently is called the *specific* card.

At each operating cycle, one or more interface card may be declared inoperational: the information it supplies is not taken into account. This corresponds to a *degenerated operating mode:* only part of the inputs are processed. A card identified as inoperational remains in that state until the next reset of the system. In the worst situation all cards are inoperational and the module delivers a *minimal service:* no measure is provided and only routine checks are carried out.

Extensive hardware self-checking is used so that errors when reading a card are unlikely. Nevertheless, for defensive programming concerns, this case is specified: the application is stopped and has to be restarted.

After acquisition, the data are checked and filtered. Three checks are carried out: the corresponding rod sensor is connected, the parity bit is correct and the data is stable (several identical values must be read before acceptance). The stringency of the third check (required number of identical values) depends on the outcome of the preceding checks of the same rod. After filtering, the measurements of the rod positions (in Gray code) are converted into a number of mechanical *steps*. The result of data conversion may be a valid number of mechanical steps or an invalid number or two special limit values.

4.2 Functional decomposition of the specification

The hierarchical decomposition involves two FSMs (M_0, M_1), and one DT (M_2). In the simplest case an FSM transition condition is the occurrence of a specific input value, e.g. a 'read error'. It may also include more complicated expressions, e.g. the current value of a rod position has to be identical to the previous one. The action resulting from a transition is either an output result or a transfer of control to lower level models until an output result be determined.

4.2.1 First level of decomposition: finite state machine M_0

The first level of decomposition consists in describing the **various operating modes,** and in identifying the conditions that make the system switch on them. Twelve operating modes are identified (Figure 1). The transition conditions relate to four factors:

- A(i):   i generic cards switch from 'operational' to 'inoperational';

- B   :   the specific card switches from 'operational' to 'inoperational';

- C   :   an error has occurred when reading an operational card
         (initiating a restart);

- D   :   a reset is forced while all the cards are inoperational.

operating mode		state label
full service	all cards operational	1
partial service	1 generic card inoperational	2
	2 generic cards inoperational	3
	3 generic cards inoperational	4
	all generic cards inoperational	5
	specific card inoperational	6
	specific and 1 generic cards inoperational	7
	specific and 2 generic cards inoperational	8
	specific and 3 generic cards inoperational	9
minimal service	all cards inoperational	10
initialization	reset	11
initialization	restart following a read error	12

Figure 1: States of the finite state machine M_0.

M_0 involves 88 transitions. By way of example, Figure 2 shows the outcoming transitions of state 8.

The information considered at this first level is not sufficient to determine an output result: when at least one card is operational and read without error, we have to proceed with the decomposition and study the processing of the measures acquired.

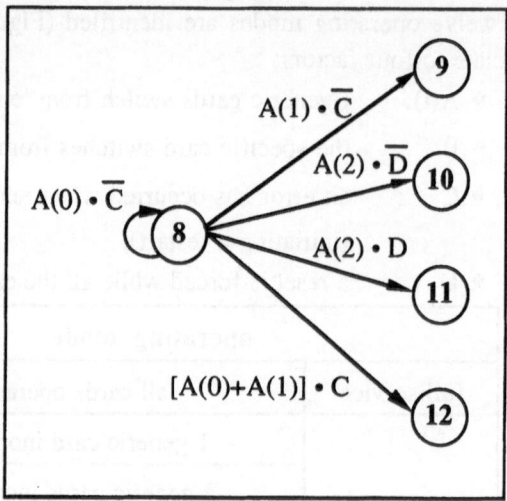

A(i) : i generic cards switch
        from 'operational' to
        'inoperational'.

C   : read error.

Ḍ   : reset forced while all the
      cards are inoperational.

•   : logical connective AND

+   : logical connective OR

Figure 2: Outcoming transitions of state 8.

4.2.2   Second level of decomposition: finite state machine M_1

The second level of decomposition M_1 models the **checks and filtering performed on one measure**. As the number of measures acquired depends on the operational cards, the number of FSM M_1 running in parallel is determined by the state occupied in M_0. The machines M_1 are created, deleted, or maintained according to the transition made at the top level. For example (Figure 2), taking the transition '8 → 9' means that one generic card less is operational: as this card contains four measures, the four corresponding machines M_1 are deleted.

M_1 has twelve states: four of them are related to the stringent filtering mode, and the eight others to the normal filtering mode. The 54 possible transitions depend on three conditions:

- the sensor is connected;

- the parity bit of the measure is correct;

- the value of the measure is identical to the one read at the preceding operating cycle for the same rod.

It is worth noting that the latter condition will imply that successive test cases will not be selected independently within a statistical test set.

Each transition induces the acceptance or rejection of the measure acquired: twelve of them correspond to the case where a value passes the checks and has to be converted.

4.2.3 Third level of decomposition: decision table M_2

The **conversion** function, invoked by twelve transitions of M_1, is described by a DT with eight rules defining the rod position to deliver. The rule conditions involve:

- the value of the measure (valid or invalid number of mechanical steps, special values);

- a boolean input value specifying whether or not a special value is expected.

4.3 Behavior analysis

The previous analysis has identified attributes of input values that are significant for exercising the module functions and their interactions. The next step **assigns input probabilities**, in order to study the dynamic behavior of the module when subjected to a test profile.

The FSMs M_0 and M_1 weighted with symbols become **stochastic graphs**, i.e. graphs weighted with probabilities. For example, consider the condition for transition '8 $\rightarrow$ 9' (Figure 2): one generic card becomes inoperational (A(1)), and the remaining card is read without error ($\overline{C}$). Let:

• q be the probability that a generic card switches to inoperational,

• r be the probability that a read error occurs.

Then, the probability weighting transition '8 $\to$ 9' is: $\binom{2}{1}$ q (1-q) (1-r).

Once the transformation is completed for all transitions, the dynamic behavior is studied through operations on the transition matrices.

As regards the DT M_2, the probability of exercising each rule is directly obtained from the probability of its condition of activation.

4.4   Design of statistical test sets

Here a partition of the three models into two subsets PS_j is suitable: PS_0 contains M_0; PS_1 groups M_1 and M_2, that may easily be studied at the same time since both relate to the processing of one measure (while M_0 relates to the acquisition of a bundle of measures). Then, two distinct test sets must be designed: the first one will ensure the coverage of the filtering/conversion functions $(M_1 + M_2)$; the second one will probe the operating modes (M_0). Each set involves its proper input distribution and test size, informations deduced from the other level (upper or lower level input parameters) being incorporated when required.

The test sizes given below are assessed for a target test quality of 0.9999 (q_N, in steady state conditions) and for the requirement that the asymptotic state probabilities are reached with a precision of 10^{-6}.

4.4.1   Input distribution and test size to cover M_1 and M_2

Upper level parameters are forced to their activating values: a full service is delivered with no read error. Hence, the maximum number of measures are processed in parallel. The probabilities of the values appearing in the rule conditions of M_2 are determined so as to ensure a good balance between the rules. Then, an input distribution appropriate to cover the FSM M_1, that is, a distribution that maximizes the asymptotic probability of the least likely state, is investigated. By considering the small number of parameters that govern the process, an approximated solution may be obtained by **simulation, by sampling the relevant probabilities over [0...1]**.

The distribution obtained supplies a sufficiently high probability for the twelve transitions invoking $M2$: each rule is exercised with the same (or higher) probability as the least likely state of $M1$, i.e., approximately 0.0087.

For a given q_N, the test size in steady state conditions is derived from relation (2), and divided by the number of measures processed in parallel: for nineteen measures, $1055/19 = 56$ inputs are required. This size is augmented with the number of initial transitions needed to ensure that the transients die down, namely 29 inputs, leading to: $N_1 = 85$ test inputs.

4.4.2 Input distribution and test size to cover M_0

As previously stated, the study of the dynamic state coverage allows us to assign optimum probabilities on the subset of inputs related to M_0. State coverage of M_0 in steady-state conditions requires $N_0 = 356$ test inputs (302 for state coverage, and 54 to reach the asymptotic state probabilities). To complete the input profile, the probabilities of lower level parameters, e.g. the values of the measures, are the same as in the preceding distribution.

5 Experiments and results

The experiments involve two versions of the module, providing us with a back-to-back testing scheme. REAL is the real version, and STU a version developed by a student from the same high-level specification; both versions are written in C language. The size of their object code approximates 20 K-bytes (a thousand lines of source code without comments). The experiment proceeds as follows: apply a test set to REAL and STU; examine the first output result for which a discrepancy is observed; identify and fix the corresponding fault(s). The process is iterated until REAL and STU agree on the whole test set.

5.1 Overview of the statistical test sets

5.1.1 Functional test sets

Section 4 aimed to show the practicality of deriving statistical test inputs from the behavior models of a non-trivial application. We now investigate the ability of these test data to expose actual faults, repeatedly whatever the particular values drawn from the defined input distributions: it is pointless to define a testing method whose efficiency depends heavily on the particular input values selected, rather than on adequate properties of the test data related to the method. Hence, in order to expose eventual disparities, **five different functional test sets F-Set$_i$** (i = 1, ..., 5) have been generated, each being composed of 85 inputs ensuring the coverage of M_1 and M_2 followed by 356 inputs ensuring the coverage of M_0.

5.1.2 Structural test sets

Although structural statistical testing is highly efficient in a unit testing phase, its relevance for larger scale programs, that is, when the complexity of the source code forces us to use only weak criteria such as branch or instruction coverage, may be questioned. One can wonder whether these criteria are sufficient to distinguish relevant input cases for a target module involving the aggregation of several functions. Another drawback of structural testing is that the selection of test data is driven by the source code, rather than by the specification: in particular, if two different versions of a same application have been designed, each of them requires its own test profile and test size.

The test sets used in the experiments are derived from the structure of the **STU version** only since few, if any faults are expected to reside in the REAL one. The complexity of the source code forces us to choose the weakest criterion, namely **instruction testing**, and to proceed empirically to derive an input distribution. Starting from a large number of input data uniformly drawn from their valid range, we progressively refine the test profile until the frequency of an instruction is deemed sufficiently high (the C-compiler supports the automatic insertion of code to count the number of times each basic block of instructions is executed). At each step, the analysis focuses on a few blocks (the "hidden" ones) and aims at determining the input conditions that force their activation. The final input distribution is very different from the uniform one. A crude estimate of the probability of the least likely block is derived; given a high test quality requirement of 0.9999, an upper bound N = 500 on the test size is drawn from relation (2). **Five structural test sets S-Set$_i$** (i = 1, ..., 5) have been generated. It has been verified a posteriori that they provide a good coverage of STU (14 executions on average for the least likely blocks).

5.1.3 Uniform test set

As the notion of random patterns is often connected to a **uniform distribution over the input domain**, it was also used experimentally for comparison purposes. Selecting uniform patterns is a *black-box strategy*, the definition of the valid input domain being derived from the program specification. Actually, "blind" uniform testing is probably the poorest test strategy, since it does not take into account information relative to the target piece of software. In [18], the unit testing of four low level functions of REAL convincingly showed that a uniform distribution was far from adequate in most cases. The results are expected to be even worse when testing the whole module. A **single uniform test set**, denoted **U-Set**, has been generated involving a large number of test data, namely 5300 inputs: this is in conformity with the foundation of uniform testing, that is, large test sets generated cheaply.

5.2 Overview of the faults uncovered

Twelve faults have been identified (Figure 3) in which 11, denoted A, B, ..., K, were found in STU; the last one, Z, resides in our test driver that provides the interface between REAL and the files containing the test sets. Faults A, G and J are **structural faults**, directly linked to the coding of STU. Faults B to F, and I, result from the **lack of understanding of the filtering check requirements** by the student, this function being at the heart of the module. The others are **initialization faults**: either an improper initial value is assigned (K) or the initialization is missing (H, Z). The initialization faults are most subtle since their activation depends on the states that follow the wrong initialization. For example, revealing H requires that G be removed and that the specific card be inoperational immediately after a restart or a reset. Finally, although Z resides in the test driver that we have developed for REAL, it has a ripple effect on the module: the simulation of hardware restart/reset fails to restore the correct initial context of REAL.

Figure 4 summarizes the results supplied by the various test sets. The succeeding sections provide the main comments and conclusions related to each type of statistical test sets: uniform, structural and functional.

A	wrong operator used in the processing of an output value
B, C, D, E, F, I	the filtering checks, as implemented, do not comply with the specification
G	wrong control flow when the specific card is inoperational
H	initialization missing for variables related to the specific card
J	a variable in a loop is initialized out of loop instead of at each iteration
K	wrong initial state for the filtering process
Z	initialization missing for a variable of our test driver

Figure 3: List of the twelve faults uncovered.

	A	B	C	D	E	F	G	H	I	J	K	Z
U-Set N = 5300	✔	—	—	✔	—	✔	✔	✔	—	—	—	—
S-Sets N = 500	✔	✔	✔	✔	✔	✔	2/5	1/5	✔	✔	✔	3/5
F-Sets N = 441	✔	✔	✔	✔	✔	✔	✔	✔	✔	✔	✔	4/5

✔	always revealed
i/j	revealed by i sets out of j
—	not revealed

Figure 4: Results supplied by the test sets.

5.3 Inadequacy of the uniform distribution

As anticipated, uniform testing provides the **poorest results**, since it reveals only five of the twelve faults identified. The U-Set poorly probes STU and REAL, from both a structural and a functional viewpoint. With respect to the structural coverage, four blocks of instructions of STU and one of REAL are never exercised; some are seldom executed (less than three times). As regards the functional coverage, some M_1 states are seldom or never reached: most faults related to the filtering checks are not revealed.

In conclusion, a test data generation based on a uniform distribution is **definitely not efficient** to design a statistical test experiment. It is often argued that uniform testing gives a best return on investment than other approaches, since a large number of test cases can be generated cheaply. But, such data are unlikely to exhibit a good fault revealing power, so that little improvement is to be expected from a *realistic* increase of the test size. Here, the uniform set is an order of magnitude larger than the other sets: the five faults are found within the first 633 executions; the remaining 4667 executions being garbage.

5.4 Weakness of instruction testing

Nine faults are repeatedly revealed by all the structural sets; as they induce a high failure rate under the structural profile, the first quarter of the sets generally suffices to expose them. The other faults (G, H, and Z) are occasionally revealed, usually late in the test experiment. The case of Z is special, because the test sets have been designed to cover STU and Z corrupts REAL from its test driver. However, the S-Sets also provide a good structural coverage of REAL (14 executions on average for the least likely blocks, as for STU).

An interesting property observed for the five S-Sets is that the high instruction coverage is preserved throughout the debugging process, despite the fact that the structure of STU is modified by successive corrections. The problem with structural criteria is that the test data are program dependent: strictly speaking, a new analysis should be conducted for each intermediate version of the program, in order to adapt the test inputs to the evolution of the source code. It seems here that the probabilistic approach allows us to circumvent the problem.

As the structural profile ensures a **suitable probe of the models M_1 and M_2**, the S-Sets are very efficient for faults in the filtering checks. On the other hand, they provide **poor coverage of M_0**: most of the time the system resides in state 1 (full service delivered); the theoretical probability of ever reaching state 10 (minimal service) is only 4.10^{-4} for N = 500. A fault linked to the degeneration of the operating mode would not be uncovered. The specificity of the fifth card is not identified in this distribution. This accounts for the poor results for G and H, while both faults are rapidly revealed by all functional sets. With 500 structural data, the probability of revealing G is about 0.49, and the exposure of H requires that G be fixed. It can be said that some major module features are hidden in the implementation structure: actually, they correspond to family of subpaths in the control flow graph that are invisible at the instruction level. This problem did not occur when testing involved small units of REAL [20]: then, selecting weak criteria was deemed as the most cost-effective approach for the design of structural statistical testing.

In conclusion, structural testing is particularly **well-suited in unit testing**, but its effectiveness diminishes as the source code under test grows. The functions supported by the software move away from the instruction level, while finer examination of the structure becomes intractable. Indeed the size of the module under test is a limit above which instruction testing itself is no longer tractable. Moreover, this module belongs to the broad class of reactive systems, and whether or not the static graph of control is a relevant model for such systems is debatable.

5.5 Promising features of functional testing

The five functional test sets yield the best results, despite the fact that they involve the smallest number of inputs (441 versus 500 and 5300). Every fault in STU exposed by some S-Set is repeatedly found by all the F-Sets. The exposure of Z is less accurate, since Z is not revealed by one functional test set, but the result is still better than the one observed for the structural sets.

The **two input distributions involved in the F-Sets exhibit complementary features**: the first subsets of 85 test data reveal the faults related to the bad processing of the measures, notably during the filtering checks, while the second subsets are more appropriate for faults related to the specific card and for initialization faults. Hence, the behavior models constructed from the specification appear to be meaningful with respect to the faults. Full instruction coverage is provided by each F-set and for each intermediate version supplied by the fixes, although slower than during structural testing.

In the light of our experiments, functional statistical testing seems to exhibit promising features that justify the continuation of the investigation. Now, the question arises as to whether the F-Sets, as designed, ensure a suitable probe of the **interactions** between high and low level functions. When defining an input distribution for the coverage of M_0, we have included probabilities deduced from the lower behavior models. But as Z is not revealed by one set this may be insufficient: the exposure of Z depends on conditions involving states of both M_0 and M_1, and these conditions are never fulfilled in F-Set$_2$. Hence, the testing of the interactions will be addressed in our new research work.

6 Conclusions

Because statistical testing is generally related to uniform test data, it is often deemed inadequate for the exposure of faults. Our study removes this preconceived idea. The random inputs have to be designed by using some model of the target program, whether structural or functional, as investigated in this paper.

For large or even medium scale programs, the model complexity forces us to use weak criteria, e.g. instruction or state coverage. The results reported here suggest that, as soon as one shifts from unit to integrated module testing, the functional approach is the most efficient: it allows us to distinguish the important features of the module early in the development process, while still providing a good coverage of the implemented code. The models used to depict the expected behavior are conventional: finite state machines and decision tables are available for any program specified according to current SA/RT techniques. Hence the approach is consistent with modern trends in software development, these trends being reflected by the increasing popularity of the CASE tools that support these techniques.

As a result the CASE tool **Statemate** [8] will be used as it offers interesting facilities that will help us to refine the approach: the behavior description is more powerful than FSMs; simulations can be programmed according to a chosen input distribution, so that statistics are easily gathered on the models; and these statistics should provide us with a significant assistance during the basic step of our approach, that is the search for a proper input distribution. Moreover, the high simulation power of Statemate should address the *oracle issue*, namely how to determine the correct output which a program should give in response to a given input [2], [23].

Emphasis will be placed on the study of the interactions between the various levels of decomposition. We will also look at practical ways to identify extremal/special values from the models. In essence such values have a low probability of occurrence during statistical testing and are more efficiently covered by deterministic inputs specifically aimed at them: the ultimate goal of the study is to define a **mixed test strategy** combining random and deterministic inputs. In our investigation, the efficiency of the test sets thus designed will be assessed relative to real faults (the twelve already found and maybe other residual ones), and a larger sample of seeded faults (mutations).

Acknowledgments

We wish to thank our colleagues Alain Costes, Yves Crouzet, Jean-Claude Laprie and David Powell for their helpful comments during the preparation of this paper. This work was supported in part by the CEC under ESPRIT Basic Research Action no. 3092: "Predictably Dependable Computing Systems (PDCS)".

References

[1] B. Beizer. *Software testing techniques.* Van Nostrand Reinhold, New York, 1983. Second Edition, 1990.

[2] D. B. Brown et al. An automated oracle for software testing. *IEEE Transactions on Reliability*, Vol. 41, No. 2, June 1992, pp. 272-280.

[3] A. M. Davis. A comparison of techniques for the specification of external system behavior. *Communications of the ACM*, Vol. 31, No. 9, September 1988, pp. 1098-1115.

[4] R. A. DeMillo, R. J. Lipton, F. G. Sayward. Hints on test data selection: help for the practicing programmer. *IEEE Computer Magazine*, Vol. 11, No. 4, April 1978, pp. 34-41.

[5] J. W. Duran, S. C. Ntafos. An evaluation of random testing. *IEEE Transactions on Software Engineering*, Vol. SE-10, No. 4, July 1984, pp. 438-444.

[6] B. Goodenough, S. L. Gerhart. Toward a theory of test data selection. *IEEE Transactions on Software Engineering*, Vol. SE-1, No. 2, June 1975, pp. 156-173.

[7] R. Hamlet . Theoretical comparison of testing methods. *Proc. 3rd IEEE Symposium on Software Testing, Analysis and Verification*, Key West, USA, December 1989, pp. 28-37.

[8] D. Harel et al. STATEMATE: a working environment for the development of complex reactive systems. *IEEE Transactions on Software Engineering*, Vol. SE-16, No. 4, April 1990, pp. 403-414.

[9] W. E. Howden. A functional approach to program testing and analysis. *IEEE Transactions on Software Engineering*. Vol. SE-12, No. 10, October 1986, pp. 997-1005.

[10] W. E. Howden. Functional program testing and analysis. *Computer Science Series*, McGraw-Hill Book Company, 1987.

[11] J-C. Laprie (Ed.). Dependability: Basic Concepts and Terminology. *Vol. 5 in the Series on Dependable Computing and Fault-Tolerant Systems*, Springer-Verlag, Austria, 1992.

[12] G. J. Myers. *The art of software testing*. Wiley, New York, 1979.

[13] S. C. Ntafos. A comparison of some structural testing strategies. *IEEE Transactions on Software Engineering*, Vol. SE-14, No. 6, June 1988, pp. 868-874.

[14] T. J. Ostrand, M. J. Balcer. The category-partition method for specifying and generating functional tests. *Communications of the ACM*, Vol. 31, No. 6, June 1988, pp. 676-686.

[15] S. Rapps, E. J. Weyuker. Selecting software test data using data flow information. *IEEE Transactions on Software Engineering*, Vol. SE-11, No. 4, April 1985, pp. 367-375.

[16] P. Thévenod-Fosse. Software validation by means of statistical testing: retrospect and future direction. *Proc. 1st IEEE Working Conference on Dependable Computing for Critical Applications*, Santa Barbara, USA, August 1989, pp. 15-22. Published in Dependable Computing for Critical Applications, *Vol. 4 in the Series on Dependable Computing and Fault-Tolerant Systems*, Springer-Verlag, 1991, pp. 23-50.

[17] P. Thévenod-Fosse. On the efficiency of statistical testing with respect to software structural test criteria. *Proc. IFIP Working Conference on Approving Software Products*, Garmisch, Germany, Elsevier Science Publishers B.V., North-Holland, 1990, pp. 29-42.

[18] P. Thévenod-Fosse, H. Waeselynck, Y. Crouzet. An experimental study on software structural testing: deterministic versus random input generation. *Proc. 21st IEEE Int. Symposium on Fault-Tolerant Computing (FTCS-21)*, Montréal, Canada, June 1991, pp. 410-417.

[19] P. Thévenod-Fosse, H. Waeselynck. An investigation of statistical software testing. *Journal of Software Testing, Verification and Reliability*, Vol. 1, No. 2, July-September 1991, pp. 5-25.

[20] P. Thévenod-Fosse, H. Waeselynck, Y. Crouzet. Software structural testing: an evaluation of the efficiency of deterministic and random test data. *LAAS Report 91.389*, December 1991.

[21] H. Ural. Formal methods for test sequence generation. *Computer Communications*, Vol. 15, No. 5, June 1992, pp. 311-325.

[22] E. J. Weyuker, T. J. Ostrand. Theories of program testing and the application of revealing subdomains. *IEEE Transactions on Software Engineering*, Vol. SE-6, No. 3, May 1980, pp. 236-246.

[23] E. J. Weyuker. On testing non-testable programs. *The Computer Journal*, Vol. 25, No. 4, 1982, pp. 465-470.

FUNCTIONAL TEST CASE GENERATION FOR REAL-TIME SYSTEMS

Dino MANDRIOLI[1], Sandro MORASCA[2], Angelo MORZENTI[1]
[1]Dipartimento di Elettronica, Politecnico di Milano,
Piazza L. Da Vinci, 32, 20133, Milano, Italy.
[2]Institute for Advanced Computer Studies, University of Maryland,
College Park, Maryland 20742.

Abstract

We address the problem of automatic derivation of functional test cases for real-time systems. We present techniques for deriving test cases from formal specification of such systems coded in TRIO, a language which extends classical temporal logic to deal explicitly with time measures. An interactive tool is outlined that implements such techniques. Essentially the tool is based on interpretation algorithms of the TRIO language. Several heuristic criteria are suggested to drastically reduce the size of generated test cases.

1 Introduction

Testing has always been a fundamental approach to system verification, despite

Work partially supported by ENEL-CRA, by Piano Finalizzato Sistemi Informatici e Calcolo Parallelo of CNR, and by NASA (Grant NSG5123-S117).

its theoretical limits–mainly with reference to the verification of software correctness–have often been emphasized [6]. There is in fact a general agreement that the use of systematic testing strategies, possibly supported by suitable tools, can strongly help the effectiveness of this activity by increasing both the probability of finding errors and the reliability of the system in the case that testing did not show the presence of errors.

It is usual and natural to divide testing strategies in *white-box testing* or *structural testing* and *black-box testing* or *functional testing*. The former drives the testing activity on the basis of the implemented system– i.e., code, if the system is implemented as software–; the latter is based on the functional specification of the system. In principle both categories can be supported by automatic or semiautomatic tools; this requires, however, that the techniques are applied to well formalized objects, what is true for programming language code but is seldom the case for specifications which are generally stated rather informally. As a consequence, it is easier to find tools supporting test case generation of structural type [8] than functional ones.

In the latter category we find compiler testing tools [3] which derive sample programs to be used as input to the compiler under testing from the grammar of the programming language; techniques to derive test cases from operational specifications [7], [12], [22], and from algebraic specifications [13], [1], [17], [18], [28], [2]. Detailed comments and comparisons on the above approaches can be found in [20]. None of them addresses, however, the main issues of this paper. In particular, no approach we are aware of, can deal explicitly with timing issues.

This paper proposes a method and a tool for the derivation of functional test cases for real-time systems. By real-time systems we mean those systems whose correct behaviour does depend on the relative speed of the involved processes–not necessarily all of them being software processes. Typical examples of such systems are plant control systems, air traffic control systems, embedded applications, etc. In general, it is widely acknowledged [31], [28] that specification, design, and verification of real-time systems are more difficult than traditional software development, due to the necessity of dealing explicitly with time, whereas traditional design techniques abstract away from this component. On the other hand real-time systems have often also very strict reliability requirements, so that much effort must be devoted to their verification.

The method and the tool proposed in this paper are based on the specification language TRIO, an extension of temporal logic aimed precisely at dealing explicitly and quantitatively with strict time requirements [15]. A key feature of TRIO is its *executability*. This means that, under suitable conditions, TRIO formulas can be automatically checked for satisfiability or validity [11]. When a formula specifying a given property of a system is interpreted, a *model* thereof is generated, i.e., a possible system behaviour that is compatible with such a specification formula. The main idea underlying this paper is to use such a model as a test set for the system implementation. In other words, the specification provided by TRIO formulas is used to generate a simulation of the specified system. Then, the implemented system is executed and its real behaviour is checked against the simulated ones.

By this way, not only possible *stimuli*–i.e., system inputs–are generated from system specifications to test the system under verification, but also system *reactions*–i.e., outputs–are provided to check whether the system behaviour really complies with the desired properties, thus solving the oracle problem. Thanks to the fact that we derive test cases from system specifications, not from the code, the method can be applied to the verification of any system, not exclusively to software verification.

The details of the idea are illustrated in the rest of this paper according to the following structure. Section 2 provides a short summary of the TRIO language, together with a sketchy illustration of a model parametric semantics motivated by the wish to achieve *executability* of specification as a basis for the construction of effective tools for their validation [19] [16]. The structure of two fundamental interpretation tools is briefly described. These are a *history generator*, i.e., an interpreter that receives as input a TRIO formula and produces as result a set of histories that are compatible with such a formula, and a *history checker*, i.e., an interpreter that receives as input a TRIO formula and a history and states whether the history is compatible with the given formula. Section 3 explains how simulations derived from TRIO specifications can be used as test cases for the implemented system, so that the two above tools can be the kernel of a tool for the generation of test cases derived from TRIO specifications. Section 4 faces the problem of distinguishing, within a sequence of events, which events are inputs and which are outputs of the system. In fact,

TRIO has no explicit notion of input and output–or, in other words, of causes and effects–in the description of system behaviour. Section 5 copes with the problem of nondeterministic behaviours. In fact TRIO, as well as other formalisms such as Petri Nets, allows nondeterministic specifications: this gives more freedom to the implementer permitting to specify only strictly necessary requirements. Nondeterminism, however, makes system verification more difficult, because several histories with the same inputs could be compatible with the same specification so that a discrepancy between the expected behaviour as produced by the system simulator and the observed behaviour of the implemented system is not necessarily a symptom of an error. Sections 6 and 7 suggest criteria to limit the testing effort (number of test cases) yet trying to obtain as much information as possible from the experimental activity. Both formal properties of test sets and heuristic criteria for test data selection are provided. Section 8 outlines the structure of an interactive tool for the derivation of functional test cases from TRIO specifications. Finally, Section 9 gives some concluding remarks and sketches the way for future work.

For reasons of brevity the present paper omits most of the technical details and proofs of theorems: the interested reader can find them in [20].

2 A brief summary of TRIO

The purpose of the following brief presentation of the language is to make the paper self-contained, *not* to provide a complete and convincing discussion of TRIO's features and of its practical utility. This was done in previous articles [23] and [14]. TRIO is a first order logical language, augmented with temporal operators that allow the specifier to express properties whose truth value may change over time. The meaning of a TRIO formula is not absolute, but is given with respect to a *current time instant* which is left implicit in the formula, much in the same way as in classical temporal logic.

2.1 Syntax: the temporal operators

The alphabet of the TRIO language includes sets of names for variables, functions, and predicates, and a fixed set of basic operator symbols. Every variable name x has an associated *type* or *domain*, called $D(x)$, which is just the set of values the variable may assume. Among the domains there is a distinguished one, required to be numeric in nature, which is called the *Temporal Domain*. Every function name has an associated arity, which is a number $n \geq 0$ (when $n=0$ the function is called a *constant*), and the indication of a type for every component of the domain and for the range. Similarly, every predicate name is associated with the number and type of its arguments. Predicates are divided into *time dependent* and *time independent* ones: time independent predicates always represent the same relation, while a time dependent predicate corresponds to a possibly distinct relation at every time instant. The predicates $<$, $\leq$, $=$, and all other usual predicates on numbers, are assumed to be time independent, so that the associated relational operations are applicable to elements of the Temporal Domain. Also, addition and subtraction are assumed as operations, with the usual properties, applicable to elements of the temporal domain.

The syntax of TRIO defines terms in the usual inductive way: terms are constants, variables, or function applications. Formulas are obtained by predicate application, or inductively by propositional composition and first-order quantifications of formulas. Besides the traditional constructs of mathematical logic, TRIO provides two basic modal operators, *Futr* and Past. If A is a formula and t a temporal term, the formulas Futr*(A, t)* and *Past(A, t)* intuitively mean that A holds at an instant laying t time units in the future (resp. in the past) with respect to the current time value, which is left implicit in the formula.

A large number of derived temporal operators may be defined by means of quantification over TI variables in the temporal terms of Futr and Past. These derived operators include all the operators of classical linear temporal logic. We mention, among others, the following ones

$$
\begin{array}{lll}
\text{AlwF(A)} & \overset{\text{def}}{=} & \forall t \ (t > 0 \rightarrow \text{Futr} \ (A, t)) \\
\text{AlwP(A)} & \overset{\text{def}}{=} & \forall t \ (t > 0 \rightarrow \text{Past} \ (A, t)) \\
\text{Alw (A)} & \overset{\text{def}}{=} & \text{AlwP} \ (A) \land A \land \text{AlwF} \ (A) \\
\text{SomF(A)} & \overset{\text{def}}{=} & \neg \ \text{AlwF} \ (\neg A) \\
\text{SomP(A)} & \overset{\text{def}}{=} & \neg \ \text{AlwP} \ (\neg A) \\
\text{Som(A)} & \overset{\text{def}}{=} & \text{SomP} \ (A) \lor A \lor \text{SomF} \ (A) \\
\text{Lasts(A, t)} & \overset{\text{def}}{=} & \forall t' \ (0 < t' < t \rightarrow \text{Futr} \ (A, t'))
\end{array}
$$

$$\text{Until}(A_1, A_2) \quad \underset{\text{def}}{=} \quad \exists t \, (t{>}0 \wedge \text{Futr}(A_2, t) \wedge \text{Lasts}(A_1, t) \,)$$
$$\text{NextTime}(A, t) \quad \underset{\text{def}}{=} \quad \text{Futr}(A, t) \wedge \text{Lasts}(\neg A, t)$$

AlwF(A) means that *A* will hold in all future time instants, while *AlwP* has the same meaning with respect to the past; *Alw(A)* means that *A* holds in every time instant of the temporal domain; *SomF(A)* means that *A* will take place sometimes in the future, and *SomP* has the same meaning in the past; *Som(A)* means that *A* takes place sometimes in the past, now or in the future; *Lasts(A,t)* means that *A* will be true in the next *t* time units; *Until(A1,A2)* means that A2 will take place sometimes in the future, and A1 holds until then; *NextTime(A, t)* means that the first time in the future when *A* occurs is *t* time units from now.

Example 1 One very simple Real-Time system is a transmission line, which receives messages at one end and transmits them unchanged to the other end with a fixed delay. The time-dependent predicate *in* means that a message has arrived to the enter-end at the current time (left implicit); the predicate *out* means that a message is emitted at the other end. The TRIO formula

$$\text{in} \leftrightarrow \text{Futr (out, 5)}$$

means that a message arrives at the current time if and only if 5 time units later a message will be emitted, so that no message gets lost and no spurious message is generated.

o

Example 2 As a second example, we consider a hydroelectric power generation plant, where the quantity of water held in the basin is controlled by means of a sluice gate. The gate is controlled via two commands, *up* and *down* which respectively open and close it, and are represented as a TRIO time dependent unary predicate named *go* with an argument in the range {*up*, *down*}. The current state of the gate is modeled by the time dependent unary predicate *position* that admits for its argument one of the four values: *up* and *down* (with the obvious meaning), and *mvUp*, *mvDown* (meaning respectively that the gate is being opened or closed). The unary time dependent predicate *signal*, whose argument can assume the values {*up*, *down*, *danger*}, represents the signaling that, respectively, the gate reaches the open or closed position, or that a dangerous situation occurred. The following formula describes the fact that it takes the sluice gate Δ time units to go from the *down* to the *up* position, after receiving a go(up) command.

$$(\text{*goUp*}) \quad \begin{pmatrix} \text{position(down)} \\ \wedge \\ \text{go(up)} \end{pmatrix} \mapsto \begin{pmatrix} \text{Lasts}(\text{position(mvUp)},\Delta) \\ \wedge \\ \text{Futr}\left(\begin{pmatrix} \text{position(up)} \\ \wedge \\ \text{signal(up)} \end{pmatrix}, \Delta \right) \end{pmatrix}$$

When a *go (up)* command arrives while the gate is not still in the *down* position, but is moving down because of a preceding *go(down)* command, then the direction of motion of the gate is not reversed immediately, but the downward movement proceeds until the *down* position has been reached. Only then the gate will start opening according to the received command.

$$(\text{*moveDown*}) \quad \begin{pmatrix} \text{position(mvDown)} \\ \wedge \\ \text{go(up)} \end{pmatrix} \to \exists t \begin{pmatrix} \text{NextTime}\left(\begin{pmatrix} \text{position(down)} \\ \wedge \\ \text{signal(down)} \end{pmatrix}, t \right) \\ \wedge \\ \text{Futr}\left(\begin{pmatrix} \text{Lasts}(\text{position(mvUp)}, \Delta) \\ \wedge \\ \text{Futr}\left(\begin{pmatrix} \text{position(up)} \\ \wedge \\ \text{signal(up)} \end{pmatrix}, \Delta \right) \end{pmatrix}, t \right) \end{pmatrix}$$

If the behaviour of the sluice gate is symmetrical with respect to its direction of motion, two similar TRIO formulas, called *goDown* and *moveUp*, will describe the commands and their effects in the opposite direction. Finally, predicate *alarm* indicates an emergency situation. The activation of the alarm signal, one time unit after the emergency occurrence, is modeled by the formula

(*danger*)   alarm $\to$ Futr(signal(danger), 1)

o

TRIO has proved to be an adequate language for real-time systems specification, but its use becomes difficult when considering large and complex systems. This is because TRIO specifications are very finely structured: the language does not provide powerful abstraction and classification mechanisms, and lacks an intuitive and expressive graphic notation. To overcome this difficulty, we enriched TRIO with concepts and constructs from object oriented methodology, yielding a language called TRIO+ [26]. Among the most important features of TRIO+ are: the ability to partition the universe of objects into classes; inheritance relations among classes, and mechanisms such as inheritance and genericity to support reuse of specification modules and their top-down, incremental development; an expressive graphic representation of classes in terms of boxes, arrows, and connections, which allow to depict class instances and their components, information exchanges and logical equivalencies among (parts of) objects. TRIO+ was used successfully in the specification of software/hardware systems of significant architectural complexity, like pondage power stations of ENEL, the Italian electric energy board [24].

2.2 TRIO's model-parametric semantics

TRIO has been given a model theoretic formal semantics [15] in a fairly natural way. The essential difference from traditional interpretation schemata of logic languages [21] [9] consists of defining an *interpretation structure* that associates a distinguished set, the temporal *domain*, denoted as T, to each term of temporal type. Any formula F is then evaluated in a generic element i of the temporal domain and its truth value is denoted as $S_i(F)$. The definition of function S_i is however more complex if we want to evaluate formulas with reference to finite (time) domains, a possibility that is of primary interest for the definition of effective methods for performing system validation and for the implementation of tools based on them. In [27] a *model-parametric semantics* for TRIO is defined, which may refer to any finite or infinite temporal domain. We omit here the detailed description of the model parametric semantics, and illustrate it on a simple example.

Example 3 The following formula specifies that if the event represented by the atomic predicate P ever occurs, it repeats itself every three time units

$$Alw (P \leftrightarrow Futr (P, 3))$$

Suppose the formula is evaluated at time instant 4 of the temporal domain shown in Figure 1.

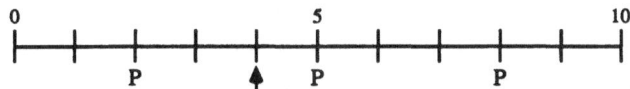

Figure 1: Structure S for the evaluation of the TRIO formula: Alw(P ↔ Futr(P, 3)). The arrow points to the instant in which the formula is evaluated. P is supposed true at times 2, 5, 8, and false elsewhere.

Intuitively, Figure 1 represents the portion of a possibly infinite behaviour that satisfies the specification, but the formula cannot be satisfied in that structure if the semantics assigns conventional values to elements of the formula outside the temporal domain. In fact, under the conventional assumption that everything is false outside the finite temporal domain, the specification requires P to be identically false in [0..10], while if P is assumed true outside the temporal domain then it is required to be identically true in [0..10]. In both cases, the finite window of Figure 1 would not describe a legal behaviour according to the given specification.

o

Model parametric semantics imposes restrictions on the evaluability of formulas implicit in the definition of the semantics with respect to the chosen domains. We define the notion of *evaluability of a formula* in a structure adequate to it. If it is not possible to evaluate the formula without any reference to time points external to the temporal domain, the formula is considered not evaluable in that domain, and its truth value is immaterial. In addition, for formulas containing quantifiers, the set of values that can be assigned to the quantified variables does not necessarily coincide with its type, but must be suitably restricted, in order to prevent exiting the temporal domain when evaluating the formula. Thus evaluability of a formula with respect to a structure does not depend just on the types of entities that the structure assigns to the components of the TRIO alphabet–i.e. on the adequacy of the structure–but also on their values.

The formula of Example 3 can be rewritten in the unabbreviated form

$$F = (P \leftrightarrow Futr\,(P,\,3)) \wedge \forall s(s>0 \rightarrow Futr\,(P \leftrightarrow Futr\,(P,\,3),\,s)) \wedge$$
$$\forall t\,(t>0 \rightarrow Past\,(P \leftrightarrow Futr\,(P,\,3),\,t)).$$

Assuming that variables s and t belong to [1..10], the formula is evaluable at time instant 4 of the temporal structure S of Figure 1, with the following evaluation domain for the quantified variables

Dom(F, 4, S) = {(s, t)∈ [1..10]² |{4, 7, 4+s, 4+s+3, 4-t, 4-t+3}⊆[0..10] }=

= { (s, t)∈ [1..10]² | 1≤s≤3 ∧ 1≤t≤4 }

Such evaluation domain is shown in Figure 2. Notice that F is satisfied in the structure, because the evaluation of subformula P ↔ Futr (P, 3) is prevented at time instants that do not belong to the interval [0..7].

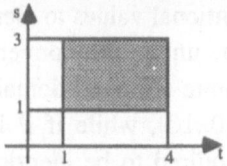

Figure 2: Evaluation domain for quantified variables of formula Alw(P ↔ Futr(P, 3)).

A specification formula F is said to be *temporally satisfiable* for a given interpretation if there exists a time instant i∈ T for which $S_i(F)$=true. F is said to be *temporally valid* in the given interpretation if and only if $S_i(F)$=true for every i∈ T; it is *valid* if it is temporally valid in every syntactically adequate interpretation.

In the remainder of the paper, when referring, even indirectly, to the evaluation of a TRIO formula *F* at instant *i* of a bounded time domain, it will be implicitly assumed that *F* is evaluable at time *i*.

3 TRIO's histories as test cases

In the present section we illustrate two basic methods useful for validating both TRIO specifications and implementations of the specified systems. These are a history generator, which verifies the satisfiability of a TRIO specification by constructing models of the formula or parts thereof, and a history checker, which determines whether a specification formula is satisfied in a given interpretation structure. Next we define a notion of test cases which is adequate to perform functional testing of real-time systems specified in TRIO, and show how the history generator and checker can be effectively used, both to generate test cases and to support the testing activity itself.

3.1 Satisfiability of a TRIO formula

In order to validate requirements expressed as TRIO specifications, one may try to prove the satisfiability of the formula by constructing a model for it. In this view, some parts of the temporal structure to be constructed are assumed to be known–namely the temporal domain T, the domains for variables, and the interpretation of time independent predicates, which describe its static configuration. Given a system specification as a TRIO formula and the static part of a structure adequate to its evaluation, the construction of the remaining parts of the structure determines the dynamic evolution of the modeled system: the events that take place and the values assumed by the relevant quantities.

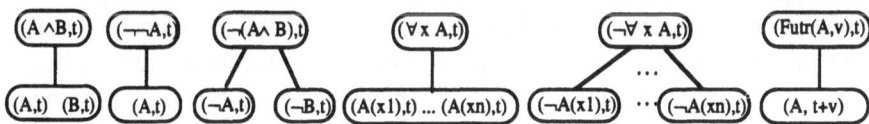

Figure 3: Pictorial description of the decomposition of formulas by the tableaux algorithm.

If the interpretation domains for variables and the temporal domain T are all finite, the satisfiability of a TRIO formula is a decidable problem and effective algorithms to solve it can be defined. In [11] we presented an algorithm that, under the hypothesis of finite domains, determines the satisfiability of a TRIO formula using a constructive method. The principal steps of the algorithm for verifying satisfiability are schematically depicted in Figure 3. The specification formula is associated with a time value t that indicates the instant where it is assumed to hold; then a decomposition process is performed which transforms a formula into a set of simpler formulas, associated to possibly different instants, whose conjunction is equivalent to it. The decomposition uses well known and intuitive properties of the propositional operators, and treats universal (resp. existential) quantifications as generalized conjunctions (resp. disjunctions); it ends when each set of subformulas, called a *tableau*, contains only literals: every tableau that does not contain any contradiction (i.e. a literal and its negation) provides a compact representation of a model for the original formula, and thus constitutes a constructive proof of its satisfiability.

The study of the complexity of the above algorithm, performed in [11], shows that it is exponential in the number of existential quantifications, with the cardinality of the domains of the quantified TI variables appearing as the base of the exponential, while it is hyper-exponential with respect to the number of universal quantifications, with the cardinality of the domain of the quantified variables appearing as the exponent.

3.2 Checking a history against a TRIO formula

Executability of TRIO formulas is also provided at lower levels of generality: the tableaux algorithm can be adapted to verify that a given temporal evolution of the system (called a history) is compatible with the specification. In the following, this operation will be called *history checking*. The history checker is implemented through a specialization of the tableaux algorithm, whose principal steps are depicted in Figure 4. Now each tableau includes only *one* formula associated with a time instant at which it must be evaluated. An and/or tree is built and the literals obtained in the leaf nodes are checked against the history.

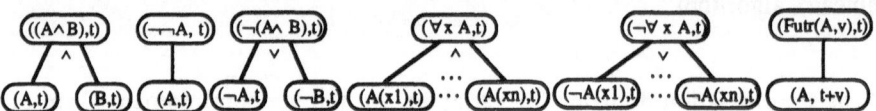

Figure 4: Pictorial description of the decomposition of formulas by the history checker algorithm.

The history checking algorithm has been shown [11] to be exponential with respect to the dimension of the formula, i.e. the number of its quantifications and binary operators. It should however be noticed that the cardinality of the domains of the quantified variables appears now as the base of the exponential, not as the exponent like in the more general algorithm for deciding satisfiability of formulas. In other words, for a given formula the complexity of history checking algorithm is a polynomial function of the cardinality of the evaluation domains. This result might be considered discouraging; however we point out that the dimension of the formula is usually relatively small with respect to the cardinality of the evaluation domains, so that, in the total complexity of checking the specification formula, the exponential factor has a limited influence with respect to the polynomial one. This was confirmed by experimental use of the prototype in cases of practical interest [4].

It should be now clear that any history can be interpreted as a test case, since it represents an evolution trace of the modeled system. In this view, the two interpreters (i.e. the algorithm for verifying satisfiability and the history checker) can become the core of a tool that allows to systematically generate test cases for the specified system and validate its responses to the provided stimuli. This intuitive and somehow idealized picture is however complicated by two fundamental facts. First, since most real-time systems are reactive in nature, a testing activity must be accomplished by providing adequate stimulations and observing the system's reactions. Yet a history, as defined so far, is just a collection of elementary events and values characterizing one dynamic evolution of the system, so the problem arises to discriminate among input and output signals in order to use a history for a testing activity. This problem is dealt with in Section 4. Second, since each leaf tableau generated by the algorithm for verifying satisfiability corresponds to a history of the specified system it is clear that such algorithm might generate, for some specification formulas, a very large number of histories, much more than those needed for performing an effective testing. As stated in the introduction, this can derive from the nondeterministic nature of the specified system: methods for dealing with this case will be provided in Section 5. Another cause of the exceedingly large number of histories could be the great variety of (not necessarily nondeterministic) behaviours exhibited by the specified system; for instance, each distinct history might correspond to a different combination of input stimuli and consequent system reactions. Sections 6 and 7 present criteria to cope with such a kind of complexity.

3.3 Test cases

Generating a test case for a TRIO specification F means building an execution H of the system to be verified, such that H satisfies F. The description of an execution H consists of the description of all events occurring in H. An event describes the fact that at a given time instant some circumstance holds.

Definition 1 Event and Execution

An event is a pair $<L,i>$, where

- L is a literal
- i is a time instant, belonging to the time domain.

An execution (or, synonymously, a history) is an event set which, for every literal L and instant i, does not include both $<L, i>$ and $<\neg L,i>$.

A history is said to be *complete* if it contains a unique truth value for each predicate of the formula at each instant of the time domain.

                                                                                    o

Thus, an event consists of a truth value for a predicate coupled with a time instant. For instance, referring to Example 1, the pair $<in,1>$ describes the event of receiving a message at time 1; the pair $<\neg out,1>$ describes the event that no message is sent at time 1. Three executions for Example 1 are in Figure 5.

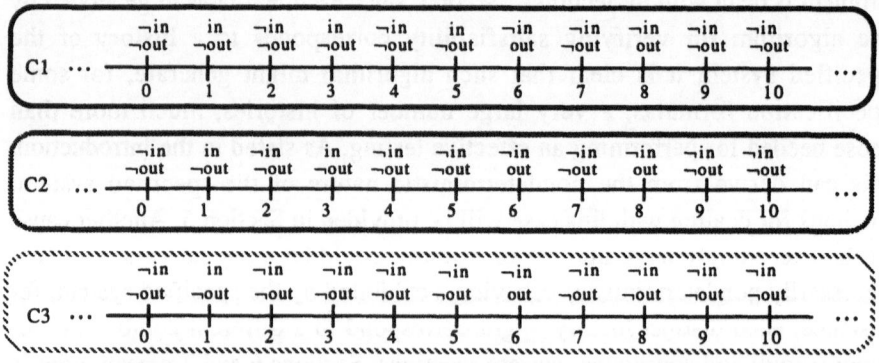

Figure 5: Three executions for the line of example 1.

An execution H satisfies a given formula F iff there exists *at least one time instant* where F is true, regardless of what happens in the other time instants. Thus, all three executions in Figure 5 satisfy the formula of Example 1. As to $C1$, F is satisfied in all time instants except $i=5$. In fact, after event $<in,5>$ (a message is input at time 5), event $<\neg out,10>$ occurs (no message is delivered at time 10). Executions $C2$ and $C3$ satisfy the formula in all time instants. This leads to the following definition.

Definition 2 Test Case

A *test case C* for a *formula F* is a complete history describing a system behaviour in an execution that satisfies *F*. A test case for a formula *F* is said to *refer* to time *i* if it makes *F* true at time *i*.

o

A test case can refer to more than one time instant. For instance, test case *C1* in Figure 5 refers to all time instants, except *i=5*; test cases *C2* and *C3* refer to all time instants.

4 Input and Output in TRIO's histories

The TRIO language *per se* does not provide a means to characterize input events (data or commands introduced into the system) and output events (data or signals generated by the system). Thus, in a test case, defined as a set of events, there is no way of discriminating input and output events. During the testing activity, however, it is essential to understand which events are actions from the external environment on the system, and which ones are actions from the system on the external environment. Consider, for instance, a modification to Example 1, where the messages on the transmission line are counted. The TRIO specification of the system is

$$\text{in} \wedge \text{count(num)} \leftrightarrow \text{Futr(out} \wedge \text{count(num+1)),5)}$$

In this formula the *in* predicate is always associated with input events, the *out* predicate is always associated with output events, while the *count* is sometimes associated with input and sometimes with output events (the value of *count* is an input when *in* occurs whereas its value is an output when *out* occurs).

Unfortunately, a test case generator is not able to perform automatically such a classification, so the problem is solved in a semiautomatic fashion. The user can partition the predicate names of the specification formula as input, output, and input/output predicate names. For graphical convenience, here we will underline input predicate occurrences (e.g. in)and outline output predicate occurrences (e.g. out). Both graphical versions (e.g. count - count) will be used for input/output predicates, according to the input vs. output nature of each predicate occurrence. The result of this operation on the counting line specification would be the following annotated formula

$$\underline{in} \wedge \underline{count}(num) \leftrightarrow Futr(out \wedge count(num+1)),5)$$

The test case generation tool applies the history generator to this formula, propagating the different marking of input and output predicate names throughout the tree, down to the terminal nodes. Figure 6 shows leaf nodes obtained for this example. In such nodes the events are marked, and can therefore be classified as either input or output events.

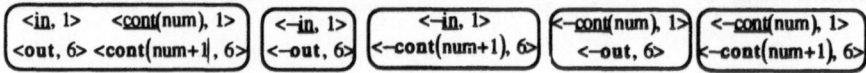

Figure 6: Input and output events for the counting line example.

In the example, *all* predicate name occurrences were classified on the original formula before running the history generator. Thus, *all* the events in the terminal nodes are classified accordingly. However, in some situations it may be more useful not to mark initially the occurrences of input/output predicates, and to mark the resulting events *after* executing the history generator. This can greatly help in reducing the presence of events classified both as input and output in the same terminal node, symptom of lack of understanding in the classification given on the formula. Should this happen, the test case generator returns such events together with a warning notice.

5 Coping with nondeterminism

A TRIO formula *F* may allow a number of different executions corresponding to the same input data, i.e. it may specify a nondeterministic system behaviour. Thus, different executions under the same input data may yield different output data, as shown in the following example.

Example 4 As a modification of Example 1, suppose that two communication lines, *l1* and *l2*, exist for transmitting the messages. The TRIO specification of the system is

$$\underline{in} \leftrightarrow (Futr(out(l1), 5) \vee Futr(out(l2), 5))$$

stating that any input message is nondeterministically output on any of the two lines

o

Nondeterminism may cause some testing problems [30]. In fact, the execution of the system under the input data of a given test case C may yield a history C', different from C, though satisfying the specification. With reference to Example 4, suppose that C contains the input event $<\underline{in},1>$ and the output event $<out(l1),6>$. If in an actual execution C' the input message at time 1 is enrouted on line $l2$, then the output event $<out(l2),6>$ occurs, instead of $<out(l1),6>$. If all other events of C and C' coincide, then C' satisfies the specification as well.

This problem can be solved in a simple and effective way by using the *history checker* (see section 3). Once the output events produced in the execution C' of the system under the input data of some test case C differ from the output events of C, then by applying of the history checker to C' we can check whether the actual execution C' still satisfies the specification.

6 Partial and elementary test cases

Building a test case requires that *every* predicate must be given a truth value at *every* time instant. This implies that if the temporal domain is unbounded, an unbounded set of events must be built. Even when the temporal domain is finite, in a real-life system specification the amount of predicates to be provided with truth values may be so huge to definitely make impractical the explicit construction of a whole test case. However, when building a test case we are often only interested in the occurrence of a *limited number of events*, in which the relevant information about a specific system behaviour of interest is embodied. Thus, to determine whether the system under study complies with the specification, we do not want to generate *all* the events of a test case, but only a *subset* thereof. This yields a dramatic reduction in the number of events to be generated, and consequently a remarkable improvement in test case generation effectiveness. The significance of the generated event sets is also increased, in that irrelevant events are not present.

To this end, recall that a test case need not satisfy a formula at *all* time instants. Actually, it is sufficient that there exists at least *one* time instant in which the specification is satisfied. For instance, with reference to Example 1, consider the following event sets

$$PC1 = \{<in, 1>, <in, 3>, <in, 5>, <out, 6>, <out, 8>\}$$

$$PC2 = \{<in, 1>, <in, 3>, <out, 6>, <out,8>\}$$

$$PC3 = \{<in, 1>, <out, 6>\}$$

A test case can be obtained from PC1 by completing it in whatever way, e.g. by adding any set of events in such away that in all time instants all predicates have a truth value. The event set *PC1* is said to be a *partial test case*, according to the following definition.

Definition 3 Partial test case

A partial test case is a history such that all complete histories that are supersets of it are test cases. A partial test case is said to refer to time *i* iff so does each test case including it.

o

For instance, *PC1* and *PC2* refer to time 1 and 3, *PC3* refers to time 1.

Generating partial test cases remarkably reduces the effort needed to build complete test cases. However, a partial test case may contain events that are not strictly necessary for it to satisfy the specification. For instance, a test case *is* a partial test case; further, consider the above reported partial test case *PC1*: the deletion of event *<in,5>* from it yields the partial test case *PC2*, which refers to the same time instants 1 and 3. But, if any of the events of *PC3* is removed, what we obtain is not a partial test case any longer. For instance, suppose to eliminate the event *<out,6>* from *PC3*, obtaining the event set *{<in,1>}*. It is now possible to complete it, i.e. to give a truth value to all remaining predicates in all time instants, in such a way to obtain an event set that is not a test case, e.g. by taking *in* as always true, and *out* always false.

This leads to the following definition.

Definition 4 Elementary test case

A partial test case *EC* is said to be an *elementary test case*, iff none of its proper subsets *ECS* (i.e. such that $ECS \subset EC$) is a partial test case.

o

The presence of elementary test cases when building partial test cases and therefore test cases is mandatory, as shown by the following theorem.

Theorem 1 Elementary and partial test cases

A partial test case referring to time *i* always contains a (possibly empty) elementary test case referring to the same time *i*.

o

Since elementary test cases are also partial test cases, in order to generate a test case it is sufficient to generate an elementary test case. Then, if necessary, the elementary test case can be extended to a test case in a trivial way, by using "no care" values. This yields a second significant reduction in the amount of events to be generated.

Next, we provide two basic operations that will be used to build complete test cases starting from a limited number of events. These notions constitute the basis for the test case generation methods and algorithms presented in Section 6.3.

6.1 Shift

In order to minimize the amount of information to be generated, when building the set of elementary test cases it is not necessary to generate them all explicitly. Instead, the knowledge of the elementary test cases referring to one time instant is sufficient to determine the elementary test cases referring to any other time instant. Thus, the number of events to be generated is further reduced. This is made possible by the Shift operation, which is defined next.

Definition 5 Shift

An event set *ES'* is said to be the Δ-*Shift* of an event set *ES"* iff for each event $<P,i> \in ES'$ there exists an event $<P,i+\Delta> \in ES"$ and vice versa. This will be denoted by $ES"=Shift(ES',\Delta)$.

o

The Shift operation transforms test cases into test cases, as stated by the following theorem.

Theorem 2 Δ-Shift of partial and elementary test cases in bounded domains.

Consider a structure S including a bounded time domain T; consider instants $i, j \in T$, and a formula F having q quantified variables $x_1, \ldots x_q$; assume that $Dom(F,i,S) = Dom(F,j,S) = D(x_1) \times \ldots \times D(x_q)$. If an event set ES1 is a partial (res., elementary) test case for F at i, then the event set

$$ES2 = Shift(ES1, j-i) - \{<L, k> \mid k \notin T\}$$

(i.e., the event set obtained by shifting ES1 and then removing the events, if any, referring to points outside T) is a partial (res., elementary) test case for F at j.

o

The above theorem has an important constructive consequence. It states that, once we know the elementary test cases referring to a given instant i, we can build the elementary test cases referring to any other time instant $i+\Delta$, by simply Δ-Shifting the former ones. Thus, assume the set of partial test cases referring to a time instant i of a bounded time domain is known. Then the set of partial test cases referring to time instant $i+\Delta$ (where F is also evaluable) is obtained by:

1. Δ-Shifting each partial test case referring to time i;

2. removing from the sets thus obtained those events that refer to a point outside the time domain.

6.2 Composition

Consider the partial test case *PC2* presented in the initial part of section 6. In all test cases that contain it the system is tested under two messages, respectively introduced at times 1 and 3. Actually, *PC2* can be seen as the union of the event sets

$$EC1 = \{<in,1>,<out,6>\}; \qquad EC2 = \{<in,3>,<out,8>\}$$

i.e. of the elementary test cases containing respectively event *<in,1>* and event

<in,3>. This suggests the use of elementary test cases, or, more generally, of partial test cases as the building blocks upon which all the partial test cases can be derived.

Definition 6 Composition

The *composition* of two event sets is their union, provided that it does not contain contradictory events, i.e., two events of the form <L, i> and <¬L, i>.

o

Thus, the partial test case *PC2* is said to be the composition of the elementary test cases *EC1* and *EC2*.

Partial test cases can be generated by composition of elementary test cases or of other partial test cases, as asserted by the following theorem.

Theorem 3 Composition of partial test cases

The composition *PC* of two partial test cases *PC1* and *PC2* is a partial test case. It refers to the union of the time instants to which *PC1* and *PC2* refer.

o

Composition allows to verify a system with respect to as many interesting situations as needed. For instance, in order to verify the communication line of Example 1 under the hypothesis that several messages are transmitted, each one at a different time instant *i*, we can generate a suitable partial test case as the composition of the elementary test cases{*<in,i>,<out,i+5>*}.

Theorem 4 Composition and formulas of type Alw(F)

The composition of a set of elementary test cases for formula *F*, one for each time instant in the time domain (where *F* is evaluable), is a partial test case such that every test case containing it makes the formula Alw(F) true.

o

This property provides a significant result when a given formula must be satisfied on all the time domain, as requested by the *Alw* construct.

Summarizing, based on the properties of elementary and partial test cases, the following procedure can be applied, in order to minimize the amount of information to be explicitly derived.

1.   Generate the elementary test cases referring to a time instant;
2.   Shift such elementary test cases, by a given set of Δ values;
3.   Compose the obtained test cases, to build partial test cases;
4.   Complete the obtained partial test cases, in order to build test cases.

6.3 Partial Test Case Generation

Based on the results of the preceding subsections, the main goal of a test case generation algorithm is to derive the set of elementary test cases for a given time instant. To this end, the history generator (see Section 3) can be significantly useful, as stated by the following theorem.

Theorem 5 history generator and partial test cases

Given a TRIO formula F and an evaluation time instant i, each unmarked leaf of the tree generated by the history generator applied to F in i contains a partial test case for F, referring to time i.

o

For instance, if we choose the evaluation time 1 for the counting line example of Section 4, we obtain the partial test cases (they are also elementary test cases) contained in Figure 6.

Actually, it is easy to see that the application of the history generator does not always yield elementary test cases in the leaves of the generated tree. In general, as stated by the above Theorem 5 the leaves of the tree contain partial test cases and they might not be elementary test cases. For instance consider the formula $A \vee (A \wedge B)$ where A and B are atomic predicates. Under the evaluation time 1 the history generator generates the tree in Figure 7.

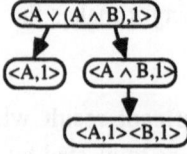

Figure 7: Non elementary test case produced by the history generator.

The event set $\{<A,1>,<B,1>\}$, generated in one of the leaves, is a partial test case, but it is not an elementary test case: it contains the elementary test case $\{<A,1>\}$, generated in the other leaf. The possibility of generating some partial test cases instead of elementary test cases has no relevant consequence on verification, except for the loss of some minimality characteristics, owned by elementary test cases only. In fact, the history generator generates a "complete" set of partial test cases for a given time instant, as the following theorem precisely states.

Theorem 6 Completeness of the generated set of partial test cases

Given an elementary test case EC for a TRIO formula F referring to time instant i, there always exists a partial test case PC referring to i, generated by applying the history generator to F with the evaluation time i, such that PC contains EC.

o

7 Test case selection criteria

In the previous sections we introduced a set of basic properties of events that permit the definition of effective methods for generating test cases. In the present section we provide test selection criteria, which allow the user to further reduce the complexity of test data generation activity by choosing those test cases that are likely to be most useful in detecting errors. Thus a test selection criterion is a *heuristic technique*, which should however provide a reasonable degree of confidence in the correctness of the system.

Test selection criteria, when applied to sequential software, exploit the structure of the programs. For instance, they provide strategies to deal with control or data flow in a program. Here, we want to define similar strategies on specifications, instead of programs. Thus, in this section, we will briefly motivate some of the test selection criteria which might be adopted in the production of test cases for non-trivial systems. The criteria refer to the various TRIO constructs used to build a system specification, so they can be inductively combined when applied to complex formulas. The presentation of the criteria will be sketchy and informal: we will rely on the reader's intuition, and present an overall example of their application at the end of the section. Formal definitions and generation procedures can be found in [20].

1 **Sometimes** An execution satisfies the TRIO formula Som(F) iff at least one time instant exists in which the formula F is satisfied. A formula like this can be used as a *part* of a system specification, when requiring that a condition eventually holds. As to test case generation, histories for *Som(F)* are a superset of those for F. Thus, the frame and the results we obtained in the previous sections can be directly exploited.

2 **Conjunction and Universal Quantification** Conjunctions and universal quantifications can be dealt with in similar ways, since a universal quantification can be interpreted as a generalized conjunction, which considers all the admissible values for the quantified variable. To achieve a complexity reduction, the subformulas deriving from the quantification can be separated. Furthermore, test case generation for a universal quantification (resp. a conjunction) can be transformed into test case generation for an existential quantification (resp. a disjunction).

3 **Always** The *Alw* construct can be treated using methods similar to those devised for universal quantification, since the construct is defined by means of a universal quantification of the temporal variable. Thus searching for partial test cases for *Alw(F)* transforms the original specification into *Som(F)* in the history generator. The conditions to satisfy the latter are far weaker than for the former: an execution might satisfy F in only one time instant. However, useful insights can be obtained from this approach. An intuitive rational for points 2 and 3 above can be deduced by a comparison with white-box testing techniques of imperative programming languages: when a *while* statement is tested, usually only one or two paths are considered, corresponding to skipping the statement or executing it once, whereas all the other cases (repeating the cycle more than once) are ignored.

4 **Implication** An implication, $A \rightarrow B$, is verified when A is false, independently of the truth value of B, or when B is true, independently of the truth value of A. Thus, if A and B are atomic predicates, $\{<\neg A,i>\}$ and $\{<B,i>\}$ are elementary test cases for $A \rightarrow B$. There is intuitive evidence, however, that the most significant test cases are obtained when A is true, i.e. the ones containing the pair $<A,i>$, $<B,i>$. Therefore, when generating test cases, an implication $A \rightarrow B$ can be usefully transformed into a conjunction $A \wedge B$.

7.1 An example of application of test case selection techniques

In this section we show the use of the above test case selection techniques, with reference to the specification of a sluice gate given in Example 2 and a time instant conventionally assumed as 0. The system specification is

Alw((*goUp*) ∧ (*moveDown*) ∧ (*goDown*)
∧ (*moveUp*) ∧ (*danger*))

A possible procedure for generating test cases is here outlined (the reader should keep in mind that test cases are selected on the basis of common sense, thus they should be evaluated according to the *likelihood*–not certainty–of showing possible errors in the implementation.

1 The *Alw* construct is transformed into the *Som* construct (criterion outlined in point 3 above)

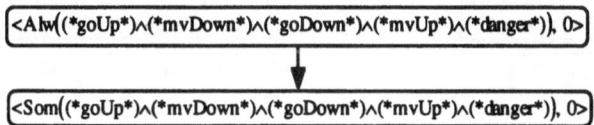

2 For each of the conjoint subformulas a different node is built (criterion 2 above)

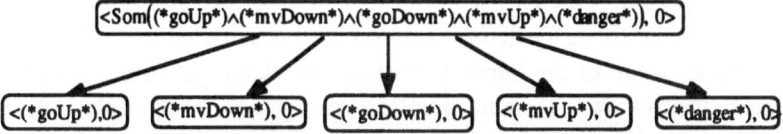

When the procedure ends, it must be checked whether the partial test cases generated for each of the new nodes are partial test cases for the subformulas contained in the other nodes; this can be done using the history checker.

3 The user chooses to transform some of the implications into conjunctions (point 4 above). For instance, the *goUp* and *danger* formula can be expanded as follows:

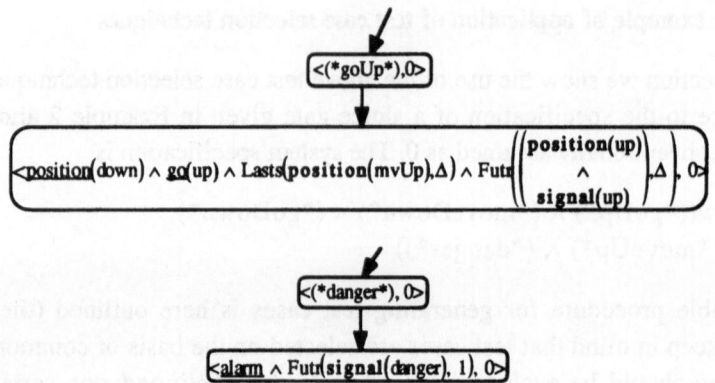

4 Developing the node derived from *(*goUp*)* by building a separate node
   for each of the subformulas (point 2) we obtain the four following nodes
   (terminal nodes are framed in bold).

<*position*(down), 0>	**<Lasts(position(mvUp),Δ), 0>**
<go(up),0>	**<Futr(position(up) ∧ signal(up), Δ), 0>**

5 As to node *<Lasts(position(mvUp),Δ),0>*, recall that the definition of the
   Lasts operator contains a universal quantification, since Lasts(A,t) is
   defined as ∀t'(0<t'<t → Futr(A,t')). Again, we drastically reduce the
   number of test cases by giving up testing all t'. For instance, here it can
   be significant to verify that *position(mvUp)* is satisfied in the first and the
   last time instant of the interval in which the sluice gate moves. Thus only
   <position(mvUp), 1> and<position(mvUp), Δ-1> are selected in the
   expansion of<Lasts(position(mvUp), Δ), 0>.

6 The node with the formula *<Futr(position(up) and signal(up),Δ)),0>* is
   developed according to the history generator method

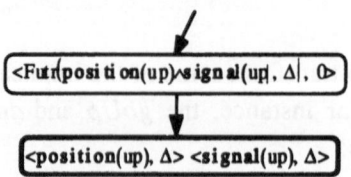

7 At last, by means of the history checker, we check that the obtained set of
   events is also a partial test case for the other nodes generated in step 2.

At the end of the procedure we obtained the partial test case

{<<u>position</u>(down),0>,<<u>go</u>(up),0>,<position(mvUp),1>, ...,

<position(mvUp),Δ-1>, <position(up),Δ>,<signal(up),Δ>}

whose significance can be evaluated on the basis of an intuitive analysis.

8 Outline of a test case generation tool based on TRIO

The final goal of the present research is to build an interactive test case generation tool. As shown in the previous sections, the activities of such a tool cannot be fully automated. The human guidance to the tool reduces the complexity inherent to test data generation and allows to generate *significant* test data. Anyway, the intervention of the user is rather restricted: the tool automatizes tedious activities, such as, for instance, creating and exploring the tree generated by the history generator. Figure 8 depicts a high-level view of the tool and its interactions towards the external world.

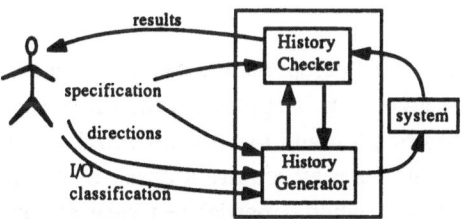

Figure 8: High–level view of the tool for test case generation.

From an external viewpoint, the test case generation tool receives the TRIO formula that specifies the system. This is essentially an input to the tool. However, it might be the case that during testing we ascertain that the specification allows behaviours that would not be allowed by the requirements. As a consequence the specification itself has to be modified.

The user interacts with the tool in two different ways.

1 The user classifies predicate names and their occurrences as inputs and outputs. The test case generation tool returns (partial) test cases, where

the events are marked according to the user's classification, possibly together with a warning notice (see Section 4).

2 The test case selection criteria are applied, according to the specification
   structure. The tool must allow the user to guide the test case generation. It
   is the user's care to interactively choose the activities that can be
   performed automatically by the tool. In particular, the user is required to
   provide choices for each step in the application of test case selection
   criteria of Section 7. This is implemented by means of *pop-up* menus, appearing when an operator (e.g. *Som*, *Alw*, etc.) is selected by means of a
   mouse. Each menu provides all the testing strategies referring to the
   selected operator.

The system under test is stimulated using the input part of the test cases
generated by the tool. The resulting executions (input and output events) are
then submitted to the tool, in order to check whether they satisfy the
specification (Sections 5 and 7).

From an internal viewpoint, the test case generation tool is composed of two
parts.

1 *History generator*. This component is based on the algorithm presented in
   section 3.1. The algorithm is applied to the given specification under the
   input/output classification and the options provided by the user. Some of
   the produced event sets may require a check by the history checker. For
   instance this happens when a partial test case for one conjoint must be
   shown to be also a partial test case for the other conjoints. The produced
   test cases are then input to the system to be verified.

2 *History checker*. Actually, this component is a tool in its own right. Here
   it is totally reused, without modifications. Its aim is to check whether the
   event sets generated by the other tool components or by system executions
   satisfy the specification. If they do not, a message is sent to the user.

Prototype versions of the history generator and of the history checker have
already been implemented. Their integration into the whole tool as outlined in
Figure 8 is presently under development.

9 Concluding remarks

We proposed a technique for the (semi)automatic derivation of test cases for real-time systems from formal specifications coded in TRIO. At the best of our knowledge, the state of the art regarding tools for the automatic generation of functional test cases of real-time systems is rather unsatisfactory. In fact, only a few examples of such tools exist, which are based on the finite automata formalism [7], which is often too simplified to deal satisfactorily with complex systems and–being operational in nature–is not suitable for the high level specification of system requirements. Thus, we are confident that our technique contributes to filling up a critical hole in the field of the rigorous verification of real-time systems.

Essentially, the proposed technique consists of generating *models* (*histories*) of the axioms specifying the system by means of a suitably specialized interpreter of the language TRIO. Such models are used as test cases by separating input and output events. When generating and checking histories we assume finite variable domains, including the time domain. The finiteness hypothesis is essential, since the TRIO language is undecidable if infinite domains are assumed. It must be pointed out, however, that this hypothesis does not impose a severe restriction on the practical usefulness of the method, since the tool allows the user to change easily the cardinality of all domains. This possibility, combined with the model parametric semantics of TRIO, provides a view of the time domain as a finite–but variable–width window on a possibly infinite behaviour of the analyzed system. The problem of testing nondeterministic systems [30] has been solved by means of an *a posteriori* verification performed by the history checker. Several techniques to drastically reduce the size of test cases have been proposed.

Our on-going research in this field is directed by the following guidelines.

- Developing more test selection criteria. In fact, the structure of our tool is *open*, in the sense that it may be complemented with new features corresponding to new techniques. These may be either of general applicability or tailored to particular classes of systems. For instance, if one developed a specialized heuristic technique that is suitable to appropriately select test cases for–say–nuclear plant control systems, this should be implemented as a new module of the whole tool to be included only when developing systems of such a category.

- The algorithms for generating and checking histories can be executed in reasonable time when applied to specifications that are manageable by an average human being, but their complexity is nevertheless exponential in the number of quantification of the analyzed formula. Hence the principal means for scaling–up the proposed method to real-life, industrial-size cases will be the development of techniques and tools for *testing in the large*. In fact, as we mentioned earlier, TRIO is just the kernel of a complete specification language, TRIO+, which allows the specification of complex systems thanks to modularization mechanisms. Accordingly, the techniques we presented in this paper are techniques for *testing in the small* in the sense that they are well suited to test systems whose properties are specified through simple TRIO formulas. The situation is much the same as in white-box testing where there are techniques that are suitable to test small pieces of code, such as the path coverage criterion, but that cannot be applied directly to the testing of large programs. Similarly, we need techniques to combine and/or simplify test cases generated for small modules into test cases for larger systems. Furthermore, since TRIO+ allows writing specifications that are only partially formalized [25], it is important to be able to integrate test cases that are edited by hand on the basis of informal requirements with test cases that have been generated automatically on the basis of formal specifications.

- Another analogy with white-box testing leads to techniques and tools to measure how much of the system has been tested so far: whereas in white-box testing we may be interested in knowing which portions of the code have been executed at least once, here we may be interested in knowing which axioms (i.e., requirements) have been experienced at least once.

The above issues are exploited in a forthcoming paper. Finally, we mention that a whole *environment* is associated with TRIO and TRIO+. Such an environment presently includes:

- A graphical, interactive editor of TRIO/TRIO+/TRIO*[1] specifications.

---

[1] TRIO* is a further language based on TRIO which is not considered in this paper [5].

- An interpreter checking the satisfiability of TRIO formulas. This is the kernel of all semantic tools associated with TRIO, included the history generator and the history checker mentioned in Section 8 [16].

For a less immediate future other tools are scheduled to help system implementation starting from TRIO specifications and formal correctness proofs of such an implementation [10]. We expect complementary benefits from tools supporting formal correctness proofs and tools supporting testing activity [16].

References

[1] G. Bernot, M. C. Gaudel, B. Marre. Software testing based on formal specifications: a theory and a tool. *Software Engineering Journal* , November 1991.

[2] L. Bougé, N. Choquet, L. Epibourg, M. C. Gaudel. Test Case Generation from Algebraic Specifications Using Logic Programming. *The Journal of Systems and Sofware* Vol. 6, 1986, pp. 343-360.

[3] A. Celentano, S. Crespi-Reghizzi, P. DellaVigna, C. Ghezzi, G. Granata, F. Savoretti. Compiler testing using a sentence generator. *Software-Testing and Experience* Vol. 10, 1980, pp.987-918.

[4] A. Coen, A. Morzenti, D. Sciuto. Specification and verification of hardware systems using the temporal logic language TRIO. *In Computer hardware description languages and their application,* D.Borrione, R. Waxman Eds., IFIP, North-Holland, Marseille, France, April 1991, pp. 43-62.

[5] E. Corsetti, A. Montanari, E. Ratto. Dealing with different time granularities in formal specifications of real-time systems. *The Journal of Real-Time Systems*, Vol. 3, 1991, pp. 191-205.

[6] O. J. Dahl, E. W. Dijkstra, C. A. R. Hoare. *Structured Programming*. Academic Press, 1972.

[7] B. Dasarathy. Timing constraints of real-time systems: constructs for expressing them, methods of validating them. *IEEE Transaction on Software Engineering*, Vol. SE-11, January 1985, pp. 80-86.

[8] R. A. DeMillo, M. W. McCracken, R. J. Martin, J. F. Passafiume. *Software Testing and Evaluation*. Benjamin/Cummings, Menlo Park, CA, 1987.

[9] H.B. Enderton. *A Mathematical Introduction to Logic*. Academic Press, New York, 1972.

[10] M. Felder, D. Mandrioli, A. Morzenti. Proving properties of Real-Time Systems through logical Specifications and Petri Nets Models. *Tech. Rept. 91-72 Dip. di Elettronica*, Politecnico di Milano, December, 1991.

[11] M. Felder, A. Morzenti. Validating Real-Time Systems by executing logic specifications in TRIO. *Proc. IEEE/ACM ICSE 14*, 1992.

[12] S. Fujiware, G. V. Bochman, F. Khendek, M. Amalou, A. Ghedamsi. Test selection based on finite automata. *IEEE Transaction on Software Engineering*, Vol. SE-17, No. 6, June 1991, pp. 591-603.

[13] J. Gannon, P. Mc Mullin, R. Hamlet. Data abstraction Implementation Specification and testing. *ACM Transactions on Programming Languages and Systems*, Vol. 3, No. 3, July 1981, pp. 211-223.

[14] C. Ghezzi, D. Mandrioli, A. Morzenti. TRIO, a logic language for executable specifications of realtime systems. *In 10th French Tunisian Seminar on Computer Science*, Tunis, May 1989, pp. 322-349.

[15] C. Ghezzi, D. Mandrioli, A. Morzenti. TRIO, a logic language for executable specifications of real-time systems. *Journal of Systems and Software*, Vol. 12, No. 2, May 1990, pp. 107-123.

[16] C. Ghezzi, M. Jazayeri, D. Mandrioli. *Fundamentals of Software Engineering*. Prentice-Hall International Editors, Englewood Cliffs, N.J., 1991.

[17] P. Jalote. Testing the Completeness of Specifications. *IEEE Transactions on Software Engineering*, Vol. SE-15, No. 5, May 1989, pp. 526-531.

[18] P. Jalote. Specification and Testing of Abstract Data Types. *Computer Languages*, Vol. 17, No. 1, 1992, pp. 75-82.

[19] R. A. Kemmerer. Testing Formal Specifications to Detect Design Errors. *IEEE Transactions on Software Engineering*, January 1985.

[20] D. Mandrioli, S. Morasca, A. Morzenti. Functional test case generation for real-time systems. *Tech. Rept. 92-8*, Politecnico di Milano, Dipartimento di Elettronica, Milano, Italy, January, 1992.

[21] E. Mendelson. *Introduction to Mathematical Logic*. Van Nostrand Reinhold Company, New York, 1963.

[22] S. Morasca, M. Pezze. Using high-level Petri nets for testing concurrent and real-time systems. *In Real-time systems: theory and applications*, H. Zendan Ed., British Computer Society, North Holland, September 1990, pp. 119-131.

[23] A. Morzenti, E. Ratto, M. Roncato, L. Zoccolante. TRIO: a Logic Formalism for the Specification of Real Time Systems. *Proc. Euromicro Workshop on Real Time*, IEEE, Como, Italy, 1989.

[24] A. Morzenti, P. SanPietro. TRIO+ an Object Oriented Logic Specification Language. *Tech. Rept. Research Report*, ENELCRA, in Italian, January, 1990.

[25] A. Morzenti, P. SanPietro. An object-oriented logic language for modular system specification. *Tech. Rept. 90-27*, Poiltecnico di Milano, Dipartimento di Elettronica, Milano, Italy, June, 1990.

[26] A. Morzenti, P. SanPietro. An Object Oriented Logic Language for Modular System Specification. *Proc. ECOOP'91*, Geneva, July 1991, Springer-Verlag.

[27] A. Morzenti, D. Mandrioli, C. Ghezzi. A Model Parametric Real-Time Logic. *ACM Transactions on Programming Languages and Systems* , 1992.

[28] D. H. Pitts, D. Freestone. The derivation of conformance tests from LOTOS specifications. *IEEE Transactions on Software Engineering*, Vol. SE-16, No. 12, December 1990, pp. 1337-1343.

[29] J. A. Stankovic. Misconcepts about real-time computing: A serious problem for next-generation computing. *IEEE Computer*, Vol. 21, No. 10, October 1988, pp. 10-19.

[30] K. C. Tai, E. E. Obaid. Reproducing testing of Ada taskprograms. *In Conference on Ada applications and environments*, IEEE, 1986.

[31] N. Wirth. Toward a Discipline of Real-Time Programming. *Communications of the ACM*, Vol. 20, No. 8, August 1977.

SPECIFICATION AND VERIFICATION
OF FAULT TOLERANCE

DESIGN FOR

DEPENDABILITY

Jens NORDAHL
Department of Computer Science, Technical University of Denmark
DK-2800 Lyngby, Denmark

Abstract

The concepts *design, correctness of design, failure mode* and *fault tolerance* are formally defined in terms of CSP. A systematic approach to verification of fault tolerance properties of designs is presented. The verification comprises a number of compositional proofs. It addresses safety and liveness properties of the design. As an example, a "cold stand by spare" fault tolerant design is formally defined and verified, using property oriented specifications of component failure assumptions and the compositional inference rules of CSP.

1 Introduction

Computers are increasingly used in critical systems, e.g., to control aircraft, monitor hospital patients, and execute banking transactions. Although the use of computers in such systems offer considerable benefits, it also poses serious risks in that we are increasingly vulnerable to their failures. Thus methodologies for analyzing dependability properties of systems, as well as for designing dependable systems, are needed.

The aim of this paper is to demonstrate how design and verification of dependable systems can utilize concepts and theories for communicating processes.

During the past fifteen years, extensive research in modelling, specifying and reasoning about the behavioural properties of communicating systems has appeared. As a result of this research, several theories have matured. We use the theory of communicating sequential processes, CSP, as described in [2].

Formal definitions of the system concepts *design* and *correctness of design* and of the dependability concepts *failure mode* and *fault tolerance* are given. The main result is a systematic approach to verifying the failure modes of a system with a given design and given component failure modes.

2   System concepts

This section defines general concepts such as system, specification, correctness and design. The definitions are based on the CSP concepts process, specification and satisfaction. The CSP processes with which we are concerned are the non-deterministic ones, as defined in chapter 3 of [2], i.e., processes represented as well formed triples (process alphabet, failure set, divergence set). The process alphabet is a set of events through which a process interacts with its environment. The failure set is a set of pairs, each consisting of a trace (a sequence of events) and a refusal set (a set of events). Each such pair describes one possible observation of the process, namely the sequence of events it has participated in, and the set of events it is currently refusing to participate in. The divergence set is a set of traces after which the system exhibits divergent, chaotic or unpredictable behaviour.

The objects whose dependability properties we want to argue about are systems. By a system we informally mean an object which interacts with its environment, and which at a given point in time bases its future pattern of interactions on the interactions of the past.

Definition 1 *A* system *is a process.*

We impose the restriction, that process alphabets must be subsets of a universal set, E, of events. This restriction is introduced only in order to allow for quantification over processes. The set E can be thought of as containing the alphabets of all processes, with which one is concerned. Let *SYSTEMS* be the set of all systems, i.e., the set of well formed triples, (process-alphabet, failure-set, divergence-set), whose alphabets are subsets of E:

$$SYSTEMS = \{(A, F, D) \mid (A, F, D) \text{ is well formed } \wedge \ A \subseteq E\}$$

A specification is a description of the intended behaviour of a system. In accordance with [2] we define

Definition 2 *A specification is a predicate with free varables tr and ref.*

A specification constrains the traces of a system by imposing constraints on the free variable *tr*. The free variable *ref* corresponds to the events refused by a system. Thus a specification prescribes readiness for certain events by imposing constraints on *ref*.

If the behaviour of a system conforms to what is prescribed in a specification, the system is said to implement the specification. In accordance with [2] we define

Definition 3 *Let P be a system and S(tr, ref) a specification. P implements S (denoted P sat S) if and only if*

$$\forall tr, ref \bullet tr \in traces(P) \wedge ref \in refusals(P/tr) \Rightarrow S(tr, ref)$$

The conceptual difference between systems and specifications in CSP is, that a system is an element in the computational model (described by *SYSTEMS*), while a specification describes properties, which may be possessed by more than one element in the computational model, i.e., a specification describes a set of systems. CSP's distinction between systems and specifications is not essential for our approach. Case studies have shown, that our approach can be based on formalisms, which do not have this distinction.

When analyzing a system, knowledge of the internal structure of the system is necessary. We envisage a system as consisting of a collection of components, and a design. The design determines how the components interact. Each component is itself a system, whose internal structure may be subject to analysis. Hence the total internal structure of a system is in this framework a hierarchy of system–component levels. We restrict ourselves to dealing with one level at a time.

A design relies on assumptions on the behaviour of the components to be used. These assumptions are an integral part of the design, and should therefore be explicitly stated (as specifications for the components).

Definition 4 *A design* is a pair

$$D = (Comb, (S_1, S_2, \ldots, S_N))$$

where Comb is a combinator of type

$$Comb: \quad SYSTEMS^N \rightarrow SYSTEMS$$

and where $(S_1, S_2, \ldots, S_N)$ *is an N-tuple of specifications.*

The combinator can be one of the operators of CSP (eg. $\|$, $\sqcap$, $;$, $\setminus$), a composition of these operators, or defined in terms of triples.

Because component specifications are considered part of a design, a design contains all the knowledge of the internal structure of a system, necessary for our analysis of the corresponding level in the system–component hierarchy.

If a system, whose internal structure is given in a design D, does not implement a given specification, we would like to know whether the incorrectness is caused by incorrect design or incorrect components. Therefore a notion of correctness of design, namely the **Dsat** relation, is introduced. Informally, a design is considered correct with respect to a specification, if the system resulting from the design and a collection of components implement the specification, assuming that the components implement the component specifications of the design.

Definition 5 *Let* $D = (Comb, (S_1, S_2, \ldots, S_N))$ *be a design and let S be a specification. The design D **Dsat** S if and only if*

$$\bigwedge_{i=1}^{N} (C_i \text{ sat } S_i) \quad \Rightarrow \quad Comb(C_1, C_2, \ldots, C_N) \text{ sat } S$$

As we expect, incorrectness can always be attributed to either incorrect components or incorrect design:

Theorem 1 *Let* $D = (Comb, (S_1, S_2, \ldots, S_N))$ *be a design and let S be a specification. Then*

$$\neg(Comb(C_1, C_2, \ldots, C_N) \text{ sat } S) \quad \Rightarrow \quad \neg(D \text{ **Dsat** } S) \vee \bigvee_{i=1}^{N} \neg(C_i \text{ sat } S_i)$$

This theorem follows from definition 5.

Figure 1: The unbounded buffer.

3 Failure modes

When reasoning about a systems behaviour, in case some of its components are behaving incorrectly, it is often necessary to restrict attention to a specific set of incorrect component behaviours, i.e., to make assumptions about how the components may fail. For example, in the case of fault tolerance, attention could be restricted to the incorrect behaviours to be tolerated. Conversely, if a system can not be guaranteed to behave correctly, it is often desirable to characterize its incorrect behaviour.

A specification S_f characterizes the systems which are correct with respect to S_f. The systems characterized by S_f might be incorrect with respect to another specification S_{org} (which is pragmatically thought of as the "original" or "correct" specification). When S_f is chosen to characterize systems, which are thought of as incorrect, S_f is pragmatically termed a *failure mode*. As an extreme case, the original specification for a system is thought of as a failure mode describing correct behaviour.

Definition 6 A failure mode *is a specification.*

Example 1 Consider an unbounded buffer, modelled in CSP as a system with communication channels i and o for input resp. output (see Figure 1). The correct behaviour is specified by $S_{BUF,org}$. BUF must only output values which have previously been input, and their ordering must be preserved, i.e., the trace restricted to the output must be a prefix of the trace restricted to the input:

$$S_{BUF,org} \triangleq tr \downarrow o \leq tr \downarrow i$$

The incorrect behaviour consisting in corrupting values, but not giving any spurious output, is characterized by the failure mode

$$S_{BUF,nso} \triangleq \#(tr \downarrow o) \leq \#(tr \downarrow i)$$

Finally a failure mode characterizing all kinds of incorrect behaviour (i.e., a *byzantine* failure mode) is

$$S_{BUF,byz} \triangleq \underline{\textbf{true}}$$

End of example.

4   Verification of failure modes

Generally several failure modes are defined for a system, and hence for components. This can be modelled as a collection, F, of system failure modes[1] and — given a design combinator, $Comb$, of N components — an N-tuple, $(F_1, F_2, \ldots, F_N)$, of collections of component failure modes, for each individual component. (Assume, that each such collection of failure modes is formed by parameterizing a specification with a parameter ranging over a given set. Then, quantification over collections of failure modes is equivalent to quantification over the corresponding parameter ranges.)

Now assume, the system is to be constructed from the components $C_1, C_2, \ldots, C_N$ using the design combinator $Comb$. In order to verify, that components with a specific combination of component failure modes

$$(S_{f1}, S_{f2}, \ldots, S_{fN}) \in F_1 \times F_2 \times \ldots \times F_N$$

implies the system satisfying a specific system failure mode, $S_f \in F$, one has to prove

$$(Comb, (S_{f1}, S_{f2}, \ldots, S_{fN})) \text{ \textbf{Dsat} } S_f$$

If this holds, $(S_{f1}, S_{f2}, \ldots, S_{fN})$ is said to *correspond* to S_f with respect to $Comb$.

Furthermore, in order to verify the defined system failure modes cover all behaviours made possible by the defined component failure modes, we must prove

---

[1]Note that the choice of specific failure modes in such a collection is a pragmatic one.

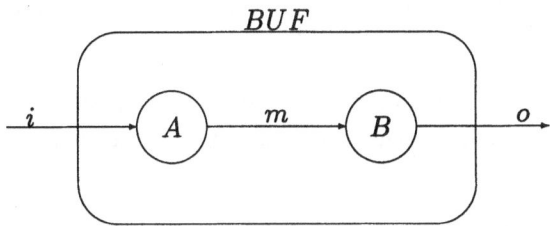

Figure 2: The two-buffer system.

$\forall (S_{f1}, S_{f2}, \ldots, S_{fN})$:
$F_1 \times F_2 \times \ldots \times F_N \bullet \exists S_f : F \bullet (Comb, (S_{f1}, S_{f2}, \ldots, S_{fN}))$ **Dsat** S_f

Example 2 Consider a system consisting of two unbounded buffers, A and B (see Figure 2) with the design combinator

$$Comb(A, B) \triangleq (A \parallel B) \backslash \{m\}$$

For A and B we define failure modes similar to those for BUF:

$$S_{A,X} \triangleq S_{BUF,X}[m/o] \quad \text{and} \quad S_{B,X} \triangleq S_{BUF,X}[m/i]$$
$$\text{for } X = org, nso, byz$$

Thus for both A, B and BUF we have defined the failure modes "correct behaviour" (org), "no spurious output" (nso) and "byzantine" (byz).

Using the proof rules of CSP we can show that

$(S_{A,org}, S_{B,org})$	Corresponds to	$S_{BUF,org}$
$(S_{A,org}, S_{B,nso})$	Corresponds to	$S_{BUF,nso}$
$(S_{A,nso}, S_{B,org})$	Corresponds to	$S_{BUF,nso}$
$(S_{A,nso}, S_{B,nso})$	Corresponds to	$S_{BUF,nso}$
All combinations containing $S_{A,byz}$ or $S_{B,byz}$	Corresponds to	$S_{BUF,byz}$

Thus, with this design combinator, the system inherits the weakest (i.e., most incorrect) of its components failure modes. **End of example.**

Fault tolerance: With the concepts defined previously we are now ready to define the concepts fault tolerance and fail softness.

Informally, the design of a system is fault tolerant if – assuming its components implement a given combination of failure modes – the system implements its specification [3].

Formally, the design combinator is fault tolerant with respect to a combination of component failure modes and a system failure mode if and only if the design comprising the combinator and the component failure modes is correct with respect to the system failure mode.

Definition 7 *Let $Comb$ be a function of type $SYSTEMS^N \rightarrow SYSTEMS$, let S be a specification, and let $(S_{f1}, S_{f2}, \ldots, S_{fN})$ be an N-tuple of failure modes. Then $Comb$ is fault tolerant with respect to S and $(S_{f1}, S_{f2}, \ldots, S_{fN})$ if and only if*

$$(Comb, (S_{f1}, S_{f2}, \ldots, S_{fN})) \text{ \textbf{Dsat} } S$$

Thus fault tolerance of a combinator is a special case of correspondence between combinations of component failure modes and system failure modes. For example, in the case of N-modular redundancy, all combinations of component failure modes for which more than $N/2$ of the failure modes for replicated components describe correct behaviour corresponds to the system failure mode describing correct behaviour.

Informally, the design of a system is fail soft if – assuming its components implement a given combination of failure modes – the system provides a degraded service, ie., the system implements a specification which is weaker than the original specification. But deciding which specification is to be considered the original one is a pragmatic issue. Therefore fail softness is considered equivalent to fault tolerance.

5   Stand by spare — an example

In order to illustrate the concepts and proof technique outlined in the previous sections, a fault tolerant system architecture, known as *stand by spare*, is analysed.

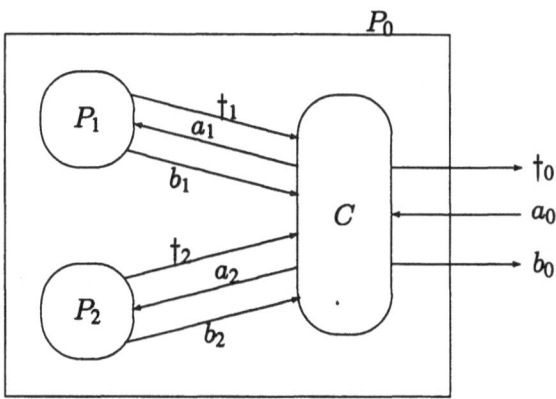

Figure 3: The internal structure of the stand by spare system.

First an informal description is given of the stand by spare architecture. This description is formalized by specifying the failure modes of the system and the system design (including component failure modes). Finally the design is proven correct wrt. different combinations of component and system failure modes.

5.1 Introductory description

A base system, P_i, inputs a value from a channel, a_i, apply a function to that value and outputs the result on the channel b_i before accepting another input. Ideally the system should behave like this, but if the system becomes unable to do so, it must notify its environment by sending a signal on an output channel, $\dagger_i$. The $\dagger_i$ channel can be thought of as carrying exception signals corresponding to the detected internal malfunction.

The task is to construct a more dependable system, P_0, from two replicas, P_1, P_2, of the base system and a coordinating component, C, such that the composite system behaves like the base system, and is tolerant to detected malfunction in one of the replica (Figure 3). If both replicas detect malfunction, P_0 must itself signal an exception to the environment.

The design of the system is the parallel composition of P_1, P_2, C with hiding of internal channels:

$$P_0 = (P_1 \parallel P_2 \parallel C) \backslash Int$$

where Int denotes the set of replicated channels:

$$Int \stackrel{\triangle}{=} \{a_1, b_1, \dagger_1, a_2, b_2, \dagger_2\}$$

The coordinating component, C, must behave as follows: 1) As long as no detection signal, $\dagger_1$, from the first replica, P_1, has occurred, C routes all input data from the environment to P_1 for processing, and all the results from P_1 are routed to the environment. 2) If $\dagger_1$ has occurred and $\dagger_2$ has not yet occurred, data are routed to and from P_2. 3) If both $\dagger_1$ and $\dagger_2$ have occurred, C signals detected malfunction on channel $\dagger_0$. Some care is needed when switching from P_1 to P_2, to ensure transparency, especially when $\dagger_1$ occurs after receiving a data object from channel a_0 but before delivering the corresponding result on channel b_0.

Replica P_1 can thus be thought of as the normally operational replica while P_2 can be thought of as the stand by spare replica.

In the above description, values are communicated over channels. However, in the formal specifications and proofs we will abstract away from values. (The specifications and proofs including value passing have been carried out. It turned out, that correctness of the stand by spare architecture is concerned mainly with sequencing and readiness properties.) Thus all communications are simple events in the following.

5.2 Specification style

The specification will be given in a constraint oriented style as a conjunction of predicates describing different properties: 1) Properties constraining the occurrence of exception signals, 2) sequencing constraints and 3) readiness constraints.

The advantage of a constraint oriented style is that the separation of properties in the specifications enable the correctness proofs to be separated into subproofs, one for each property. Each subproof for a property of a composed system is mainly based on the corresponding properties in the component specifications.

5.3 Auxiliary functions

The function *last* giving the last event of a nonempty trace is made total by returning the special value **nil** for the empty trace:

$$last(t) \triangleq (\quad t = <> \quad \rightarrow \quad \textbf{nil},$$
$$t \neq <> \quad \rightarrow \quad t[\#t] \quad)$$

In order to specify switching of the data flow from P_1 to P_2 after detected malfunction of P_1, the partial function $After\dagger_1$ is introduced on traces. Given a trace containing one occurrence of $\dagger_1$, it yields the trace after $\dagger_1$. However, if the trace before the dagger contains an input (event a_0) from the environment to which no output (event b_0) has been given, the last such input is prefixed to the trace after the dagger.

$$After\ \dagger_1\ (p\char`^ < \dagger_1 > \char`^ s) \triangleq (\quad \#(p \downarrow a_0) = \#(p \downarrow b_0) \quad \rightarrow \quad s,$$
$$\#(p \downarrow a_0) \neq \#(p \downarrow b_0) \quad \rightarrow \quad < a_0 > \char`^ s \quad)$$

5.4 System specification

The system is specified as a collection of failure modes. Each failure mode specifies that the system must process at least k pairs of input and output before internal malfunction occurs and $\dagger_0$ is communicated. The failure modes differ only in the value of k. Thus k is a measure for how long the system will behave properly. All failure modes for the system are specified by the specification $S_{i,k}$ and the alphabet αP_i (for $i = 0$) below.

The alphabet of the system is

$$\alpha P_i \triangleq \{a_i, b_i, \dagger_i\}$$

The structure of the system specification with respect to occurrence of exception signals is given by

$$S_{i,k} \triangleq (\dagger_i \notin elems(tr) \Rightarrow (Seq_i \wedge Ready_i))$$
$$\wedge Limit_{i,k} \wedge OnceDetect_i$$

When no exception has occurred, the trace is an alternation of a_i and b_i

$$Seq_i \triangleq tr \leq < a_i, b_i >^*$$

When no exception has occurred, the system is always ready for some event.

$$Ready_i \overset{\triangle}{=} \quad (last(tr) \in \{b_i, \underline{nil}\} \Rightarrow a_i \notin ref \vee \dagger_i \notin ref)$$
$$\wedge \quad (last(tr) = a_i \qquad \Rightarrow b_i \notin ref \vee \dagger_i \notin ref)$$

No exception occurs for the first k pairs of input and output

$$Limit_{i,k} \overset{\triangle}{=} \#tr \leq 2k \Rightarrow \dagger_i \notin elems(tr)$$

Only one exception signal is allowed

$$OnceDetect_i \overset{\triangle}{=} \#(tr \downarrow \dagger_i) \leq 1$$

5.5 Design of the system

Formally, the design of the system comprises a design function, $Comb$, mapping the components (i.e., the two replicas P_1, P_2 and the coordinator C) into a system, and for each component a collection of failure modes. We have

$$Comb(P_1, P_2, C) \overset{\triangle}{=} (P_1 \parallel P_2 \parallel C)\backslash Int$$

The components P_1 and P_2 have failure modes similar to those for the system, given by $S_{i,k}$ and αP_i for $i \in \{1, 2\}$.

For the coordinating component C, only one failure mode is specified (ie., this analysis does not take failure of C into account). The alphabet of C is

$$\alpha C \overset{\triangle}{=} \alpha P_0 \cup \alpha P_1 \cup \alpha P_2$$

The specification predicate for C is structured according to the following conditions on occurrence of exception signals: 1) $\dagger_1$ has not occurred, in which case events are routed to and from P_1. 2) $\dagger_1$ has occurred but $\dagger_2$ has not occurred, in which case events are routed to and from P_2. 3) Both $\dagger_1$ and $\dagger_2$ have occurred in which case an exception is signalled to the environment of P_0. The occurrence of

$\dagger_2$ before $\dagger_1$ is inhibited by C, because C will not communicate with P_2 before the occurrence of $\dagger_1$.

$$
S_C \triangleq \quad
\begin{aligned}
&(\dagger_1 \notin elems(tr) && \Rightarrow CSeq_1 \wedge CReady_1) \\
\wedge\ &(\dagger_1 \in elems(tr) \wedge \dagger_2 \notin elems(tr) && \Rightarrow CSeq_2 \wedge CReady_2) \\
\wedge\ &(\dagger_1 \in elems(tr) \wedge \dagger_2 \in elems(tr) && \Rightarrow CSeq_3 \wedge CReady_3)
\end{aligned}
$$

Before occurrence of $\dagger_1$, the trace is an alternation of events a_0, a_1, b_1, b_0. Furthermore, C must always be ready for $\dagger_1$ and either a_0, a_1, b_1 or b_0:

$$
\begin{aligned}
CSeq_1 &\triangleq tr \leq\ < a_0, a_1, b_1, b_0 >^* \\
CReady_1 &\triangleq \quad
\begin{aligned}
&(last(tr) \in \{b_0, \mathbf{nil}\} && \Rightarrow a_0 \notin ref \wedge \dagger_1 \notin ref) \\
\wedge\ &(last(tr) = a_0 && \Rightarrow a_1 \notin ref \wedge \dagger_1 \notin ref) \\
\wedge\ &(last(tr) = a_1 && \Rightarrow b_1 \notin ref \wedge \dagger_1 \notin ref) \\
\wedge\ &(last(tr) = b_1 && \Rightarrow b_0 \notin ref \wedge \dagger_1 \notin ref)
\end{aligned}
\end{aligned}
$$

After occurrence of $\dagger_1$ and before occurrence of $\dagger_2$, the trace is continued with alternation of events a_0, a_2, b_2, b_0. Furthermore, C must always be ready for $\dagger_2$ and either a_0, a_2, b_2 or b_0:

$$
\begin{aligned}
CSeq_2 &\triangleq After\ \dagger_1\ (tr) \leq\ < a_0, a_2, b_2, b_0 >^* \\
CReady_2 &\triangleq \quad
\begin{aligned}
&(last(After\ \dagger_1\ (tr)) \in \{b_0, \mathbf{nil}\} && \Rightarrow a_0 \notin ref \wedge \dagger_2 \notin ref) \\
\wedge\ &(last(After\ \dagger_1\ (tr)) = a_0 && \Rightarrow a_2 \notin ref \wedge \dagger_2 \notin ref) \\
\wedge\ &(last(After\ \dagger_1\ (tr)) = a_2 && \Rightarrow b_2 \notin ref \wedge \dagger_2 \notin ref) \\
\wedge\ &(last(After\ \dagger_1\ (tr)) = b_2 && \Rightarrow b_0 \notin ref \wedge \dagger_2 \notin ref)
\end{aligned}
\end{aligned}
$$

After the occurrence of both $\dagger_1$ and $\dagger_2$, C must send a single exception signal to the environment:

$$
\begin{aligned}
CSeq_3 &\triangleq tr = s^\frown < \dagger_1 >^\frown t^\frown < \dagger_2 >^\frown u \ \Rightarrow\ u \leq < \dagger_0 > \\
CReady_3 &\triangleq \dagger_0 \notin elems(tr) \Rightarrow \dagger_0 \notin ref
\end{aligned}
$$

5.6 Correctness of the stand by spare design

The stand by spare mechanism is considered correct if the system satisfies $S_{0,m}$ assuming its components satisfy $S_{1,k}, S_{2,l}, S_C$. Thus correctness of the mechanism is equivalent to correctness of a design. The relation between m and (l, k) is

a measure for the degree of dependability improvement offered by the mechanism.
We can prove correctness for $m = k + l - 2$. Formally, this is stated as

Theorem 2

$$(Comb, (S_{1,k}, S_{2,l}, S_C)) \textbf{ Dsat } S_{0,k+l-2}$$

The correctness proof for theorem 2 will be based on proof rules from chapter 3
of [2]. Due to our specification style, some derivatives of the proof rules will be
useful. Below these derivatives as well as a logic rule are presented.

Derived proof rules: From L3 of [2] we get

DPR 1   *If* P **sat** $S(tr)$
              and Q **sat** $T(tr)$
              and   *neither P nor Q diverges*
              then $P \parallel Q$ **sat** $S(tr \downarrow \alpha P) \wedge T(tr \downarrow \alpha Q)$

From L3 of [2] we get

DPR 2   *If* P **sat** $tr = t_p \Rightarrow e \notin ref$
              and Q **sat** $tr = t_q \Rightarrow e \notin ref$
              and   *neither P nor Q diverges*
              then $P \parallel Q$ **sat** $(tr \downarrow \alpha P = t_p \wedge tr \downarrow \alpha Q = t_q) \Rightarrow e \notin ref$

From the definition of **sat** we get

DPR 3   *If* $e \notin \alpha Q$
              then Q **sat** $e \notin ref$

From **DPR 3** and **DPR 2** we get

DPR 4   *If* P **sat** $tr = t_p \Rightarrow e \notin ref$
              and $e \notin \alpha Q$
              and   *neither P nor Q diverges*
              then $P \parallel Q$ **sat** $tr \downarrow \alpha P = t_p \Rightarrow e \notin ref$

From L7 of [2] we get

DPR 5 *If* P *sat* $S(tr) \wedge NODIV$
and $S(tr) \Rightarrow S'(tr \downarrow (\alpha P - C))$
then $P \backslash C$ *sat* $S'(tr)$

where $NODIV$ states that the number of hidden symbols that can occur is bounded by some function of the non-hidden symbols.

A logic rule:

LR 1 *If* $\exists x : \{x \mid A(x)\} \bullet (A(x) \Rightarrow C)$
and x *not free in* C
then C

The proof for theorem 2 is based on lemmas: First the behaviour of the parallel composition of the components is treated (lemma 1). Secondly the application of hiding is treated, based on assumptions on the parallel composition (lemma 2). Finally correctness of the entire design is proven, based on the lemmas.

Lemma 1 *If* $P = P_1 \parallel P_2 \parallel C$
And P_1 *sat* $S_{1,k}$, P_2 *sat* $S_{2,l}$, C *sat* S_C
Then P *sat* $S_{P,k,l}$
where

$$S_{P,k,l} \triangleq \begin{aligned} &(\dagger_1 \notin elems(tr) &&\Rightarrow PSeq_1 \wedge PReady_1) \\ &\wedge (\dagger_1 \in elems(tr) \wedge \dagger_2 \notin elems(tr) &&\Rightarrow PSeq_2 \wedge PReady_2) \\ &\wedge (\dagger_1 \in elems(tr) \wedge \dagger_2 \in elems(tr) &&\Rightarrow PSeq_3 \wedge PReady_3) \\ &\wedge PLimit_{k,l} \end{aligned}$$

$$PSeq_1 \triangleq tr \leq < a_0, a_1, b_1, b_0 >^*$$

$$PReady_1 \triangleq \begin{aligned} (\ &(last(tr) \in \{\underline{nil}, b_0\} &&\Rightarrow a_0 \notin ref) \\ &\wedge (last(tr) = a_0 &&\Rightarrow (a_1 \notin ref \vee \dagger_1 \notin ref)) \\ &\wedge (last(tr) = a_1 &&\Rightarrow (b_1 \notin ref \vee \dagger_1 \notin ref)) \\ &\wedge (last(tr) = b_1 &&\Rightarrow b_0 \notin ref) \ ) \end{aligned}$$

$$PSeq_2 \triangleq After \ \dagger_1 \ (tr) \leq < a_0, a_2, b_2, b_0 >^*$$

$$PReady_2 \;\triangleq\; (\;\; (last(After\; \dagger_1\; (tr)) \in \{\underline{nil}, b_0\} \;\Rightarrow\; a_0 \notin ref)$$
$$\wedge\; (last(After\; \dagger_1\; (tr)) = a_0 \qquad \Rightarrow\; (a_2 \notin ref \vee \dagger_2 \notin ref))$$
$$\wedge\; (last(After\; \dagger_1\; (tr)) = a_2 \qquad \Rightarrow\; (b_2 \notin ref \vee \dagger_2 \notin ref))$$
$$\wedge\; (last(After\; \dagger_1\; (tr)) = b_2 \qquad \Rightarrow\; b_0 \notin ref)\;\;)$$

$$PSeq_3 \;\triangleq\; tr = s^\wedge < \dagger_1 > \,\hat{}\, t^\wedge < \dagger_2 > \,\hat{}\, u \;\Rightarrow\; u \leq\, < \dagger_0 >$$

$$PReady_3 \;\triangleq\; \dagger_0 \notin elems(tr) \Rightarrow \dagger_0 \notin ref$$

$$PLimit_{k,l} \;\triangleq\; \exists s, t, u \bullet (\;\; tr \leq s^\wedge < \dagger_1 > \,\hat{}\, t^\wedge < \dagger_2 > \,\hat{}\, u$$
$$\wedge\; \{\dagger_1, \dagger_2\} \cap elems(s\,\hat{}\,t\,\hat{}\,u) = \{\}$$
$$\wedge\; \#s \geq 4k - 1$$
$$\wedge\; \#t \geq 4l - 2 \;\;)$$

Proof of lemma 1

Components P_i and C satisfying $S_{i,k}$ resp. S_C are divergence free. Thus we can safely use proof rules having freedom from divergence as an assumption.

From

$$C\; \mathbf{sat}\; (\dagger_1 \notin elems(tr) \Rightarrow CSeq_1)$$

and **DPR 1** we get

$$P\; \mathbf{sat}\; (\dagger_1 \notin elems(tr) \Rightarrow PSeq_1)$$

From $PSeq_1, \alpha C, \alpha P_1$ we get

$$SYS_\| \; \mathbf{sat}$$
$$(\dagger_1 \notin elems(tr) \Rightarrow \tag{1}$$
$$(\; tr \downarrow \alpha C = tr$$
$$\wedge\; (last(tr) \in \{b_0, \underline{nil}, b_1, a_0\} \;\Rightarrow\; last(tr \downarrow \alpha P_1) \in \{b_1, \underline{nil}\})$$
$$\wedge\; (last(tr) = a_1 \qquad\qquad \Rightarrow\; last(tr \downarrow \alpha P_1) = a_1)\;\;))$$

From **DPR 3**, $\alpha P_1, \alpha P_2$ we get

$$P_1\; \mathbf{sat}\; ref \cap \{a_0, b_0, \dagger_0, a_2, b_2, \dagger_2\} = \{\}$$
$$P_2\; \mathbf{sat}\; ref \cap \{a_0, b_0, \dagger_0, a_1, b_1, \dagger_1\} = \{\} \tag{2}$$

From $CReady_1, Ready_1$, (1),(2) and **DPR 2** we get

$$P\; \mathbf{sat}\; (\dagger_1 \notin elems(tr) \Rightarrow PReady_1)$$

From

$$C \text{ sat } ((\dagger_1 \in elems(tr) \land \dagger_2 \notin elems(tr)) \Rightarrow CSeq_2)$$

and **DPR 1** we get

$$P \text{ sat } ((\dagger_1 \in elems(tr) \land \dagger_2 \notin elems(tr)) \Rightarrow PSeq_2)$$

From $OnceDetect_1$ and $PSeq_1$ we get

> **P sat**
> $$((\dagger_1 \in elems(tr) \land \dagger_2 \notin elems(tr)) \Rightarrow$$
> $$tr \downarrow \alpha P_2 = After \dagger_1 (tr) \downarrow \alpha P_2) \qquad (3)$$

From $PSeq_2, \alpha C, \alpha P_2$ and (3) we get

> **P sat**
> $$(\ (\dagger_1 \in elems(tr) \land \dagger_2 \notin elems(tr)) \Rightarrow$$
> $$(After \dagger_1 (tr \downarrow \alpha C) = After \dagger_1 (tr)$$
> $$\land (last(After \dagger_1 (tr)) \in \{b_0, \underline{\text{nil}}, b_2, a_0\} \Rightarrow$$
> $$last(tr \downarrow \alpha P_2) \in \{b_2, \underline{\text{nil}}\})$$
> $$\land (last(After \dagger_1 (tr)) = a_2 \Rightarrow$$
> $$last(tr \downarrow \alpha P_2) = a_2))\ ) \qquad (4)$$

From $CReady_2, Ready_2$, (2,4) and **DPR 2** we get

$$P \text{ sat } ((\dagger_1 \in elems(tr) \land \dagger_2 \notin elems(tr)) \Rightarrow PReady_2)$$

From

$$C \text{ sat } ((\dagger_1 \in elems(tr) \land \dagger_2 \in elems(tr)) \Rightarrow CSeq_3)$$

and **DPR 1** we get

$$P \text{ sat } ((\dagger_1 \in elems(tr) \land \dagger_2 \in elems(tr)) \Rightarrow PSeq_3)$$

From

$$C \text{ sat } ((\dagger_1 \in elems(tr) \wedge \dagger_2 \in elems(tr)) \Rightarrow C\,Ready_3)$$

and (2), **DPR 2** we get

$$P \text{ sat } ((\dagger_1 \in elems(tr) \wedge \dagger_2 \in elems(tr)) \Rightarrow P\,Ready_3)$$

In $PLimit_{k,l}$ the decomposition

$$tr \leq s\char`\^ < \dagger_1 > \char`\^ t \char`\^ < \dagger_2 > \char`\^ u \quad \wedge \quad \{\dagger_1, \dagger_2\} \cap elems(s\char`\^ t\char`\^ u) = \{\}$$

is possible, since $PSeq_1$ implies that $\dagger_2$ never occurs before $\dagger_1$, and since $OnceDetect_1$ and $OnceDetect_2$ implies that $\dagger_1$ and $\dagger_2$ each occur at most once. Now, $\#s \geq 4k - 1$ follows from $PSeq_1$ and $Limit_{1,k}$. Finally $\#t \geq 4l - 2$ follows from $PSeq_2$ and $Limit2, l$, because $\dagger_1 \notin elems(tr) \Rightarrow tr \downarrow \alpha P_2 = <>$. Therefore

$$P \text{ sat } PLimit_{k,l}$$

End of proof.

Lemma 2   *If* $P_0 = P\backslash Int$
             And $P \text{ sat } S_{P,k,l}$
             Then $P_0 \text{ sat } S_{0,k+l-2}$

Proof of lemma 2
First it is proven, that the number of hidden events that can occur in P is bounded by some function of the nonhidden events. This enables the use of proof rule L7 of [2] and derivatives thereof.

From $S_{P,k,l}, PSeq_1, PSeq_2, PSeq_3$ we get

$$P \text{ sat } \exists s, t \bullet Sub(s, t, tr) \tag{5}$$

where

$$Sub(s, t, tr) \triangleq$$
$$(\quad tr \leq s \char`^ < \dagger_1 > \char`^t\char`^ < \dagger_2, \dagger_0 > \qquad\qquad (6)$$
$$\wedge \quad s \leq < a_0, a_1, b_1, b_0 >^*$$
$$\wedge \quad (last(s) \in \{b_0, \underline{\mathbf{nil}}\} \Rightarrow t \leq < a_0, a_2, b_2, b_0 >^*)$$
$$\wedge \quad (last(s) \notin \{b_0, \underline{\mathbf{nil}}\} \Rightarrow< a_0 > \char`^t \leq < a_0, a_2, b_2, b_0 >^*)$$
$$\wedge \quad SubLimit_{k,l}(s, t) \,)$$

where $SubLimit_{k,l}$ is to be defined later.

From (5,6) we see that P can at most perform contiguous sequences of length six of events to be hidden. Therefore the above mentioned bound exists.

From (6) we have

$$Sub(s, t, tr) \Rightarrow$$
$$(\quad tr \downarrow Ext \leq (s \downarrow Ext)\char`^(t \downarrow Ext)\char`^ < \dagger_0 > \qquad\qquad (7)$$
$$\wedge \quad (s \downarrow Ext) \leq < a_0, b_0 >^*$$
$$\wedge \quad (last(s \downarrow Ext) \in \{b_0, \underline{\mathbf{nil}}\} \Rightarrow (t \downarrow Ext) \leq < a_0, b_0 >^*)$$
$$\wedge \quad (last(s \downarrow Ext) = a_0 \Rightarrow< a_0 > \char`^(t \downarrow Ext) \leq < a_0, b_0 >^*) \,)$$

where

$$Ext \triangleq \alpha P - Int$$

From (5,7) and **DPR 5** follows

$$P_0 \ \mathbf{sat} \ (\dagger_0 \notin elems(tr) \Rightarrow Seq_0)$$

By L7 we have

$$P_0 \ \mathbf{sat} \ \exists s \bullet (tr = s \downarrow Ext \wedge S_{P,k,l}[s/tr, ref \cup Int/ref])$$

We want to prove

$$P_0 \ \mathbf{sat} \ (\quad (P'_{tr}(tr) \Rightarrow (a_0 \notin ref \vee \dagger_0 \notin ref)) \qquad\qquad (8)$$
$$\wedge \quad (P''_{tr}(tr) \Rightarrow (b_0 \notin ref \vee \dagger_0 \notin ref)) \,)$$

where

$$P'_{tr}(tr) \triangleq \dagger_0 \notin elems(tr) \wedge last(tr) \in \{b_0, \mathbf{nil}\}$$
$$P''_{tr}(tr) \triangleq \dagger_0 \notin elems(tr) \wedge last(tr) = a_0$$

The first line of (8) is proved by showing

$$(P'_{tr}(tr) \wedge \exists s \bullet (tr = s \downarrow Ext \wedge S_{P,k,l}[s/tr, ref \cup Int/ref]))$$
$$\Rightarrow (a_0 \notin ref \vee \dagger_0 \notin ref)$$

We have

$$(P'_{tr}(tr) \wedge \exists s \bullet (tr = s \downarrow Ext \wedge S_{P,k,l}[s/tr, ref \cup Int/ref])) \Rightarrow \qquad (9)$$
$$\exists s : \{s \mid P'_{tr}(s \downarrow Ext) \wedge SeqConstr(s)\} \bullet S_{P,k,l}[s/tr, ref \cup Int/ref]$$

Where the sequencing constraints on s are given by

$$SeqConstr(s) \triangleq \exists ref \bullet S_{P,k,l}[s/tr]$$

The set over which s is quantified in the consequent of (9) can be classified according to different cases in $S_{P,k,l}$:

$$\{s \mid P'_{tr}(s \downarrow Ext) \wedge SeqConstr(s)\} = \bigcup_{i=1}^{3}\{s \mid A'_i(s) \wedge SeqConstr(s)\}$$

where

$$A'_1(s) \triangleq \dagger_1 \notin elems(s) \wedge last(s) \in \{b_0, \mathbf{nil}\}$$
$$A'_2(s) \triangleq \dagger_1 \in elems(s) \wedge \dagger_2 \notin elems(s) \wedge last(After \dagger_1(s)) \in \{b_0, \mathbf{nil}\}$$
$$A'_3(s) \triangleq \dagger_1 \in elems(s) \wedge \dagger_2 \in elems(s) \wedge \dagger_0 \notin elems(s) \wedge$$
$$last(s \downarrow Ext) \in \{b_0, \mathbf{nil}\}$$

The connection between $A'_i(s)$ and the readiness properties of $S_{P,k,l}$ are given by:

$$S_{P,k,l}[s/tr, ref \cup Int/ref] \Rightarrow \quad (\quad (A'_1(s) \Rightarrow a_0 \notin ref) \qquad (10)$$
$$\wedge \ (A'_2(s) \Rightarrow a_0 \notin ref)$$
$$\wedge \ (A'_3(s) \Rightarrow \dagger_0 \notin ref))$$

The consequent of (9) implies

$$\bigvee_{i=1}^{3} \exists s : \{s \mid A_i'(s)\} \bullet S_{P,k,l}[s/tr, ref \cup Int/ref]$$

By **LR 1** and (10) this implies

$$a_0 \notin ref \vee t_0 \notin ref$$

The proof for the second line of (8) is similar, only the subdivision of the set, over which s is quantified, changes:

$$\{s \mid P_{tr}''(s \downarrow Ext) \wedge SeqConstr(s)\} = \bigcup_{i=1}^{5} \{s \mid A_i''(s) \wedge SeqConstr(s)\}$$

where

$$A_1''(s) \triangleq t_1 \notin elems(s) \wedge last(s) \in \{a_0, a_1\}$$
$$A_2''(s) \triangleq t_1 \notin elems(s) \wedge last(s) = b_1$$
$$A_3''(s) \triangleq t_1 \in elems(s) \wedge t_2 \notin elems(s) \wedge last(After\ t_1\ (s)) \in \{a_0, a_2\}$$
$$A_4''(s) \triangleq t_1 \in elems(s) \wedge t_2 \notin elems(s) \wedge last(After\ t_1\ (s)) = b_2$$
$$A_5''(s) \triangleq t_1 \in elems(s) \wedge t_2 \in elems(s) \wedge t_0 \notin elems(s) \wedge$$
$$last(s \downarrow Ext) = a_0$$

We have

$$S_{P,k,l}[s/tr, ref \cup Int/ref] \Rightarrow (\ (A_1''(s) \Rightarrow \textbf{\underline{false}}) \tag{11}$$
$$\wedge\ (A_2''(s) \Rightarrow b_0 \notin ref)$$
$$\wedge\ (A_3''(s) \Rightarrow \textbf{\underline{false}})$$
$$\wedge\ (A_4''(s) \Rightarrow b_0 \notin ref)$$
$$\wedge\ (A_5''(s) \Rightarrow t_0 \notin ref)\ )$$

Thus

$$(P_{tr}''(tr) \wedge \exists s \bullet (tr = s \downarrow Ext \wedge S_{P,k,l}[s/tr, ref \cup Int/ref]))$$
$$\Rightarrow \bigvee_{i=1}^{5} \exists s : \{s \mid A_i''(s)\} \bullet S_{P,k,l}[s/tr, ref \cup Int/ref]$$

By **LR 1** and (11) this implies

$$b_0 \notin ref \lor \dagger_0 \notin ref$$

Thus

$$P_0 \textbf{ sat } (\dagger_0 \notin elems(tr) \Rightarrow Ready_0)$$

From (5,7) and **DPR 5** we get

$$P_0 \textbf{ sat } OnceDetect_0$$

Because of $PLimit_{k,l}$ we can define

$$SubLimit_{k,l}(s,t) \triangleq \#s \geq 4k - 1 \land \#t \geq 4l - 2$$

With this definition of $SubLimit_{k,l}$ we get

$$Sub(s,t,tr) \Rightarrow (\quad tr \downarrow Ext \leq (s \downarrow Ext)\hat{\ }(t \downarrow Ext)\hat{\ } < \dagger_0 >$$
$$\land \quad \dagger_0 \notin elems(s\hat{\ }t)$$
$$\land \quad \#(s \downarrow Ext) \geq 2k - 1 \land \#(t \downarrow Ext) \geq 2l - 2 \quad)$$

Then by **DPR 5**

$$P_0 \textbf{ sat } \exists s \bullet (\quad tr \leq s\hat{\ } < \dagger_0 >$$
$$\land \quad \dagger_0 \notin elems(s)$$
$$\land \quad \#s \geq 2(k + l - 2) \quad)$$

Thus

$$P_0 \textbf{ sat } Limit_{0,k+l-2}$$

End of proof.

Proof of theorem 2

From lemma 1 and lemma 2 we get

If $\quad P_0 = (P_1 \parallel P_2 \parallel C) \backslash Int$

And $\quad P_1$ sat $S_{1,k}$, $\quad P_2$ sat $S_{2,l}$, $\quad C$ sat S_C

Then $\quad P_0$ sat $S_{0,k+l-2}$

from which theorem 2 follows. **End of proof.**

6 Conclusion

A systematic approach to verifying the failure modes of a system with respect to a given design and given component failure modes has been outlined. It is based on the concepts *design, correctness of design* and *failure mode*, which have been formally defined. The approach seems well suited for analyzing how incorrectness of components propagates from one level of design to the level above it. Furthermore it seems suitable for analyzing or synthesizing designs for fault tolerance wrt. specific component failure modes. The approach has been applied to the analysis of a stand by spare mechanism.

Assumptions about the behaviour of a failing component have been described by a single specification – a failure mode – instead of calculating the behaviour of a failing component as a function of its correct behaviour and a specification of its failure assumptions. This approach has been shown to work, and although it produces fairly complex specifications it has the benefit of eliminating the calculations. Furthermore, in calculation based approaches, the behaviour of a failing component is often partly determined by the calculation chosen by the inventor of the approach, and thereby concealed from the designer using the approach. In our approach, specification of failing behaviour is left completely to the designer, thereby giving the designer the highest possible degree of freedom. Similar approaches to specifying failing behaviour may be found in [1], [5]. In contrast to these, the calculation based approach is used in [4], [6] and also in the state machine approach [8], although the calculations are not explicitly carried out.

Scalability is a prime issue for any verification method. Therefore we have given priority to the ability to perform compositional proofs. To achieve this, we have advocated the use of CSP's compositional proof rules. The global assumptions inherent in many fault tolerant designs are an impediment to compositionality. By global assumptions we understand such component assumptions which cannot be expressed as a conjunction of single component assumptions. To circumvent this problem we have split the verification of designs with global assumptions into several subproofs without global assumptions. Each subproof is based on a conjunction of single component assumptions satisfying the global assumptions. As shown in the stand by spare example, such subproofs can be conducted as a single generalized proof by parameterizing the failure mode specifications.

A contrasting approach is found in [6], where global assumptions are expressed as an equivalence between a fault free system and a fault affected system. The verification with respect to global assumptions consists of proving this equivalence by algebraic calculations on models of the components.

Our approach also have some similarities with the D-calculus [7] used for generating test patterns for revealing faults in digital circuits. In the D-calculus, an expression describing a fault of one of the circuits components together with an input test pattern results in a particular output. If the output is faulty, the input pattern is a test for the fault. If no input pattern results in faulty outputs, the circuit tolerates the fault.

In order to make our approach more feasible for engineers, specification styles for failure modes (including repair and transient failures) as well as specialized proof rules are fields of current research. Also, a library of verified off–the–shelf fault tolerant designs and tools for verification support would be useful.

For specification purposes it is not sufficient to define the set of failure modes relevant for the system under consideration. The failure modes should be accompanied by some quantitative probabilistic measures specifying the likelihood for the system being in each of the defined failure modes. Furthermore, for verification purposes, these measures should apply to component failure modes too, and ideally a compositional method for verification of these measures should be devised.

Acknowledgments

Anders P. Ravn, Hans Rischel and E. V. Sørensen are gratefully acknowledged for their constructive criticism, inspiring discussions and thorough reviews of this document. Furthermore thanks to Tom Anderson, Flaviu Cristian, C.A.R. Hoare, Jean Claude Laprie, Kim Guldstrand Larsen, Brian Randell, IFIP Working Group 10.4 and the Esprit BRA (3104) project ProCoS for helpful suggestions and fruitful discussions.

References

[1] F. Cristian. A rigorous approach to fault-tolerant programming. *IEEE Transaction on Software Engineering*, SE-11(1), 1985.

[2] C. A. R. Hoare. *Communicating Sequential Processes*. Prentice Hall, 1985.

[3] J-C. Laprie. Dependable computing and fault tolerance: concepts and terminology. *Proc. 15th International Symposium on Fault-tolerant Computing*, IEEE, 1985.

[4] Z. Liu. Transformation of programs for fault tolerance. *Formal Aspects of Computing*, 4(5), 1992.

[5] L. V. Mancini, G. Pappalardo. Towards a theory of replicated processing. *Proc. Symposium on Formal Techniques in Real-time and Fault Tolerant Systems (Lecture Notes in Computer Science 331)*, Springer Verlag, September 1988.

[6] J. Peleska. Design and verification of fault tolerant systems with csp. *Distributed Computing*, 5(2), 1991.

[7] J. P. Roth. Diagnosis of automata failures : a calculus and a method. *IBM J. Res. Development*, 10(4), 1966.

[8] F. B. Schneider. Implementing fault tolerant services using the state machine approach: a tutorial. *ACM Computing Surveys*, 22(5), 1990.

TRACING

FAULT TOLERANCE

Henk SCHEPERS
Department of Mathematics and Computing Science
Eindhoven University of Technology
P.O. Box 513, 5600 MB Eindhoven, The Netherlands

Abstract

A fault may cause a process to behave abnormally, and a fault hypothesis divides such abnormal behaviour into exceptional and catastrophic behaviours. The set of normal and exceptional behaviours can be considered the set of *acceptable* behaviours. In this report traces, or communication histories, are used to denote the behaviour of a process. The semantic function $\mathcal{H}[\![P]\!]$ defines the set of possible communication sequences that can be observed up to any point in an execution of process P. A fault hypothesis is defined as a predicate representing a reflexive relation between the normal and acceptable histories of a process. Such relations enable one to abstract from the precise nature of a fault and to focus on the exceptional behaviour it causes. For a fault hypothesis χ the construct $(P \wr \chi)$ indicates execution of process P under the assumption of χ. Then, the set $\mathcal{H}[\![(P \wr \chi)]\!]$ is the set of acceptable histories of P with respect to χ.

Supported by the Dutch NWO under grant number NWI88.1517: 'Fault Tolerance: Paradigms, Models, Logics, Construction'.

1   How to characterize fault tolerance

According to Laprie [7] fault tolerance is the property of a system 'to provide, by redundancy, service complying with the specification in spite of faults occurred or occurring'. In this report we present a formal method to specify and verify a fault tolerant system based on a set of traces, or histories, that capture the behaviour of the system in terms of the messages that are communicated.

In fault tolerant systems, three domains of behaviour are usually distinguished: normal, exceptional and catastrophic (see for instance [8]). Normal behaviour is the behaviour when no faults occur. The discriminating factor between exceptional and catastrophic behaviour is the fault hypothesis which stipulates how faults affect the normal behaviour. An example is the hypothesis that a communication medium may lose messages. Relative to the fault hypothesis an exceptional behaviour exhibits an abnormality which should be tolerated (to an extent that remains to be specified), and a catastrophic behaviour has an abnormality that was not anticipated (cf. [10]). Thus, for this communication medium the corruption of messages is catastrophic. In general, the catastrophic behaviour of a component cannot be tolerated by a system.

We consider networks of sequential processes that communicate synchronously via directed channels. To formalize the behaviour of a process we use traces, or histories, which record the communications along the observable channels of the process. For instance, a possible history h of process $Square$, which alternately inputs an integer via the observable channel in and outputs its square via the observable channel out, may be $\langle (in, 1), (out, 1), (in, 3), (out, 9) \rangle$. The semantic function $\mathcal{H}[\![P]\!]$ determines the set of possible communication sequences that can be observed up to any point in an execution of process P. The set $\mathcal{H}[\![P]\!]$ is prefix closed which means that, for instance, also $\langle (in, 1) \rangle \in \mathcal{H}[\![Square]\!]$ and $\langle (in, 1), (out, 1) \rangle \in \mathcal{H}[\![Square]\!]$. A fault hypothesis is formalized as a predicate representing a reflexive relation between the normal and acceptable histories of a process. Such relations enable one to abstract from the precise nature of a fault and to focus on the exceptional behaviour it causes. For a fault hypothesis χ the construct $(P \wr \chi)$ indicates execution of process P under the assumption of χ. The extent to which the exceptional behaviour of a process must be tolerated is specified additionally. The exceptional behaviour resulting from $Square$'s output channel becoming stuck at zero can be defined using a predicate $StuckAtZero$ such that, for instance, $\langle (in, 1), (out, 1), (in, 3), (out, 0) \rangle \in \mathcal{H}[\![(Square \wr StuckAtZero)]\!]$. The set $\mathcal{H}[\![S]\!]$ of behaviours of a system S can be expressed, using a simple composition rule, in terms of those of its components. On the basis of $\mathcal{H}[\![S]\!]$ it can be verified that S indeed tolerates the exceptional behaviour of its components to the desired extent.

Our method, in which we represent the effects of faults by relations on the communication histories, is illustrated by applying it to two examples. The first concerns a communication medium that may corrupt or insert messages. The second deals with the design of a stable virtual disk which is a highly reliable store of data implemented using multiple unreliable physical disks.

The remainder of this report is organized as follows. Section 2 introduces the notation used and presents a simple composition rule. In Section 3 various types of communication media are modeled in terms of relations between histories restricted, on the one hand, to the communication medium's input channel, and, on the other hand, to the output channel. In Section 4 we show that a similar approach can be followed to verify a stable disk. Section 5 presents the conclusions and suggests some directions for future research.

2 Notation for specification

Let $chan(P)$ be the set of visible, or observable, channels of process P. The channels we use are unidirectional and connect exactly two processes, or a process and its environment. Communication along a channel is synchronous. Let $in(P)$ be the set of observable input channels of process P, and $out(P)$ the set of observable output channels of process P.

- $in(P) \cup out(P) = chan(P)$

- $in(P) \cap out(P) = \emptyset$

For a channel c the set αc denotes the alphabet of channel c, that is, the values communicable along channel c. We represent the synchronous communication of value v on channel c by a pair (c, v), $v \in \alpha c$, whose parts are obtained using the functions $ch((c, v)) = c$ and $val((c, v)) = v$. In this report we use the set $\Lambda c = \{(c, v) \mid v \in \alpha c\}$ to denote the set of possible communications along channel c. To denote the observable behaviour of process P we use a history h

which is a finite sequence (also called a trace) of the form $\langle(c_1, v_1), \ldots, (c_n, v_n)\rangle$ of length $len(h) = n$, where $n \in \mathbb{N}$, $c_i \in chan(P)$, and $v_i \in \alpha c_i$, for $1 \leq i \leq n$. Such a history denotes the communications of P along its observable channels up to some point in an execution.

Let $\langle\rangle$ denote the empty history, i.e. the sequence of length 0. The concatenation of two histories $h_1 = \langle(c_1, m_1), \ldots, (c_k, m_k)\rangle$ and $h_2 = \langle(d_1, n_1), \ldots, (d_l, n_l)\rangle$, denoted $h_1{}^\wedge h_2$, is defined as $\langle(c_1, m_1), \ldots, (c_k, m_k), (d_1, n_1), \ldots, (d_l, n_l)\rangle$. We use $h^\wedge(c, m)$ as an abbreviation of $h^\wedge\langle(c, m)\rangle$. We say that h_1 is a prefix of h_2, notation $h_1 \preceq h_2$, if there exists a t such that $h_1{}^\wedge t = h_2$. If $h_1 \preceq h_2$ and $h_1 \neq h_2$ then h_1 is a strict prefix of h_2, notation $h_1 \prec h_2$. A set of histories is prefix closed if every prefix of an element of the set is in itself an element of that set. Let $h(i)$ denote the ith element of history h. For $1 \leq l < u \leq len(h)$, $h[l, u]$ denotes the subsequence $h(l)^\wedge \ldots {}^\wedge h(u)$ of a history h. $h[l, u] = \langle\rangle$ if $l > u$, and $h[l, u] = h(l)$ if $l = u$. In particular, $h[1, u]$, abbreviated as $h[u]$, is h's prefix of length u.

An important operation on histories is the channel projection $h{\uparrow}cset$ denoting the sequence consisting of the elements of h which are communications on the channels in $cset$. We abbreviate $h{\uparrow}\{c\}$ as $h{\uparrow}c$. A related operation is the channel hiding $h \setminus cset$ denoting the sequence obtained from h by removing all $cset$ records. Renaming a channel c into a channel d in a history h, notation $h{<}d/c{>}$, means replacing all (c, v) records by (d, v).

In this report we use expressions of the form $(\Lambda c)^{\cdot}$ to denote the set of non-empty histories with elements in Λc; $(\Lambda c)^* = (\Lambda c)^{\cdot} \cup \{\langle\rangle\}$. The expression $(\Lambda c_1, \Lambda c_2)^{\cdot}$ generates sequences of, alternately, an element of Λc_1 and an element of Λc_2, starting with a Λc_1 element and ending with a Λc_2 element. The function PC returns the prefix closure of a set of histories.

For a concurrent programming language, e.g. CSP [4], it is possible to determine the prefix closed set $\mathcal{H}[\![P]\!]$ of histories that can be observed up to any point in some execution of process P. The definition of a trace-based semantics for the sequential fragment of a concurrent programming language is rather standard (see e.g. [4], [14]). In this report our main concern is the formalization of fault hypotheses in relation to concurrency. Therefore we do not give a syntax for sequential programs, but we characterize a sequential process P by giving its semantics $\mathcal{H}[\![P]\!]$.

A *fault hypothesis* χ of a process P is a predicate, whose only free variables are h_{old} and h, representing a reflexive relation on histories. The intuition is that h_{old} represents a normal history of process P, whereas h is an acceptable history of P with respect to χ. For a fault hypothesis χ the construct $(P \wr \chi)$ indicates execution of process P under the assumption of χ. Then, this construct enables us to specify failure prone processes. An obvious restriction on the fault hypothesis χ of a process P is that χ may only refer to the channels of P.

The set $\mathcal{H}[\![(P \wr \chi)]\!]$, that is, the set of acceptable behaviours of P with respect to χ, can be defined as follows:

$$\mathcal{H}[\![(P \wr \chi)]\!] \triangleq \{\, h \mid \exists t \in \mathcal{H}[\![P]\!] : \chi[t/h_{old}] \,\}$$

where $\chi[t/h_{old}]$ is χ with t substituted for h_{old}.

Because of the reflexivity of χ we have that $\mathcal{H}[\![P]\!] \subseteq \mathcal{H}[\![(P \wr \chi)]\!]$. The fault hypotheses χ we consider ensure that $\mathcal{H}[\![(P \wr \chi)]\!]$ is prefix closed. Notice that, because of its prefix closeness, $\mathcal{H}[\![P]\!]$ already contains the fail-silent behaviour of P. The operator $\wr$ is left-associative. Consequently, we write $(P \wr \chi_1 \wr \chi_2)$ instead of $((P \wr \chi_1) \wr \chi_2)$.

Given the histories of the failure prone processes P and Q, we are interested in the histories of $P\|Q$, the system composed of P and Q in parallel. Communication via channels is synchronous and a history only records communications that have been performed. Thus, a history h of $P\|Q$ has the property that $h{\uparrow}chan(P)$ and $h{\uparrow}chan(Q)$ match histories of P and Q respectively. This leads to the following definition:

$$\mathcal{H}[\![P\|Q]\!] \triangleq \{\, h \mid h{\uparrow}chan(P) \in \mathcal{H}[\![P]\!] \wedge h{\uparrow}chan(Q) \in \mathcal{H}[\![Q]\!] \wedge \\ h{\uparrow}chan(P\|Q) = h \,\}$$

In the above definition the last conjunct assures that a history of $P\|Q$ only contains channels of P or Q.

In case there is a conflict in the names and directions of the channels when combining failure prone processes, the channel renaming construct $P{<}d/c{>}$ can be used.

$$\mathcal{H}[\![P<d/c>]\!] \triangleq \{\, h \mid \exists t \in \mathcal{H}[\![P]\!] : h = t<d/c> \,\}$$

It is sometimes convenient to hide, or internalize, certain channels of a failure prone process. For $cset \subseteq chan(P)$,

$$\mathcal{H}[\![P\backslash cset]\!] \triangleq \{\, h \mid \exists t \in \mathcal{H}[\![P]\!] : h = t\backslash cset \,\}$$

Since the histories introduced in this section denote finite observations, they can only be used to verify safety properties [14]. Liveness properties are out of the scope of this report.

3   Modeling communication media

In this section we model communication media in terms of their histories. A communication medium M is a component with two observable channels, $chan(M) = \{in, out\}$, such that $\alpha in = \alpha out$. The history h of a communication medium can be characterized using a relation between the restriction of h to, respectively, *out* and *in*, i.e. $h{\uparrow}out$ and $h{\uparrow}in$. We start by determining the set of histories of a perfect communication medium, which is a communication medium that always behaves normally. Then, we consider media that corrupt and insert messages.

3.1   Perfect communication media

A perfect communication medium passes the messages received via its input channel to its output channel in the same order as received. Using the channel renaming operator we can characterize a perfect communication medium M by defining $\mathcal{H}[\![M]\!]$ as follows:

$$\mathcal{H}[\![M]\!] \triangleq \{\, h \in (\Lambda in \cup \Lambda out)^* \mid \forall t \preceq h : t{\uparrow}out<in/out> \preceq t{\uparrow}in \,\}$$

In this definition the use of t is necessary to exclude, for instance, histories that contain an output communication before an input communication.

3.2   Communication media that corrupt messages

To model the effects of corruption we define fault hypothesis *Cor* as follows:

$$Cor \triangleq \quad len(h) = len(h_{old})$$
$$\wedge \, h \backslash out = h_{old} \backslash out$$
$$\wedge \, \forall i : 1 \le i \le len(h_{old}) : ch(h_{old}(i)) = out \rightarrow ch(h(i)) = out$$

In the above definition the first two conjuncts are necessary to exclude undesired histories. The third conjunct captures the fact that the communicated value may be an arbitrary element of αout.

In practice, an encoding function *enc* maps the set of messages into a much larger set of codewords. Only a small fraction of the codewords are valid, that is, mapped into by *enc*. It is very unlikely that due to corruption one valid codeword is changed into another. Thus, the probability that corruption is not detected is made so small that it can be neglected. Let the function $Enc(V) = \{enc(v) \mid v \in V\}$ denote the set of encodings of values from V, and let its complement, $Enc'(V)$, denote the set of invalid codewords. The decoding function *dec* returns, for a valid codeword, the decoded counterpart; for an invalid codeword *dec* returns the distinguished value '†'.

A mechanism to deal with corruption of messages is to have a receiver acknowledge each message. If the received message is valid then the sender sends a positive acknowledgement, if not then it sends a negative acknowledgement. We use two (unidirectional) communication media, one — *MES* — to send messages and the other — *ACK* — to send acknowledgements, that interconnect two processes S and R, with $chan(S) = \{in, mes_{in}, ack_{out}\}$, and $chan(R) = \{mes_{out}, ack_{in}, out\}$, as exemplified in Figure 1.

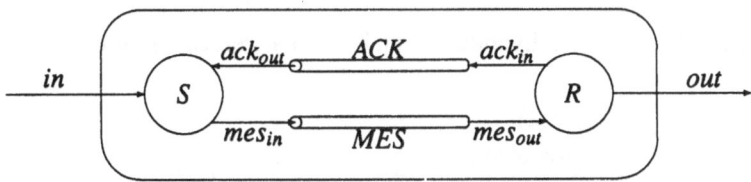

Figure 1: Duplex communication medium.

The sender process S encodes the messages received via channel *in* and relays them to channel mes_{in}. It then awaits an acknowledgement from R; if S receives a negative or a corrupted acknowledgement it retransmits the message. To deal with the duplication that is caused by the corruption of a positive acknowledgement, each message and each acknowledgement should be uniquely identifiable. To do so, we implement the indices of $h{\uparrow}in$ as sequence numbers which are attached to a message. Positive acknowledgements consist of the sequence number of the message they acknowledge; negative acknowledgements consist of a zero. Consequently, $\alpha mes_{in} = Enc(\alpha in \times \mathbf{N})$, and $\alpha ack_{in} = Enc(\mathbf{N} \cup \{0\})$. This is, in effect, a generalization of the alternating bit protocol (cf. [1]).

Let seq and msg be functions which, for an element of $\alpha in \times \mathbf{N}$, return the sequence number and the message, respectively. The set $\mathcal{H}[\![S]\!]$ can be specified as follows:

$$\mathcal{H}[\![S]\!] \triangleq \{ h \in PC((\Lambda in, (\Lambda mes_{in}, \Lambda ack_{out})^*)^*) \mid$$

$$\forall i : ch(h(i)) = in : i = 1 \vee i > 1 \rightarrow h(i-1) = (ack_{out}, enc(len(h[i-1]{\uparrow}in)))$$
$$\wedge \ \forall i : ch(h(i)) = mes_{in} : \quad dec(val(h(i))) = (len(h[i]{\uparrow}in), val(h{\uparrow}in(len(h[i]{\uparrow}in))))$$
$$\wedge \quad ch(h(i-1)) = in$$
$$\vee \ ch(h(i-1)) = ack_{out} \wedge \quad dec(val(h(i-1))) = \text{`†'}$$
$$\vee \ dec(val(h(i-1))) = 0 \ \}$$

The receiver process R properly acknowledges the messages it receives via channel mes_{out}. It sends original valid messages to the environment:

$$\mathcal{H}[\![R]\!] \triangleq \{ h \in PC(((\Lambda mes_{out}, \Lambda ack_{in})^*, \Lambda out)^*) \mid$$

$$\forall i : ch(h(i)) = ack_{in} : \quad dec(val(h(i-1))) = \text{`†'} \rightarrow h(i) = (ack_{in}, enc(0))$$
$$\wedge \ dec(val(h(i-1))) \neq \text{`†'} \rightarrow$$
$$h(i) = (ack_{in}, enc(seq(dec(val(h(i-1))))))$$
$$\wedge \ (len(h) > i \wedge seq(dec(val(h(i-1)))) > len(h[i]{\uparrow}out)) \rightarrow$$
$$h(i+1) = (out, msg(dec(val(h(i-1))))) \ \}$$

The detectable corruption hypothesis can be formalized as follows:

$$DetCor \triangleq \quad len(h) = len(h_{old})$$
$$\wedge h \backslash mes_{out} = h_{old} \backslash mes_{out}$$
$$\wedge \forall i : 1 \leq i \leq len(h_{old}) : ch(h_{old}(i)) = mes_{out} \rightarrow \quad ch(h(i)) = mes_{out}$$
$$\wedge \quad val(h(i)) = val(h_{old}(i))$$
$$\vee val(h(i)) \in Enc'(\alpha in \times \mathbf{N})$$

Let $CMES \triangleq (MES \wr DetCor)$ be the communication medium that carries and possibly corrupts messages, and let
$CACK \triangleq (ACK \wr DetCor[\ ack_{out}/mes_{out}\ ,\ Enc'(\mathbf{N} \cup \{0\})/Enc'(\alpha in \times \mathbf{N})\ ])$ be the one that does so with acknowledgements. As S does not take another message

from *in* before the previous message has been positively acknowledged, it is possible to show that

$$\mathcal{H}[\![(S\|CMES\|CACK\|R)\setminus\{ack_{in}, ack_{out}, mes_{in}, mes_{out}\}]\!] = \mathcal{H}[\![M]\!]$$

which proves tolerance.

3.3 Communication media that insert messages

Assuming that due to insertion no valid messages are generated, the phenomenon of insertion can be modeled by defining:

$$
\begin{aligned}
DetIns \triangleq \quad & h = h_{old} \\
& \vee \; \exists i : 1 \le i \le len(h) \wedge (\exists v \in Enc'(\alpha in \times \mathbb{N}) : h(i) = (mes_{out}, v)) : \\
& \qquad\qquad\qquad\qquad h[1, i-1]\char94 h[i+1, len(h)] = h_{old} \\
& \vee \; \exists i_1, \ldots, i_n : \quad 1 \le i_1 < \ldots < i_n \le len(h) \\
& \qquad\qquad \wedge \; \forall j : (\exists v \in Enc'(\alpha in \times \mathbb{N}) : h(i_j) = (mes_{out}, v)) : \\
& \qquad\qquad\qquad\qquad h[1, i_1-1]\char94 \ldots \char94 h[i_n+1, len(h)] = h_{old}
\end{aligned}
$$

Let $ICMES \triangleq (M \wr DetCor \wr DetIns)$ be the medium that inserts and corrupts messages, and let
$ICACK \triangleq (CACK \wr DetIns[\, ack_{out}/mes_{out} \,,\; Enc'(\mathbb{N} \cup \{0\})/Enc'(\alpha in \times \mathbb{N})\,])$
be the medium that does so with acknowledgements. R sending an extra negative acknowledgement upon reception of an inserted message will only result in S resending the most recent unacknowledged message. Notice that this kind of duplication does not differ from the duplication that results from the corruption of positive acknowledgements. Hence,

$$\mathcal{H}[\![(S\|ICMES\|ICACK\|R)\setminus\{ack_{in}, ack_{out}, mes_{in}, mes_{out}\}]\!] = \mathcal{H}[\![M]\!]$$

from which we conclude tolerance.

4 Designing a stable disk

A stable disk is a store of data which is highly reliable. It has a capacity of *MaxSec* sectors, each identified by a number from a set *SecNum*, and is capable of storing a data item with a value from a set *Val*. There is an operation *SWrite(sn, v)* to store

v in sector sn, and an operation $SRead(sn)$ to read the value previously stored in sector sn.

The disk receives $SRead$ and $SWrite$ requests from the environment via observable channel in:

$$\alpha in = \{\text{`SWrite'}\} \times SecNum \times Val \cup \{\text{`SRead'}\} \times SecNum$$

Define, for an element of Λin, the auxiliary functions op, sn, and v which return, respectively, the operation type, the sector number, and, where appropriate, the value.

Let '$\sqrt{}$' $\notin Val$ denote successful termination of an $SWrite$ operation. The disk sends the responses via observable channel out to the environment. Each response contains an index identifying the originating request, i.e. the index of the originating request relative to channel in. Consequently,

$$\alpha out = (Val \cup \{\text{`}\sqrt{}\text{'}\}) \times \mathbb{N}$$

For an element of Λout the functions req and res return, respectively, the element's request number field and the response part. We use function $org(h(i))$ to return the index j of h such that $ch(h(j)) = in$ and $len(h[j] \uparrow in) = req(h(i))$. Consequently, $h(org(h(i))) = h \uparrow in(req(h(i)))$.

Informally, the stable disk can be specified as follows:

> *if S^1 and S^2 are successive operations on the same sector, then they should satisfy the following properties:*
>
> 1. *if S^1 is an SWrite operation and S^2 is an SRead operation, then the value read by S^2 is the same as the value written by S^1, and*
>
> 2. *if S^1 and S^2 are both SRead operations, then they read the same value.*

4.1 A perfect disk

A perfect disk always has the value that was last written to a particular sector available. Let, for $s \in SecNum$ and $h \in (\Lambda in)^*$, the function $SN(s, h)$ return the sequence of input requests of h relating to sector s. Furthermore, let, for $ot \in \{\text{'SRead', 'SWrite'}\}$ and $h \in (\Lambda in)^*$, the function $Op(ot, h)$ return the sequence of input requests of h with operation type ot. Then, the set $\mathcal{H}[\![D]\!]$ of histories of a perfect disk D can be specified as follows:

$$
\begin{aligned}
\mathcal{H}[\![D]\!] \triangleq \{ h \in PC((\Lambda in \cup \Lambda out)^*) \mid \\
\forall i : ch(h(i)) = out : \\
op(h(org(h(i)))) = \text{'SWrite'} \rightarrow res(h(i)) = \text{'}\sqrt{\text{'}} \\
\wedge\, op(h(org(h(i)))) = \text{'SRead'} \rightarrow \\
res(h(i)) = LastWrite(sn(h(org(h(i)))), h[org(h(i))]{\uparrow}in) \}
\end{aligned}
$$

Here $LastWrite(s, h)$ abbreviates $v(Last(Op(\text{'SWrite'}, SN(s, h))))$, using $Last(h)$ to denote $h(len(h))$.

4.2 An unreliable disk

An unreliable physical disk may not always have the last written value available. If for some sector one of the previously specified properties is violated the stable disk is said to have lost that sector's value. In such a case we want an *SRead* operation to return the distinguished value '†' $\notin Val$. This constitutes a reliable *SRead* operation, and, hence, $\alpha out = (Val \cup \{\text{'}\sqrt{\text{'}}, \text{'}\dagger\text{'}\}) \times \mathbb{N}$.

In this section we confine ourselves to transient faults. We consider wipe outs and decay of the information magnetically recorded on the disk, and malfunctions of the disk's head positioning circuitry. It should be noted that, because of the transientness of the faults, not all of them lead to an *SRead* failure: a successful *SWrite* operation undoes the effects of preceding faults.

4.2.1 Disks that corrupt sectors after a successful write

Informally, a disk that corrupts sectors can be specified as follows:

> *if S^1 and S^2 are successive operations on the same sector, then they should satisfy the following properties:*
>
> 1. *if S^1 is an SWrite operation and S^2 is an SRead operation, then the value read by S^2 is the same as the value written by S^1, and*
>
> 2. *if S^1 and S^2 are both SRead operations, then they read the same value, and*

> 3. *if 1. and 2. cannot be satisfied, because in between S^1 and S^2 the*
> *value stored in the sector has been corrupted, then S^2 returns '†'.*

Analogous to our treatment of communication media we use an encoding algorithm, for instance a cyclic redundancy check (CRC) mechanism, so that we can
neglect the possibility that corruption is not detected. In particular, we use the
functions *enc* and *dec* which were introduced in Section 3.2. Doing so, the domain
of sector values that are actually stored on the disk, say *LVal* ('$\sqrt{}$' $\notin$ *LVal*), is
divided into a set *Enc(Val)* of valid values and a set *Enc'(Val)* of invalid values:

$$LVal = Enc(Val) \cup Enc'(Val)$$

The use of this CRC mechanism results in low level variants of the *SRead* and
SWrite operations, denoted *LRead* and *LWrite* respectively. Let *DISK* be the low
level variant of *D*. A process *S* relates the original input channel to channel low_{in}
(by replacing the *Val* elements by their CRC encoded counterparts), and channel
low_{out} to the original output channel (by applying a CRC check to the output
values), respectively. Consequently,

$$\alpha low_{in} = \{\text{'}LWrite\text{'}\} \times SecNum \times Enc(Val) \cup \{\text{'}LRead\text{'}\} \times SecNum$$

and

$$\alpha low_{out} = (LVal \cup \{\text{'}\sqrt{}\text{'}\}) \times \mathbf{N}$$

The functions *lop*, *lsn*, *lv*, *lreq*, *lres*, *lorg*, *LOp* and *LSN* are the low level counterparts of *op*, *sn*, *v*, *req*, *res*, *org*, *Op* and *SN*, respectively. We define the operation
LastLWrite, *LastWrite*'s internal counterpart, as follows:

$$LastLWrite(s, h) \triangleq lv(Last(LOp(\text{'}LWrite\text{'}, LSN(s, h))))$$

We model corruption as the insertion of an *LWrite* request with the *LVal* field set
to an *Enc'(Val)* element. This insertion can be regarded as initiated by the adverse
environment [2].

$DCor \triangleq$
$\quad len(h) = len(h_{old})$
$\quad \wedge\ h \backslash low_{out} = h_{old} \backslash low_{out}$
$\quad \wedge\ \forall i : 1 \leq i \leq len(h) : ch(h_{old}(i)) = low_{out} \to$
$\qquad lres(h_{old}(i)) = `\sqrt{}` \to h(i) = h_{old}(i)$
$\qquad \wedge\ lres(h_{old}(i)) \neq `\sqrt{}` \to\quad ch(h(i)) = low_{out}$
$\qquad\qquad\qquad\qquad\qquad\quad \wedge\ lreq(h(i)) = lreq(h_{old}(i))$
$\qquad\qquad\qquad\qquad\qquad\quad \wedge\ \exists t : DIns[\, t/h\, ,\ h[lorg(h(i))]/h_{old}\,] :$
$\qquad\qquad\qquad\qquad\qquad\qquad lres(h(i)) = LastLWrite(lsn(h(lorg(h(i)))), t\mathord{\uparrow}low_{in})$

where

$DIns \triangleq$
$\quad h = h_{old}$
$\quad \vee\ \exists i : 1 \leq i < len(h) \wedge (\exists s \in SecNum, v \in Enc'(Val) : h(i) = (low_{in}, (`LWrite', s, v))) :$
$\qquad\qquad\qquad\qquad\qquad\qquad\qquad\qquad\qquad\quad h[1, i-1]{}^{\wedge}h[i+1, len(h)] = h_{old}$
$\quad \vee\ \exists i_1, \ldots, i_n :\quad 1 \leq i_1 < \ldots < i_n < len(h)$
$\qquad\qquad\qquad\quad \wedge\ \forall j : (\exists s \in SecNum, v \in Enc'(Val) : h(i_j) = (low_{in}, (`LWrite', s, v))) :$
$\qquad\qquad\qquad\qquad\qquad\qquad\qquad\qquad\quad h[1, i_1-1]{}^{\wedge} \ldots {}^{\wedge}h[i_n+1, len(h)] = h_{old}$

This method of specification fits the fact that the exact moment at which corruption occurs is not observable. Notice that the intermediate history t contains the original and the inserted input communications, but still the original output communications.

The specification of the set of histories of S reads:

$\mathcal{H}[\![S]\!] \triangleq \{\, h \in PC((\Lambda in, \Lambda low_{in}, \Lambda low_{out}, \Lambda out)^*) \mid$
$\quad \forall i : ch(h(i)) = low_{in} :$
$\qquad lsn(h(i)) = sn(h(i-1))$
$\qquad \wedge\ op(h(i-1)) = `SRead' \to lop(h(i)) = `LRead'$
$\qquad \wedge\ op(h(i-1)) = `SWrite' \to (lop(h(i)) = `LWrite' \wedge lv(h(i)) = enc(v(h(i-1))))$
$\quad \wedge\ \forall i : ch(h(i)) = out :$
$\qquad req(h(i)) = lreq(h(i-1))$
$\qquad \wedge\ lres(h(i-1)) = `\sqrt{}` \to res(h(i)) = `\sqrt{}`$
$\qquad \wedge\ lres(h(i-1)) \in LVal \to res(h(i)) = dec(lres(h(i-1))) \,\}$

The histories of $\mathcal{H}[\![(DISK \wr DCor) \| S]\!]$ satisfy the requirements stated above.

4.2.2 Disks that misplace sectors

Informally, a disk that corrupts or misplaces sectors can be specified as follows:

if S^1 and S^2 are successive operations on the same sector, then they should satisfy the following properties:

1. *if S^1 is an SWrite operation and S^2 is an SRead operation, then the value read by S^2 is the same as the value written by S^1, and*

2. *if S^1 and S^2 are both SRead operations, then they read the same value, and*

3. *if 1. and 2. cannot be satisfied, because in between S^1 and S^2 the value stored in the sector has been corrupted or overwritten by a misplaced SWrite operation for another sector, or if either S^1 or S^2 has been misplaced, then S^2 returns '†'.*

This is the case where an operation is performed on a wrong sector, which we formalize as the alteration of the operation's set sector number. We could define a fault hypothesis *DMisplace* analogous to *DCor* but because of the use of intermediate histories, which is necessary to achieve transparency, the histories in $\mathcal{H}[\![(DISK \wr DMisplace \wr DCor)]\!]$ would show the effects of corruption or misplacement, but not both. Hence, we define fault hypothesis *DMisCor* such that it captures both.

DMisCor $\triangleq$

$$
\begin{aligned}
&len(h) = len(h_{old}) \\
&\wedge\ h \setminus low_{out} = h_{old} \setminus low_{out} \\
&\wedge\ \forall i : 1 \le i \le len(h) : ch(h_{old}(i)) = low_{out} \rightarrow \\
&\qquad lres(h_{old}(i)) = \text{'}\sqrt{}\text{'} \rightarrow h(i) = h_{old}(i) \\
&\qquad \wedge\ lres(h_{old}(i)) \neq \text{'}\sqrt{}\text{'} \rightarrow \\
&\qquad\qquad ch(h(i)) = low_{out} \\
&\qquad\qquad \wedge\ lreq(h(i)) = lreq(h_{old}(i)) \\
&\qquad\qquad \wedge\ \exists t_1, t_2 : DIns[\ t_2/h\ ,\ t_1/h_{old}\ ] \wedge Misplace[\ t_1/h\ ,\ h[lorg(h(i))]/h_{old}\ ] : \\
&\qquad\qquad\qquad lres(h(i)) = LastLWrite(lsn(h(lorg(h(i)))), t_2 \uparrow low_{in})
\end{aligned}
$$

where

$$
\begin{aligned}
Misplace\ \triangleq\ &len(h) = len(h_{old}) \\
&\wedge\ h \setminus low_{in} = h_{old} \setminus low_{in} \\
&\wedge\ \forall i : 1 \le i \le len(h) : ch(h_{old}(i)) = low_{in} \rightarrow \\
&\qquad ch(h(i)) = low_{in} \\
&\qquad \wedge\ lop(h(i)) = lop(h_{old}(i)) \\
&\qquad \wedge\ lop = \text{'}LWrite\text{'} \rightarrow lv(h(i)) = lv(h_{old}(i))
\end{aligned}
$$

In analogy with message passing systems, detection is possible if the destination, in this case the sector number, is encoded in the sector value. However, this method is not capable of detecting obsolescence: if after a misplaced write operation for a sector a subsequent read operation is executed correctly then the read value is valid, yet obsolete. Detection of obsolescence is possible by implementing the number of write operations on a particular sector as incarnation number in that sector, and maintaining an incarnation number table. Then obsolescence manifests itself as incarnation number inconsistency. Notice that misplacement does not affect the updating of the incarnation number table. Furthermore, the sector number is checked before the incarnation number.

Introduce the set $SIVal = SecNum \times \mathbb{N} \times Val$ plus the functions esn, inc, and siv that return, for an element of $SIVal$, the encoded sector number, the encoded incarnation number, and the value, respectively. Together with the CRC encoding discussed in the previous section, this results in a set $LSIVal = Enc(SIVal) \cup Enc'(SIVal)$ of values actually stored on the disk. As a result,

$$\alpha low_{in} = \{`LWrite'\} \times SecNum \times Enc(SIVal) \cup \{`LRead'\} \times SecNum$$

and

$$\alpha low_{out} = (LSIVal \cup \{`\sqrt{}'\}) \times \mathbb{N}$$

Adapt the functions lop, lsn, lv, $lreq$, $lres$, LOp and LSN to these new domains. Let $AddSI$ and $RemSI$ denote the functions that add, respectively remove the sector number and the incarnation number. For an element $h(i) \in \Lambda low_{out}$ which is a valid response to an $LRead$ request the Boolean function $Pass$ checks the encoded sector number and the incarnation number, i.e.

$Pass(h(i)) \triangleq$
$esn(dec(lres(h(i)))) = lsn(h(lorg(h(i))))$
$\wedge\ inc(dec(lres(h(i)))) = len(LOp(`LWrite', LSN(lsn(h(lorg(h(i)))),$
$$h[lorg(h(i))]\uparrow low_{in})))$$

with the intuition that if $Pass(h(i))$ does not hold, then $RemSI(dec(lres(h(i))))$ returns '†'.

The set of histories of process S_2 can be specified as follows:

$$\mathcal{H}[S_2] \triangleq$$
$$\{\, h \in PC((\Lambda in, \Lambda low_{in}, \Lambda low_{out}, \Lambda out)^*) \mid$$
$$\forall i : ch(h(i)) = low_{in} :$$
$$lsn(h(i)) = sn(h(i-1))$$
$$\wedge\, op(h(i-1)) = \text{'SRead'} \rightarrow lop(h(i)) = \text{'LRead'}$$
$$\wedge\, op(h(i-1)) = \text{'SWrite'} \rightarrow$$
$$(lop(h(i)) = \text{'LWrite'} \wedge lv(h(i)) = enc(AddSI(v(h(i-1)))))$$
$$\wedge\, \forall i : ch(h(i)) = out :$$
$$req(h(i)) = lreq(h(i-1))$$
$$\wedge\, lres(h(i-1)) = \text{'}\surd\text{'} \rightarrow res(h(i)) = \text{'}\surd\text{'}$$
$$\wedge\, lres(h(i-1)) \in LSIVal \rightarrow$$
$$res(h(i)) = RemSI(dec(lres(h(i-1)))) \,\}$$

Then it can be shown that the histories of $\mathcal{H}[(DISK \wr DMisCor) \| S_2]$ satisfy the above stated requirements.

4.3 Design of the stable disk

We want the disk to be stable, ideally with a guarantee that a stored value is never lost. In this section we describe how the stability of the disk originates from the multiple disk devices from which it is constructed (see Figure 2).

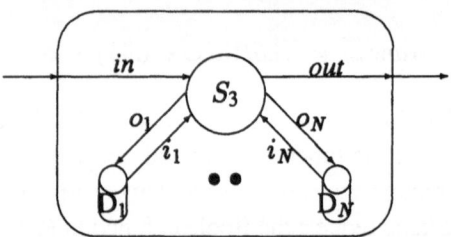

Figure 2: Design of the stable disk.

Each disk contains the same data; the intuition is that data lost on one disk is usually still available on another. For the individual disks we consider the same fault models as elaborated before, that is

$$\forall j : 1 \leq j \leq N : \mathcal{H}[D_j{<}low_{in}/i_j\,,\ low_{out}/o_j{>}] = \mathcal{H}[(DISK \wr DMisCor)]$$

Because each disk should contain the same data, an *SWrite* request has to be executed on each disk. A simple strategy to service read requests is to start on disk D_1, try disk D_2 if that is not successful, and so on. If some disk cannot service a read request, because the sector's data are lost, then the data will have to be retrieved on another disk. This creates internal requests. If an internal read request has been serviced successfully, internal write requests for the disks that need recovery are generated. The following example illustrates this scenario.

Example Consider $\langle (in, (W, 1, 3)), (out, (1, \checkmark)), (in, (R, 1)), (out, (2, 3)) \rangle$ as high level history, where 'R' and 'W' denote '*SRead*' and '*SWrite*', respectively. Using 'r' and 'w' to denote '*LRead*' and '*LWrite*', and '$\star$' to denote a response containing an invalid value, the low level history for an implementation consisting of 2 disk devices may look like: $\langle (in, (W, 1, 3)), (o_1, (w, 1, enc(AddSI(3)))), (i_1, (1, \checkmark)), (o_2, (w, 1, enc(AddSI(3)))), (i_2, (1, \checkmark)), (out, (1, \checkmark)), (in, (R, 1)), (o_1, (r, 1)), (i_1, (2, \star)), (o_2, (r, 1)), (i_2, (2, enc(AddSI(3)))), (o_1, (w, 1, enc(AddSI(3)))), (i_1, (3, \checkmark)), (out, (2, 3)) \rangle$.

5 Conclusions and future research

A number of formal methods for fault tolerance have been proposed in the literature. Much of the, by now classical, work on the formalization of fault tolerance is state based. In the state machine approach the output of several instantiations of a program, each running on a distinct processor, is compared. Lamport's original description [6] dealt with fault-free environments only; for a survey of the efforts to generalize the state machine approach to deal with faults see [12]. A well-known application of the state machine approach is the implementation of fail-stop processors [11].

Cristian [2] uses Hoare logic to make the normal and exceptional domain explicit by partitioning the initial state space into disjoint subspaces for normal and exceptional behaviour, and providing a separate specification for each part. Started in the normal subspace the program terminates normally, but started in the exceptional subspace the program terminates exceptionally, i.e by raising an exception. This formalism is, however, not natural in case there may arise a non-deterministic situation from which both normal and exceptional behaviour can result. Such situations occur, for instance, when modeling faults affecting communication.

The more recent literature proposes trace based formalisms. Weber (cf. [13]) uses fault scenarios which are traces that include records of the system's input and output operations and also a description of the faults that have occurred. A fault tolerance property is expressed as an equivalence between a fault scenario from which the fault events are removed and a fault free history. This equivalence relation, called tolerance relation, is, however, not elaborated in [13].

More CSP-based are the formalisms of Joseph, Moitra, and Soundarajan [5] and Peleska [9]. Like Weber's formalism, these formalisms treat faults explicitly.

In the formalism presented in this report the focus is not on the faults that cause exceptional behaviour, but rather on the effects these faults have on the observable behaviour of a component. The behaviour of a process is denoted by a trace which is called a history. A fault hypothesis is formalized as a predicate representing a reflexive relation between the normal and acceptable histories of a process.

The use of the method was illustrated by applying it to two examples: a communication medium and a stable disk. To model the effects of faults on the observable behaviour of the two exemplary systems different approaches were taken. For the communication medium the relation between the normal and acceptable behaviour concentrated on the sequence of output communications; for the stable disk that relation concentrated on the sequence of input communications. To achieve transparency in the stable disk case intermediate histories had to be used, which prohibited the composition of fault hypotheses that was possible for the fault hypotheses of the communication medium. To overcome this problem the formalism must deal with such intermediate histories explicitly.

A natural continuation is the definition of a programming language, an assertion language and a compositional proof theory. This proof theory enables one to prove the validity of correctness formulae of the form P **sat** ϕ, where P is a failure prone process and ϕ a safety property.

Time has not been included in this formalism. Extension to real-time is necessary to treat the important — though not mentioned here — class of timing faults. Let **T** be a time domain, i.e. a structure of instants with a least element (0), no greatest element, and an ordering ($<$). Then we define a timed history (cf. [3]) as a finite sequence of the form $\langle (c_1, v_1, T_1), \ldots, (c_n, v_n, T_n) \rangle$, with $n \in \mathbf{N}$, $T_i \in \mathbf{T}$ ($T_{i+1} \geq T_i$), and $v_i \in \alpha c_i$, for $1 \leq i \leq n$. We allow $T_{i+1} = T_i$ only for communications on channels of different processes. Note that in the timed framework verification no longer means proving equality of sets of histories, since the tolerant system may and will be slower.

In this report nothing has been said about graceful degradation. For example, the failure of one of the three components in a triple-modular-redundancy configuration usually results in the degraded duplication-with-comparison configuration. We currently investigate modeling graceful degradation as switching to another, less ambitious, set of acceptable histories.

Acknowledgments

I am very grateful for the helpful comments of Willem-Paul de Roever, Jozef Hooman, Jos Coenen, Jos Baeten, and, especially, Mathai Joseph on earlier drafts of this report. I would like to thank the referees for their useful remarks.

References

[1] K. A. Bartlett, R. A. Scantlebury, P. T. Wilkinson. A note on reliable full-duplex transmission over half-duplex links. *Communications of the ACM*, Vol. 12, No. 5, 1969, pp. 260-261.

[2] F. Cristian. A rigorous approach to fault-tolerant programming. *IEEE Transaction on Software Engineering*, Vol. SE-11, No. 1, pp. 23-31, 1985.

[3] T. A. Henzinger, Z. Manna , A. Pnueli. Timed transition systems. *Lecture Notes in Computer Science*, Vol. 600, Springer-Verlag, 1992, pp. 226-251.

[4] C. A. R. Hoare. *Communicating Sequential Processes*. Prentice-Hall International, 1985.

[5] M. Joseph, A. Moitra , N. Soundararajan. Proof rules for fault tolerant distributed programs. *Science of Computer Programming*, Vol. 8, 1987, pp. 43-67.

[6] L. Lamport. Time, clocks, and the ordering of events in a distributed system. *Communications of the ACM*, Vol. 21, No. 7, 1978, pp. 558-565.

[7] J. C. Laprie. Dependable computing and fault tolerance: concepts and terminology. *Proc. 15th IEEE Int. Symp. on Fault Tolerant Computing*, Ann Arbor, Mich., 1985, pp. 2-11.

[8] P. A Lee , T. Anderson. *Fault tolerance: principles and practice*. Springer-Verlag, 1990.

[9] J. Peleska. Design and verification of fault tolerant systems with CSP. *Distributed Computing*, Vol. 5, 1991, pp. 95-106.

[10] H. Schepers. Terminology and paradigms for fault tolerance. *Report CSN 91–08*, Eindhoven University of Technology, 1991. Also to appear in: J. Vytopil (ed.). *Formal Techniques in Real-Time and Fault Tolerant Systems*. Kluwer Academic Publishers, 1993.

[11] R. D. Schlichting , F. B. Schneider. Fail-stop processors: an approach to designing fault tolerant computing systems. *ACM Transaction on Computer Systems*, Vol. 1, No. 3, 1983, pp. 222-238.

[12] F. B. Schneider. Implementing fault tolerant services using the state machine approach: a tutorial. *ACM Computing Surveys*, Vol. 22, No. 4, 1990, pp. 299-319.

[13] D. G. Weber. Formal specification of fault-tolerance and its relation to computer security. *ACM Software Engineering Notes*, Vol. 14, No. 3, 1989, pp. 273-277.

[14] J. Zwiers. Compositionality, concurrency and partial correctness. *Lecture Notes in Computer Science*, Vol. 321, Springer-Verlag, 1989.

DEPENDABILITY AND PERFORMANCE

EVALUATION OF

FAULT-TOLERANT SOFTWARE:

A PERFORMABILITY MODELING

APPROACH

Ann T. TAI[1], *Algirdas AVIŽIENIS*[2], *John F. MEYER*[3]
[1]*Computer Science Program, University of Texas at Dallas*
Dallas, Texas 75083, USA
[2]*Department of Computer Science, University of California at Los Angeles*
Los Angeles, California 90024, USA
[3]*Department of Electrical Engineering and Computer Science*
University of Michigan
Ann Arbor, Michigan 48109, USA

Abstract

A comparative evaluation of recovery blocks and N-version programming ($N = 3$) is accomplished by means of performability modeling. For each scheme, a corresponding stochastic process model is constructed by employing a hierarchical modeling framework. Comparison is based on a performability measure that quantifies software "effectiveness" in a designated operational environment. The evaluation results reveal some interesting differences between the two schemes; in addition, they point to certain inadequacies in the use of computational redundancy which could serve as the basis for design modification.

1   Introduction

Investigations of fault tolerance techniques for software have been conducted since the early 1970's [1]. Much of this effort has focused on two well-documented approaches, namely, recovery blocks (RB) [2] and N-version programming (NVP) [3]. Included here is the important issue of how such schemes are evaluated, either experimentally or by model-based techniques. Research in the latter regard has concerned both *performance* [4], and *dependability* [4],[5], [6], [7]. However, since fault tolerance mechanisms usually have a simultaneous effect on both performance and dependability, combined assessment of the two attributes, via measures of *performability* [8], is also desirable.

Since the late 1970's, work on the development and application of performability models in a computing context (see a recent survey [9]) has been primarily oriented toward hardware systems. Some studies have considered the relationships between performance and dependability in software. Gelenbe and Mitrani studied the effect of errors and recovery on the execution time of Algol-like programs [10]. Chimento and Trivedi analyzed the execution time distribution of block structured programs run on processors subject to failure and repair [11]. Hsueh and Iyer's measurement-based performability model considers both hardware and software errors and reflects the interaction between system components [12].

Motivated by the prior work cited above, we have developed a performability modeling framework that is suited to the particular needs of software evaluation [13]. The model consists of a representation of the total system, together with a performance variable that quantifies a program's value (worth) in a specified operational environment. The model emphasizes the representation of interactions between performance attributes and dependability attributes and, moreover, captures the behavior of the software in its operational environment. The central purpose of this paper is to demonstrate the feasibility of software performability modeling via a comparative evaluation of RB and NVP. Informally, the performability measure is defined with respect to the number of "successful" iterations of the software's execution that benefit the user during a bounded time period of duration t. Letting M_t denote this variable, the measure in question is taken to be its expected value $E[M_t]$, referred to as the software's "effectiveness."

Section 2 presents some additional background for the investigation, followed by Section 3 which provides a more precise definition of the variable M_t. Section 4 describes the general modeling framework, with Sections 5 and 6 being devoted to the specific RB and NVP models, respectively. In conclusion, Section 7 presents the comparative evaluation results and discusses their implications.

2   Background

The fault-tolerant software systems evaluated here are for real-time applications of an open loop type. At the beginning of an iteration, the program accepts input; at the end of the iteration, it provides output that is a function only of the most recently accepted input value. If the output of an iteration is determined to be erroneous, it will be suppressed and/or some default value will be provided, and the system then resumes normal operation with the next iteration. However, an undetected error will cause a catastrophic failure which rules out all the benefit from prior normal operation. Further, each iteration is under a real-time constraint which takes the form of an upper bound on the time to complete an iteration. We assume that a fault-tolerant software system includes an error-free watchdog timer capable to detect violations of the real-time constraint. If the execution of an iteration does not complete before the deadline is reached, it will be aborted by the watchdog timer and a new iteration then begins. The treatment for such a detected timing error is the same as for a detected value error.

The use of redundancy in software fault tolerance introduces "similarity" considerations for results and errors [1]. The results of individual components may differ within a certain range when design diversity is used. Similar results are defined to be two or more results that are within the range of variation allowed by a decision algorithm or logically agree. When similar results are erroneous, they are called similar errors. Accordingly, two or more design faults considered to be "related" if they cause similar errors. On the other hand, such faults are regarded as being "independent" if the errors they cause are distinct. The manifestation of related faults (in the form of similar errors), is triggered by certain input data, and is a consequence of the interaction between the program deficiencies and environmental conditions. In the interest of simplifying terminology, the phrase *probability of a related fault* will mean "probability of a manifestation of a related fault"; likewise, *probability of an independent fault* means "probability of a manifestation of an independent fault."

3 The performance variable

We presume that the accomplishment of software in its operational context is a direct function of the number of *successful* iterations during some bounded time interval $[0, t]$ that *benefit* the user, where t expresses the duration of the mission. Specifically, a successful iteration refers to one that provides correct and timely output; further, it benefits the user provided there are no catastrophic failures during the mission. Accordingly, we define a performance variable M_t which quantifies the software's accomplishment as follows. An iteration which generates a correct and timely result is counted as one unit by M_t, an iteration which results in a detectable error (value or timing) is considered to have no contribution to the mission and thus it is not accounted for by M_t, and an iteration which results in an undetected erroneous result corresponds to a catastrophic failure. The latter causes the mission to be worthless, thus resulting in zero-level accomplishment ($M_t = 0$). We also define a set of random variables which describe the underlying stochastic process and support the evaluation of M_t:

I_t = number of iterations which generate correct results during $[0, t]$.

D_t = number of iterations which generate detected errors during $[0, t]$.

N_t = number of iterations which generate undetected errors during $[0, t]$.

It is noteworthy that these random variables are system-oriented and application-independent while the performance variable M_t is user-oriented and application-dependent. For example, if N_t is positive implying catastrophic failure, M_t definitely becomes zero but I_t can still be positive. On the other hand, the performance variable can be formulated via its relationships with I_t and N_t, that is

$$M_t = \begin{cases} I_t & \text{if } N_t = 0 \\ 0 & \text{otherwise} \end{cases}$$

As mentioned in the opening section, our particular interest is in the quantity of benefit which can be expected by the user. Accordingly, we define the *effectiveness* of the object software be the expected value of M_t ($E[M_t]$).

4 A hierarchical approach

To support the evaluation of effectiveness, we construct a base model (the underlying stochastic process) in a hierarchical manner. The base model is implemented

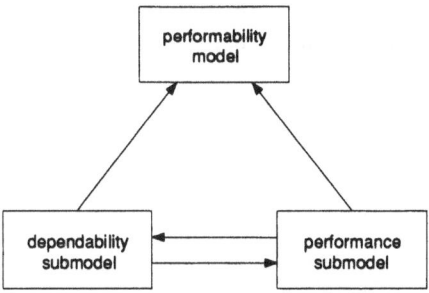

Figure 1: A hierarchical representation.

in two layers, as shown in Figure 1. The lower layer consists of a dependability submodel and a performance submodel. The dependability submodel is a discrete-time, finite-state Markov process describing the software's failure behavior. The dependability model is responsible for supplying the probabilities that a single iteration will succeed, fail detectably, or cause a catastrophic failure. The performance model is a renewal (counting) process representing the iterative nature of the software's execution, in which each program iteration is represented by a renewal cycle. This model is responsible for supplying the mean and variance of the renewal cycle time. The arrows between the lower layer submodels in Figure 1 indicate the interactions between the two attributes. Examples are 1) the probability of a detected value error, as evaluated by the dependability submodel, increases the probability of activating the use of the computational redundancies and thus increases the mean program iteration time, and 2) the probability of real-time constraint violation, as evaluated by the performance submodel, contributes to the probability of a degraded output due to a detected timing error. The information supplied by the lower layer of the base model determines its upper layer which, in turn, directly supports the evaluation of effectiveness. This hierarchical modeling framework enables us to represent different fault-tolerant software systems in a similar manner, as illustrated in the following sections.

5 The RB model

Figure 2 shows the operation of the RB scheme. The system has two alternative programs, namely, the primary and the secondary. The system also includes an acceptance test which checks the correctness of the outputs of the alternatives, and a watchdog timer which detects real-time constraint violations. The system

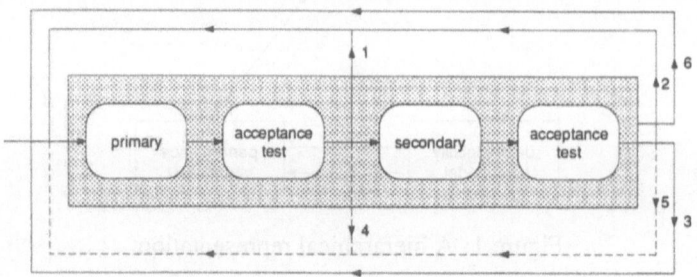

Figure 2: Recovery blocks operation.

operates as follows. An iteration begins with the execution of the primary (P); the acceptance test (AT) executes upon the completion of P. If P executes correctly and the AT accepts its result, the current iteration completes and the next one begins (Path 1 in Figure 2). If the AT rejects the result of P for any reason, the secondary (S) executes and the AT subsequently checks the result of S. If S computes correctly and the AT accepts the result, the iteration completes and the next one begins (Path 2); if the AT rejects the result of S for any reason, the result is suppressed and the next iteration begins (Path 3). Path 4 corresponds to the case where P generates an erroneous result and the AT subsequently accepts it; similarly, Path 5 corresponds to the case where S generates an erroneous result and AT subsequently accepts it. From a dependability model perspective, an undetected error leads to an absorbing state representing catastrophic failure. Therefore, Paths 4 and 5 are represented by dashed lines. Finally, Path 6 corresponds to the situation where the execution of an iteration exceeds its real-time deadline; in this case, the iteration is aborted and the next one begins immediately. Since this scenario may occur at any stage of an iteration, Path 6 starts from the shaded rectangle instead of any component encapsulated inside.

5.1   The dependability submodel for RB

The dependability submodel is a fault-manifestation model. Arlat *et al.* developed rather complete fault classifications and fault-manifestation models for RB and NVP [5]. With some minor adjustments, these serve as the dependability submodels at the lower layer. Generally, the ability of a testing procedure to detect errors, as inherent in its functional specification, is referred to as its "coverage."

Accordingly, when such a procedure is implemented (e.g., by a computer program), coverage quantifies the fault-free behavior of that implementation. With this assumption and, specifically, for an AT of the type considered here, we define its *coverage* to be the conditional probability that it rejects an erroneous result (per its specification), given that this result is incorrect. For simplicity in the evaluation that follows, we assume perfect coverage for the AT. On the other hand, we do account for errors caused by faults in the AT, i.e., erroneous judgments it makes with respect to the result under test. The fault classification and notation for probabilities of fault manifestation are illustrated in Table 1.

Fault Types	Probability of Manifestation
Related fault in P and S	q_{ps}
Related fault in P and AT or P, S and AT	q_{pa} or q_{psa}
Related fault in S and AT	q_{sa}
Independent fault in P or S	q_p or q_s
Independent fault in AT causing it to reject a result, given the result is correct	q_a

Table 1: Fault types and notation for RB.

Finally, as reflected in the construction that follows, we presume that the events i) the AT rejects an acceptable result provided by P and ii) it subsequently accepts an acceptable result from S, have a negligible probability of joint occurrence.

The detailed model, based on the above fault classification and assumptions, is shown in Figure 3. The definitions of the states are given in Table 2.

The states TP_i correspond to different types of faults in P, according to their manifestation of errors (during some iteration), i.e.,

TP_1 : no fault manifested in P.
TP_2 : manifestation of an independent fault in P.
TP_3 : manifestation of a related fault between P and S.
TP_4 : manifestation of a related fault between P and the AT.

Distinctions within states S_i and TS_i are made in a similar fashion. In particular, after an independent fault becomes manifest in P, the AT's operation causes a transition to state S_2 with probability 1. Note that this transition is certain with or

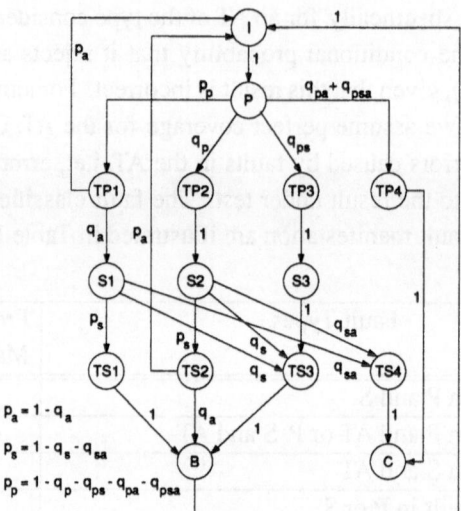

Figure 3: The dependability submodel for RB.

States	Definition
I	initial state of an iteration
P	execution of P
$\{TP_i \mid i \in \{1, 2, 3, 4\}\}$	execution of AT after P
$\{S_i \mid i \in \{1, 2, 3\}\}$	execution of S
$\{TS_i \mid i \in \{1, 2, 3, 4\}\}$	execution of AT after S
B	benign failure (caused by a detected value error)
C	catastrophic failure (caused by an undetected value error)

Table 2: State definitions for RB dependability model.

without a manifestation of an independent fault in the AT. If there is no such error, the AT correctly rejects the erroneous result provided by P (with probability 1 due to the coverage assumption). If there is a fault-caused error in the AT, because this fault is not related to the one in P (due to its independence), the errors in P and the AT are dissimilar; accordingly, the AT must take an action that does not correlate with the mistake made by P, i.e., as in the first case, it rejects the result. Manifestation of a related fault between P and S (state TP_3) corresponds to a detected error and leads through S_3 and TS_3 to state B. An error due to a related fault between P and the AT (state TP_4) is undetected and, hence, results in a catastrophic failure (state C).

Suppose now that p_c denotes the probability of a catastrophic failure and p_{bv} the probability of benign failure due to result suppression. (The other type of benign failure is a detected real-time deadline violation; this is accounted for via the analysis presented in the next section.) Then, from the state-transition diagram, we have

$$p_c = p_p \cdot q_a \cdot q_{sa} + q_p \cdot q_{sa} + q_{pa} + q_{psa} \tag{1}$$

$$p_{bv} = p_p \cdot q_a \cdot (p_s + q_s) + q_p \cdot p_s \cdot q_a + q_p \cdot q_s + q_{ps} \tag{2}$$

5.2 The performance submodel for RB

As noted earlier, we take the performance submodel to be a renewal process that represents the iterative nature of the software's execution. More precisely, if we let

$$K_t = \text{number of iterations during } [0, t]$$

then the submodel in question is the stochastic process $K = \{K_t \mid t \geq 0\}$. Note that, in terms of the variables defined in Section 3, for any time t, K_t coincides with the sum $I_t + D_t + N_t$, since every iteration is classified as being either successful, degraded (computational result suppression due to a detected value or timing error), or the cause of a catastrophic failure. Our principal interest here, in support of the performability measure $E[M_t]$, is the distribution of the time Y between successive renewals of K, i.e., the duration of an iteration of the program's execution.

To this end, we assume that the execution times of P, S and the AT are statistically independent; this is based on an assumption of enforced diversity, i.e., diverse structures and algorithms are used in the alternatives and acceptance test. Further, we take them to be exponentially distributed (as assumed by others in similar settings; see [11], for example) with parameters λ_p, λ_s, and λ_a, respectively. Figure 4 depicts this renewal process, which is derived from Figure 2. By definition, the random variable K_t is independent of outcome type. Hence Paths 1 and 4 in Figure 2 aggregate to a single path marked p_1 in Figure 4, where p_1 is the probability that an iteration completes upon a successful run of P and the AT, or fails due to the related faults between P and the AT. Likewise, Paths 2, 3 and 5 in Figure 2 aggregate to a single path marked $(1 - p_1)$. From the dependability submodel (Figure 3), it follows that

$$p_1 = p_p \cdot p_a + q_{pa} + q_{psa}.$$

Let the Laplace transform of the probability density functions (pdf's) of the execution times of P, S and the AT be denoted as F_p^*, F_s^* and F_a^*, respectively. Then the transform of the pdf of Y_c, the combined time for execution of the alternatives and acceptance test, is given by

$$F_c^*(s) = p_1 \cdot F_p^*(s) \cdot F_a^*(s) + (1 - p_1) \cdot F_p^*(s) \cdot F_s^*(s) \cdot (F_a^*(s))^2.$$

Through an inverse Laplace transform we then obtain f_c, the pdf of Y_c. Due to the

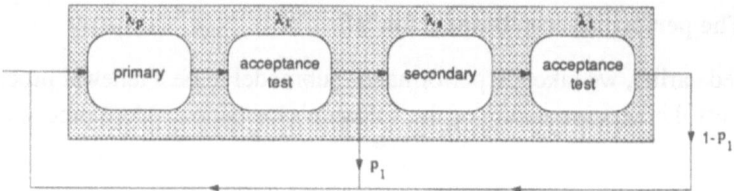

Figure 4: The performance submodel for RB.

real-time constraint, characterized as a fixed bound $\tau > 0$ on the time to complete an iteration, the pdf f of the iteration time Y coincides with f_c for $y < \tau$ (i.e., the iteration completes before timing out); otherwise it has the form of an impulse at $y = \tau$. More precisely,

$$f(y) = \begin{cases} f_c(y) & y < \tau \\ p_{bt} \cdot \delta(y - \tau) & y \geq \tau \end{cases}$$

where $\delta(y - \tau)$ is the unit impulse function, and $p_{bt} = 1 - \int_0^\tau f_c(y)dy$ is the probability that an iteration violates the real-time constraint. We can then compute the mean μ and variance σ^2 of the renewal cycle time (duration of an iteration) Y.

5.3 The performability model for RB

Figure 5 depicts the performability model at the upper layer. The shaded rectangular box corresponds to the one in Figure 4, which encapsulates the operational details of a renewal cycle. Since the mission durations t we consider (several

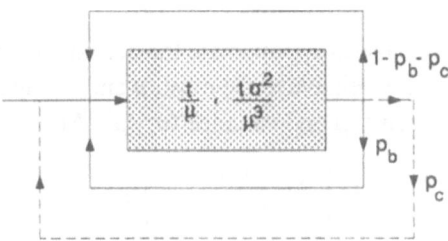

Figure 5: Performability model for fault-tolerant software.

hours) are much greater than the renewal cycle times Y (with averages in the millisecond range) and the mean and variance of Y are finite, we can approximate the distribution of K_t by its limit as $t \to \infty$. By the central limit theorem for renewal processes this asymptotic distribution is normal. The two expressions marked in the box are the mean and variance of the this distribution. They are functions of μ and σ^2, the mean and variance of the renewal cycle times, supplied primarily by the performance submodel. The outward arrows from the box denote the different outcomes of a renewal cycle, each of which is associated with a probability which is primarily supplied by the dependability submodel. Relating Figure 2 to Figure 5, Paths 1 and 2 in Figure 2 aggregate to the path marked $(1 - p_b - p_c)$ in Figure 5, Paths 4 and 5 aggregate to the path marked p_c, and Paths 3 and 6 aggregate to the path marked p_b, which is the probability of degradation, i.e., the probability that a computational result is suppressed due to either a value or timing error. Accordingly,

$$p_b = p_{bv} + p_{bt}.$$

Intuitively speaking, the performance variable M_t corresponds to the number of cycles that are fed back from the upper path during the period $[0, t]$, conditioned by the event that no cycle goes to the dashed path during this period.

The moment generating function of the performance variable M_t can then be derived (see [13]) in the following form.

$$E[e^{sM_t}] \;=\; E\left[(1 - (1 - p_c)^{K_t}) + (p_b + (1 - p_b - p_c) \cdot e^s)^{K_t}\right]. \qquad (3)$$

In turn, the measure we seek to evaluate can be expressed as

$$
\begin{aligned}
E[M_t] &= \left. \frac{d\,E[e^{sM_t}]}{d\,s} \right|_{s=0} \\
&= \frac{1 - p_b - p_c}{1 - p_c} \cdot E[K_t \cdot (1 - p_c)^{K_t}]
\end{aligned} \tag{4}
$$

As noted above, we are able to approximate the distribution of K_t by its limiting normal distribution where, relative to the mean μ and variance σ^2 of the iteration time Y, K_t then has mean t/μ and variance $t\sigma^2/\mu^3$. More precisely,

$$
\lim_{t \to \infty} P\left\{ \frac{K_t - t/\mu}{\sqrt{t\sigma^2/\mu^3}} < x \right\} = \Phi(x). \tag{5}
$$

If further we let $\hat{\mu} = t/\mu$, $\hat{\sigma} = \sqrt{t\sigma^2/\mu^3}$, and $\alpha = \log(1 - p_c)$, then

$$
\begin{aligned}
&E[K_t \cdot (1 - p_c)^{K_t}] \\
&= E[(\hat{\sigma}X + \hat{\mu})e^{\alpha(\hat{\sigma}X + \hat{\mu})}] \\
&= e^{\alpha\hat{\mu}}(\hat{\sigma}E[Xe^{\alpha\hat{\sigma}X}] + \hat{\mu}E[e^{\alpha\hat{\sigma}X}]) \\
&= e^{\alpha\hat{\mu}}\left(\hat{\sigma}\int_{-\frac{\hat{\mu}}{\hat{\sigma}}}^{\infty} xe^{\alpha\hat{\sigma}x}\varphi(x)dx + \hat{\mu}\int_{-\frac{\hat{\mu}}{\hat{\sigma}}}^{\infty} e^{\alpha\hat{\sigma}x}\varphi(x)dx\right)
\end{aligned} \tag{6}
$$

An analytic solution of the effectiveness $E[M_t]$ is thus obtained by substituting the above in equation (4).

6 The NVP model

Figure 6 shows the operation of the NVP scheme. The system has three functionally equivalent but independently developed programs — versions, a decision function which determines a consensus result from the results delivered by the versions, and a watchdog timer which detects violations of the real-time constraint. The system operates as follows. The three versions begin to execute at the same time. When all three complete their execution, the decision function votes on their results. If there exists a majority representing a correct computation, the output will be the correct result (Path 1 in Figure 6). If majority does not exist, the result will be suppressed (Path 3). The next iteration begins after either of the cases stated above. If there exists a majority representing similar errors, the output will be erroneous and undetected (Path 2). As in the RB model, an undetected

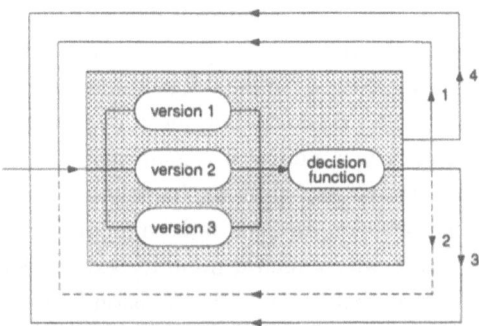

Figure 6: N-Version programming operation.

error leads to an absorbing state representing catastrophic failure. Therefore, Path 2 is represented by the dashed line. Finally, Path 4 corresponds to the situation where the duration of an iteration exceeds a specified upper bound (the real-time constraint); in this case, the execution is aborted and the next iteration begins immediately thereafter.

6.1   The dependability submodel for NVP

In a manner analogous to the construction of the RB model, we adapt the fault-manifestation model of [5] (in this case, the one for NVP) to fit the needs of the dependability submodel at the lower layer. The fault classification and notation for probabilities of fault manifestation are illustrated in Table 3. Because the implementation of the decision function is normally application-independent, we omit the category "related fault in versions and decision function."

Fault Types	Probability of Manifestation
Related fault in the 3 versions	q_{3v}
Related fault in the 2 versions	q_{2v}
Independent fault in a version given no related fault between the versions	q_{iv}
Independent fault in the decision function	q_d

Table 3: Fault types and notation for NVP.

To condition the independent fault probability q_d of the decision function on input events, we assume that 1) it is very unlikely that a majority exists among the version results but the decision function delivers a non-majority result, and 2) the probability that no majority exists, but the decision function delivers a good result from versions, is negligible. By excluding these rare events, we can condition q_d simply on the presence or absence of a majority. In other words, we need only consider two conditional probabilities, namely, q_{d1} and q_{d2}. The former is the probability that the decision function fails to recognize an existing majority and suppresses the result, given such a majority exists (decision function too "strict"); the latter is the probability that the decision function fails to recognize the discrepancies among the versions and delivers an erroneous result, given such discrepancies exist (decision function too "loose"). In [5], the term $(1 - q_{iv})$ is assumed to have an approximate value of 1. However, a principal reason for employing fault-tolerant software of this type is to enhance dependability through design diversity, even in cases where the probability q_{iv} of an independent fault (given no related fault between versions) is appreciable relative to the probability of a related fault. Thus the term $(1 - q_{iv})$ is included in our model. The detailed model, based on the above assumptions, is shown in Figure 7. The definitions of the states are shown in Table 4. The states D_i correspond to the different outcomes from the three versions' execution. That is,

D_1 : there exists a majority representing a correct result.
D_2 : no majority exists.
D_3 : there exists a majority representing an erroneous result.

States	Definition
I	initial state of an iteration
V	execution of versions
$\{D_i \mid i \in \{1, 2, 3\}\}$	execution of decision function
B	benign failure (caused by a detected value error)
C	catastrophic failure (caused by an undetected value error)

Table 4: State definitions for NVP dependability model.

As in the RB model, we let p_c denote the probability of a catastrophic failure and p_{bv} the probability of a benign failure due to result suppression. Then, from the

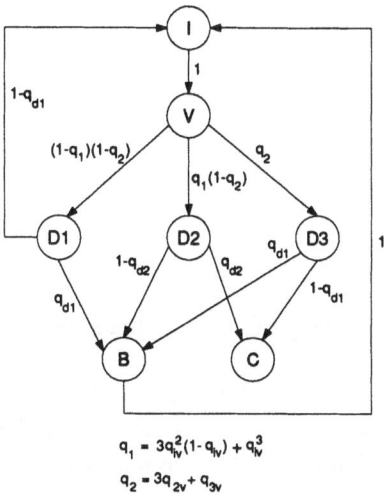

$$q_1 = 3q_{tv}^2(1-q_{tv}) + q_{tv}^3$$
$$q_2 = 3q_{2v} + q_{3v}$$

Figure 7: The dependability submodel for NVP.

state-transition diagram, it follows that

$$p_c = q_1 \cdot (1 - q_2) \cdot q_{d2} + q_2 \cdot (1 - q_{d1}) \tag{7}$$
$$p_{bv} = (1 - q_1)(1 - q_2) \cdot q_{d1} + q_1 \cdot (1 - q_2)(1 - q_{d2}) + q_2 \cdot q_{d1} \tag{8}$$

6.2 The performance submodel for NVP

To model the performance of NVP, we employ a construction similar to that of the RB performance submodel and, in particular, assume that the execution times of the first, second, and third versions in the NVP scheme are independently and exponentially distributed with parameters λ_1, λ_2 and λ_3, respectively. Figure 8 depicts this performance submodel, i.e., the renewal process $K = \{K_t \mid t \geq 0\}$ which is derived from Figure 6. By definition, for any choice of $t > 0$, the random variable K_t is independent of outcome type. Hence, Paths 1, 2 and 3 in Figure 6 aggregate to a single path in Figure 8. Let Y_v be the random variable that represents the time for parallel version execution and suppose the execution times of the first, second, and third versions are represented by the variables Y_1, Y_2, and Y_3, respectively. Then $Y_v = max\{Y_1, Y_2, Y_3\}$ and the probability distribution function (PDF) of Y_v is

$$G(y) = \begin{cases} (1 - e^{-\lambda_1 y})(1 - e^{-\lambda_2 y})(1 - e^{-\lambda_3 y}) & \text{if } y \geq 0 \\ 0 & \text{otherwise} \end{cases}$$

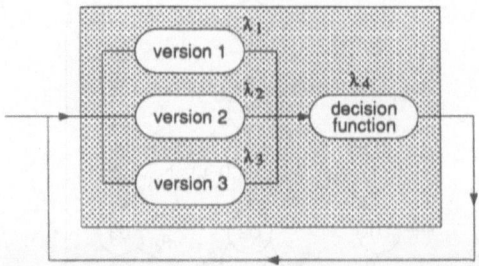

Figure 8: The performance submodel for NVP.

The pdf of Y_v, call it g, can then be obtained by differentiating G. In turn, suppose that G^* denotes the Laplace transform of g and H^* the Laplace transform of the pdf of the decision function execution time (also distributed exponentially, with parameter λ_4). Since parallel execution of the versions is followed by execution of the decision function, the transform of the pdf of Y_c (the combined time for execution of both) is $F_c^*(s) = G^*(s) \cdot H^*(s)$. Taking the inverse transform of the above, we thus obtain the pdf f_c of Y_c. The pdf f of the iteration time Y, which accounts for the real-time constraint imposed by the upper bound τ, is then derived in a manner analogous to that of the RB solution (Section 5.2).

6.3   The performability model for NVP

The resulting performability model at the upper layer is likewise similar to the corresponding RB model (see Section 5.3), as depicted in Figure 5. Relating the NVP operational model in Figure 6 to this performability model, Path 1 in Figure 6 corresponds to the path marked $(1 - p_b - p_c)$ in Figure 5, Path 2 corresponds to the path marked p_c, and Paths 3 and 4 aggregate to the path marked p_b. Thus, after integrating the information supplied by the lower layer models, the moment generating function can be derived in a manner identical to that of the RB scheme (Section 5.3).

7   Comparative evaluation results

Assuming a 10 hour mission (t = 10 hours), the effectiveness $E[M_t]$ of both RB and NVP are evaluated for fixed values of the lower layer submodel parameters. Because the intent of this study is to demonstrate the feasibility of performability modeling, as opposed to providing conclusive evaluation results, we emphasize consistency in our choice of parameter values, as opposed to, say, a match with some set of experimentally determined values. The assumptions underlying these choices are the following.

1) For the RB system, the coverage of an AT is an indicator of its complexity, where an increase in coverage generally requires a more complicated realization. Since a program's execution time and fault (manifestation) probabilities likewise increase with growing complexity, it is reasonable to assume that the mean execution time $(1/\lambda_a)$ and fault probabilities $(q_a, q_{pa}$ and $q_{sa})$ of the AT vary directly as functions of coverage. Due to this consideration, coupled with the perfect coverage assumption (Section 5.1), the values assigned to the execution time and fault probabilities of the AT are comparable to those of an alternate (primary or secondary).

2) The decision function in the NVP system, which is a reusable application-independent component, has a lower fault probability than other components. In accordance with our earlier remarks concerning the term $(1 - q_{iv})$ (see Section 6.1), probabilities of independent faults in components are allowed to be significantly higher than those of related faults between components.

3) It is also assumed that the mean execution times of components in the RB and NVP schemes differ moderately because of the assumption of enforced diversity and equivalent alternative/version functionality. Further, since it is anticipated for RB that the secondary will be utilized much less than the primary and acceptance test, the slower alternative (the one whose mean execution time is greater) is chosen for the secondary.

The resulting parameter value assignments are shown in Table 5, where the execution rates (λ's) have per-millisecond dimensions. The probabilities of related faults between two components are retained as variable parameters, in order to compare their relative influence on the effectiveness of the two schemes. Actual computations, including the Laplace transforms and derivations of density functions, were realized using Mathematica [1].

Figure 9 compares the performability of the RB and NVP schemes according to the measure $E[M_t]$ (effectiveness for a 10 hour mission, in units of 10^6). Resulting values of this measure are plotted as a function of the "probability of

---

[1] Mathematica is a registered trademark of the Wolfram Research Inc.

Parameter	Value
λ_p	1/5
λ_s	1/8
λ_a	1/5
q_{ps}	variable
q_{pa}	variable
q_{sa}	variable
q_{psa}	10^{-10}
q_p	0.0001
q_s	0.0001
q_a	0.0001
t	$3.6\ 10^7$ (ms)
τ	30 (ms)

(a) RB

Parameter	Value
λ_1	1/5
λ_2	1/6
λ_3	1/8
λ_4	2.0
q_{2v}	variable
q_{3v}	10^{-10}
q_{iv}	0.0001
q_{d_1}	10^{-9}
q_{d_2}	10^{-9}
t	$3.6\ 10^7$ (ms)
τ	30 (ms)

(b) NVP

Table 5: Parameter values.

related faults"; as noted above, this is the probability of (a manifestation of) a related fault between two components, i.e., the value of the parameters q_{ps}, q_{pa}, and q_{sa} in the case of RB and the value of q_{2v} in the case of NVP. For fixed assignments of the remaining parameter values (according to Table 5), the results show that RB is more effective than NVP throughout the considered domain of related-fault probabilities.

Some insight as to why this is so can be gained by first examining distinctions with regard to the relative dependability of the two schemes, as revealed by their respective dependability submodels. Here we see that, for both RB and NVP, the probability of a catastrophic failure (equations (1) and (7)) is dominated by the probability of a related fault between the components. In the RB scheme, where this probability coincides with the probabilities q_{ps}, q_{pa}, and q_{sa}, only q_{pa} contributes directly to the probability p_c of a catastrophic failure. This is because i) an error due to a related fault in P and S (having probability q_{ps}) cannot result in catastrophic failure and ii) an error due to a related fault in S and the AT (having probability q_{sa}) can result in a catastrophic failure only if the AT rejects P. In the NVP scheme, on the other hand, the probability q_{2v} of a related fault between

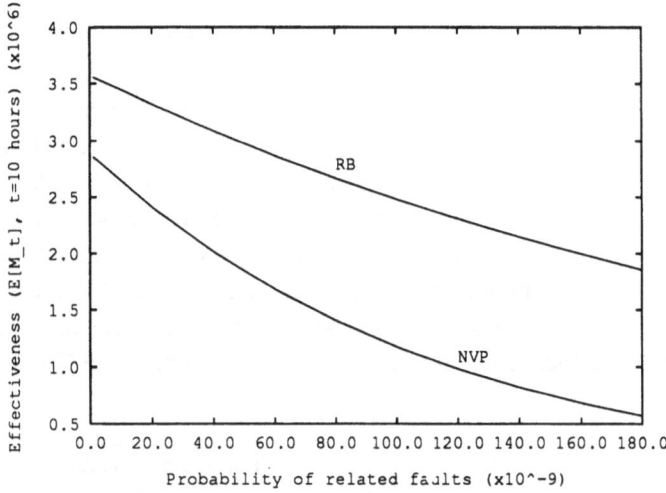

Figure 9: Comparison of RB and NVP (I).

any two versions contributes directly to p_c. Therefore, when compared to RB, occurrence of a catastrophic failure (during an iteration) for NVP is approximately three times more likely.

The performability advantage of RB, per Figure 9, is also due, in part, to distinctions in (strict) performance. Specifically, from the lower layer performance submodels we observe that i) for RB, the mean iteration time is dominated by the mean combined execution time of P and the AT, while ii) for NVP, the mean iteration time is lengthened due to the fact that version synchronization requires the system to wait for the slowest version. In the case of exponentially distributed version execution times (as assumed here), version synchronization translates into a relatively severe penalty on NVP performance.

Another thing worth noting in Figure 9 is that the difference between RB (upper curve) and NVP (lower curve) effectiveness becomes greater as the probability of a related fault increases. This can be explained via a combination of the effects cited above. If the probability of a related fault is low, the effectiveness difference between RB and NVP is due mainly to the performance cost imposed by 3-version synchronization. However, as the related-fault probability increases, this difference is amplified by the fact that NVP is more vulnerable to a catastrophic failure caused by a related fault between two components (in which case, $M_t = 0$).

Additional evaluations have been conducted with respect to other choices of parameter values that comply with the assumptions stated in the beginning of this
section. In the discussion that follows, two such cases are considered and their
corresponding results are compared. Although the assumption that mean execution times of components differ moderately is certainly reasonable in the light
of enforced diversity and equivalent (alternative/version) functionality, additional
insight can be gained by letting them differ significantly (while maintaining the
assumption that P is more efficient than S in the RB scheme). For example, if the
component execution rates are chosen according to the values in Table 6 then the
corresponding effectiveness results become those displayed in Figure 10.

Parameter	Value
λ_p	1/5
λ_s	1/18
λ_a	1/5

(a) RB

Parameter	Value
λ_1	1/5
λ_2	1/6
λ_3	1/18

(b) NVP

Table 6: Modified performance parameter values.

Specifically, when compared with the results of the previous evaluation (Figure
9), the upper curve demonstrates that the effectiveness of RB remains similar
to the case where the mean execution times differ moderately. On the other
hand, the effectiveness of NVP becomes significantly poorer in the low and
moderate related-fault probability domain. This is because the mean iteration
time for the RB scheme is dominated by the faster components (P and AT)
whereas that for the NVP scheme is bounded below by the slowest version.
As related faults become more probable, however, we note that this distinction
begins to disappear. This is due to the interaction between the performance and
dependability attributes. More precisely, for the case where the mean execution
times differ moderately (Figure 9), as the related-fault probability becomes greater,
NVP's higher (compared to this case) iteration rate results in a greater likelihood
of a catastrophic failure caused by a related fault between two components, thus
negating the performance benefit. Stating this observation in more general terms,
when the dependability deficiency of a fault-tolerant software system becomes
severe, it appears to dominate the system's effectiveness (as quantified by the
measure $E[M_t]$). In other words, its effectiveness becomes relatively insensitive

to variations in performance attributes.

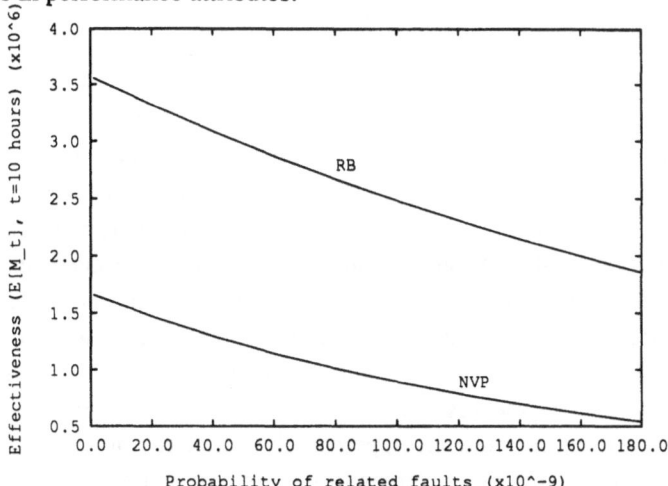

Figure 10: Comparison of RB and NVP (II).

Finally, let us consider the other extreme where the mean execution times of the components are identical. In particular, suppose that the mean execution times of P, S and AT for RB and those of the three versions for NVP all equal to 5 milliseconds. In this case we obtain the results depicted in Figure 11. Here we

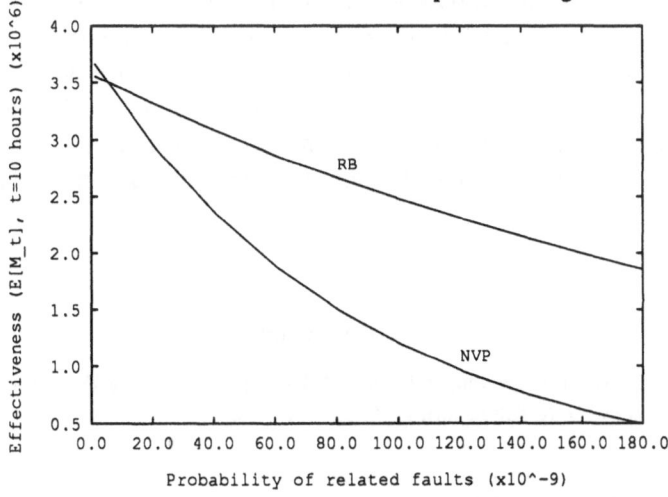

Figure 11: Comparison of RB and NVP (III).

see that the effectiveness of NVP is slightly better than that of RB in the very low related-fault probability domain. However, as the related-fault probability increases, NVP becomes increasingly inferior to RB.

Further explanation of what has been observed in these three cases (Figures 9, 10, and 11), resides in the lower layer performance submodels. For NVP, the mean iteration time is dominated by the mean of the maximum execution time for the three versions ($max\{Y_1, Y_2, Y_3\}$). Since these times are exponentially distributed, the mean of the maximum is always greater than any individual mean. Moreover, it can be shown that when the mean version execution times $1/\lambda_1, 1/\lambda_2$ and $1/\lambda_3$ differ only moderately or are all identical, the mean of the maximum is significantly greater than that of any individual. This fact, which does not seem to be intuitively obvious, is responsible for the severe reduction in performance suffered by NVP in the first and third cases. In the second case, where the component execution rates have appreciably different values, the mean of the maximum execution time converges to the greatest among them. Hence, once again, there is a severe performance penalty for version synchronization.

It should be recalled that the intent of this investigation is not that of providing definitive evaluation results based on realistic estimations of parameter values. Instead, its purpose is to demonstrate the feasibility and utility of model-based performability evaluation when applied to fault-tolerant software. In particular, we believe that the above observations lend testimony to how a measure such as $E[M_t]$, when based on a hierarchically constructed model, is indeed able to capture the collective effect of lower level performance and dependability attributes. The information supplied by such evaluations is also suggestive of design modifications conducive to performability enhancement. This possibility was explored to some extent in [13] and is currently the subject of more intensive study.

References

[1] A. Avižienis. Software fault tolerance. *Information Processing 89: Proc. of the IFIP Congress 89* (G. X. Ritter, ed.), 1989, pp. 491–498.

[2] B. Randell. System structure for software fault tolerance. *IEEE Trans. Software Engineering*, Vol. SE-12, June 1975, pp. 220–23.

[3] A. Avižienis, L. Chen. On the implementation of N-Version Programming for software fault-tolerance during program execution. *Proc. of COMPSAC-77*, 1977, pp. 149–155.

[4] A. Grnarov, J. Arlat, A. Avižienis. On the performance of software fault-tolerance strategies. *Proc. Int. Symposium on Fault-Tolerant Computing*, (Kyoto, Japan), October 1980, pp. 251–253.

[5] J. Arlat, K. Kanoun, J-C. Laprie. Dependability modeling and evaluation of software fault-tolerance systems. *IEEE Trans. Computers*, Vol. C-39, April 1990, pp. 504–512.

[6] S. Leu, E. B. Fernandez, T. Khoshgoftaar. Fault-tolerant software reliability modeling using petri-nets. *Microelectronics and Reliability*, Vol. 31, No. 4, 1991, pp. 645–667.

[7] G. Pucci. A new approach to the modeling of recovery block structure. *IEEE Trans. Software Engineering*, Vol. SE-18, February 1992, pp. 159–167.

[8] J. F. Meyer. On evaluating the performability of degradable computing systems. *IEEE Trans. Computers*, Vol. C-29, August 1980, pp. 720–731.

[9] J. F. Meyer. Performability: a retrospective and some pointers to the future. *Performance Evaluation*, Vol. 14, 1992, pp. 139–156.

[10] E. Gelenbe, I. Mitrani. Modeling the execution of block structured processes with hardware and software failures. *Mathematical Computer Performance and Reliability* (G. Iazeolla *et al.*, eds.), Elsevier Science Publishers B. V., North-Holland, 1984, pp. 329–339.

[11] P. F. Chimento, K. S. Trivedi. The performance of block structured programs on processors subject to failure and repair. *High Performance Computer Systems* (E. Gelenbe, ed.), Elsevier Science Publishers B. V., North-Holland, 1988, pp. 269–280.

[12] M. C. Hsueh, B. K. Iyer. A measurement-based performability model for a multiprocessor system. *Computer Performance and Reliability*. (G. Iazeolla *et al.*, eds.), Elsevier Science Publishers B. V., North-Holland, 1988, pp. 337–351.

[13] A. T. Tai. Performability concepts and modeling techniques for real-time software. *Ph.D. dissertation*, UCLA Computer Science Department, Los Angeles, CA, December 1991.

ON THE TRANSIENT ANALYSIS OF

STIFF MARKOV CHAINS

Jürgen DUNKEL, Harald STAHL
Universität Dortmund, Informatik IV
P.O. Box 500500, D-4600 Dortmund 50, Germany

Abstract

Dependability and performability analysis commonly requires the transient analysis of Markov chains. Because most of these models involve rates of different orders of magnitude, they lead to *stiff* Markov chains, which are ill-conditioned in a computational sense for conventional numerical methods. In this paper the well-known randomization technique is adapted to cope with a special class of stiff models. Then we present a class of models wich remain computational intractable. This leads to an appropriate new characterization of stiff Markov chains. For this model class the recently proposed implicit ODE-solvers are also computational infeasible, if they use iterative numerical techniques. A modified step size control and iterative aggregation/disaggregation techniques are proposed to improve the solver performance. The composite usage of both techniques yields large computational gains, especially for higher order methods.

1 Introduction

Most dependability and performability models of fault tolerant computing systems are based on continuous time Markov chains (CTMC). The computation of many measures like the instantaneous availability or the probability distribution of mission time calls for the transient analysis of the

Markov model. There are a lot of different numerical methods for finding the transient solution of Markov chains: the randomization, and conventional (explicit) differential equation solution methods (e.g. Runge-Kutta) [19] are well-known.

A key-problem of dependability models is that they usually lead to so-called *stiff* problems, which are ill-conditioned in a computational sense for conventional numerical methods. However, there are special (viz. implicit) numerical methods for solving stiff ordinary differential equations (ODE) [14], [2], which have been applied on continuous-time Markov chains recently [3], [19], [15]. Unfortunately, all these implicit ODE-methods require a lot of computational effort. This paper is concerned with the efficient transient analysis of stiff CTMC's.

The next section gives an introduction to stiff Markov chains. In section 3 we propose an improved randomization algorithm which works very efficiently for a special class of problems satisfying the stiffness condition given in [19] and [14]. This experience motivates a discussion of an appropriate characterization of stiff Markov chains. We will see that our modified randomization only fails, if the Markov model satisfies the stiffness condition of [19], and has the NCD-property according to Courtois [5]. In section 4 we investigate the performance of implicit ODE-methods on models satisfying the new stiffness characterization. Unfortunately, implicit ODE-methods become infeasible for those models also, if they employ standard iterative solvers. To cope with this problem we propose a modified step size control, taking the number of required Gauss-Seidel steps into account (section 5.1), and the usage of iterative aggregation/disaggregation techniques (section 5.2).

From practical experience we see, that the composite application of modified step size control and iterative A/D solvers yields large computational gains, in particular for high order implicit Runge-Kutta methods.

2 Stiff Markov chains

In this section we present the characterizations of stiff Markov chains proposed in the literature. First we will introduce some notation.

Let $\{Y(t), t \geq 0\}$ be a homogeneous finite-state continuous time Markov chain (CTMC) with state space Ω of cardinality $|\Omega|=n$. The corresponding generator matrix is given by $Q=(q_{ij})$, where q_{ij} denotes the transition rate from state i to state j, and the diagonal elements are $q_{ii}=-\sum\limits_{j \neq i}^{k} q_{ij}$. Let $P_i(t)$ be the unconditional probability of the CTMC being in state i, and the row vector $P(t)$ is the transient state probability vector of the CTMC.

The time dependent behavior of the CTMC can be described by the Kolmogorow differential equations:

$$\dot{P}(t) = P(t)\, Q \quad P(0) = P_0 , \tag{1}$$

where P_0 represents the initial state probability vector of the CTMC. A solution is given by

$$P(t) = P_0\, e^{Qt}, \tag{2}$$

where the exponential of a matrix can be formally defined by the Taylor power series $e^{Qt} = \sum\limits_{i=0}^{\infty} \frac{(Qt)^i}{i!}$.

There are a lot of different ways to compute the solution of (1), see e.g. [10], [12], [19]. For benign problems useful methods are e.g. randomization or Runge-Kutta. Stiff problems are just those which are not benign: "stiff differential equations are equations which are *ill-conditioned* in a computational sense" [14]. But we may ask, how ill-conditioning can be expressed in a formal definition. In the literature there are two different attempts to define stiffness: In the "classical" definition, see e.g. [13], p. 231 an ODE-system according to (1) is said to be stiff, if

$$\max_i |Re(\lambda_i)| \gg \min_i |Re(\lambda_i)| , \tag{3}$$

where λ_i are the (complex) eigenvalues of the generator matrix Q, and $Re(\lambda_i)$ denotes the corresponding real part. Obviously it is very difficult to obtain the eigenvalues of the generator matrix, and so the problem arises to find a stiffness definition, which is easy to check. A common assumption is that stiffness in Markov models is given, when there are transition rates of greatly different orders of magnitude [19]. An appropriate characterization is the *stiffness ratio*:

$$\frac{q_{max}}{q_{min}} \gg 1, \quad \text{with} \quad q_{max} = \max_i |q_{ii}|, \quad q_{min} = \min_i |q_{ii}| \quad (4)$$

However, the computational effort for the transient analysis of CTMC's depends obviously on the length of the solution interval [0,t], too. But in definition (4) the time scale (or mission time) t is omitted. Therefore in [14], [19] and [15] stiffness is characterized by a large value of the *stiffness index*::

$$qt \gg 1, \quad \text{with} \quad q = \max_i |q_{ii}|. \quad (5)$$

In the next section we will demonstrate that none of the definitions (4) and (5) characterize stiffness exactly in the sense of computational intractability.

Finally, we will give some examples for stiff Markov models: Many dependability and performability models satisfy the classical stiffness definition (4) involving transition rates of different orders of magnitude. Usually failure rates are orders of magnitude smaller than repair rates. In gracefully degrading systems the rates of the events within a degradation mode are much greater than the event rates due to occurrences of failures.

Note, that in all these models the length of the solution interval usually depends on the small rates: the modeler studies a relative large misson time t to take the effect of rare events (here: failures) into account.

3 Adaptation of randomization to stiff Markov chains

A lot of empirical studies demonstrate that randomization (sometimes also called uniformization) is the method of choice for non-stiff Markov chains [12], [19], [15]. It is more accurate and efficient than ODE solution techniques. Other advantages include accurate error control and ease of implementation.

The basic idea of randomization (see e.g. [11], [18]) is to consider the discrete-time Markov chain (DTMC) which is embedded in the CTMC $\{Y(t), t \geq 0\}$. The corresponding transition probability matrix is then given by $Q^* = Q/q + I$ with $q > \max_i |q_{ii}|$. Returning to the CTMC, we obtain the transient state probability vector P(t):

$$P(t) = \sum_{k=0}^{\infty} \pi(k) \ e^{-qt} \ \frac{(qt)^k}{k!} \ . \tag{6}$$

In (6) $\pi(k)$ is the state probability vector of the DTMC after k jumps, which is given by successive vector-matrix products:

$$\pi(k+1) = \pi(k) \ Q^*, \quad \pi(0) = P(0) \ . \tag{7}$$

The term $e^{-qt} \frac{(qt)^k}{k!}$ in (6) is the probability that there are k jumps of the DTMC in the interval [0,t] (Poisson probabilities). In an implementation the infinite series (6) has to be truncated after the (m+1)th term. Obviously the corresponding truncation error ε is then upper-bounded by

$$\varepsilon \le 1 - e^{-qt} \sum_{k=0}^{m} \frac{(qt)^k}{k!} \ .$$

Two problems arise applying the randomization on stiff Markov chains with a great stiffness index qt:

- First, we get a pragmatic implementation problem, because on standard work stations the computation e^{-qt} yields the result 0.0 for great qt (underflow) [9]. For example this is the case if qt > 750 on a SUN work station (SPARC-ELC). Thus the standard randomization formula yields an all-zero-vector as its result.

- Secondly, it is a crucial problem that the number of required vector-matrix products according to (7) grows strongly with qt. This is obvious, because the probability that there are many jumps of the embedded DTMC in the interval [0,t] increases with qt. The truncation point m is of the order O(qt) [19].

To cope with the first problem one can omit the factor e^{-qt} in (6) and consider the series $y_m = \sum_{k=0}^{m} \pi(k) \ \frac{(qt)^k}{k!} \ .$

A normalization is required to transform the vector y_m into a probability vector. However for great qt the term $z_n = \frac{(qt)^n}{n!}$ also becomes too large. A solution of this problem is an appropriate normalization during the randomization as described by the following algorithm:

while $(n < m)$

    1. with $z_n = \dfrac{(qt)^n}{n!}$, we obtain $y_n = y_{n-1} + \pi(n)\, z_n$;

    2. let r_n be the L_1 norm of y_n , then $r_n = r_{n-1} + z_n$;

    3. if r_n exceeds a specified threshold (e.g. $r_n > 10^{20}$)

        $y_n = y_n / r_n$;

        $z_n = z_n / r_n$;

        $r_n = 1.0$.

Finally y_m must be normalized to get a probability vector. The algorithm terminates if z_n is smaller than a user-defined threshold or if $y_n = y_{n-1}$ due to machine round-off.

To reduce the number of vector-matrix products in [9] it is proposed to truncate the series (6) at the left side, i.e. to start the series at some integer $l > 0$. Then the number of required terms is of order $O(\sqrt{qt})$. Unfortunately the problem remains to determine the corresponding first significant vector $\pi(l)$ of the DTMC. In [18] it is suggested to compute $\pi(l) = \pi(0)Q^l$ by successive squaring of Q. But in practice this approach is infeasible, because matrix squaring produces fill-in. Additionally matrix squaring often wastes a lot of CPU time in some sparse matrix schemes. Note, that even if the series is truncated at the left side, the basic problem remains: the computational effort grows strongly with the stiffness index qt.

If, however, stiffness is solely caused by a large t, then it is possible that the embedded DTMC reaches steady-state after its s-th jump, i.e.

    $\pi(k+1) = \pi(k)$ for $k > s$.

In this case the steady state vector π_s satisfies $\pi_s = \pi_s\, Q^*$ according to (7), and we can save further vector-matrix products, and our algorithm yields for $m > s$:

$$y_m = y_s + \pi(s) \sum_{i=s+1}^{m} z_i \, . \tag{8}$$

Note that the consideration of the DTMC steady state to improve randomization is also suggested in the Ph.D. thesis of Muppala [16], [17], recently.

Remark:

An approriate test on stationarity of the DTMC is crucial. We use the approaches suggested in [20]. The main idea is, to test some norm of the difference between two iteration vectors spaced further apart: $\|\pi(k) - \pi(k-m)\|$ $< \varepsilon$, where m depends on the convergence speed. For details see [20].

To demonstrate the usefulness of this approach we consider the M/M/1/K-system with the following basic parameters: Assume that there are K = 10 buffers, arrival rate $\lambda = 1.0$, service rate $\mu = 1000.0$, and mission time t = 10.0. In our model stiffness can be increased by increasing the mission time t or by increasing the service rate μ. For standard randomization and for the modified randomization algorithm (random.-stat) we obtain the following results.

mission time t	1.0		10.0		100.0		1000.0		10000.0	
randomization	0.1	1263	0.8	10804	8.0	10^5	78.4	10^6	-	-
random.-stat.	0.1	361	0.1	361	0.1	361	0.1	361	0.1	361

Table 1: CPU-time and number of required vector-matrix products vs. mission time.

The first entry in the Table denotes the CPU-time on a SUN-4 workstation, the second is the number of required vector-matrix products. Using standard randomization we can observe the effects described in [19] and [15] for the same model: the number of required vector-matrix products grows linear with the stiffness index qt. So for large mission times t this approach becomes infeasible.

The modified randomization algorithm shows that the DTMC reaches equilibrium after 361 vector-matrix products. Thus the computational effort is independent of the mission time t and therefore independent of the stiffness index qt as defined in [14] and [19].

In a next step (see Table 2) we vary the "classical" stiffness index by increasing the service rate μ. Also in this case the DTMC reaches stationarity after a few vector-matrix products and thus the improved algorithm remains efficiently, too.

service rate μ	100.0		1000.0		10000.0		100000.0		1000000.0	
randomization	0.1	1263	0.8	10804	8.0	10^5	80.5	10^6	-	-
random.-stat.	0.1	41	0.1	361	0.1	21	0.1	21	0.1	21

Table 2: computational effort of randomization vs. service rate.

Note that both examples are presented in [19] and [15] to demonstrate that randomization cannot be used for problems with great stiffness index. However, our slight modification of the randomization algorithm shows that the stiffness indices proposed in the literature are not an appropriate characterization of "computational intractability" in this case.

But how can we characterize stiff Markov chains which are intractable for this modified randomization technique? Obviously our algorithm yields no improvement, if the embedded DTMC doesn't reach stationarity during the solution interval. Thus those Markov chains are critical, which reach their stationary phase very slow. It is well-known that this type of models satisfies the *nearly completely decomposability* (NCD)-property [5].

In NCD-models there are rates of different orders of magnitude as required in the classical stiffness definition, but additionally the state space can be partitioned into sets which are "loosely" coupled by small transition rates. The small rates between the state subsets prevent that stationarity can be reached fast.

As an example of a NCD-model we consider an extended machine-repairmen model. There are k identical components which suffer a failure with rate λ. With probability c (coverage factor) this is a soft failure which is recovered with rate μ, and with probability (1-c) it is a hard failure, which is repaired with rate ν (we assume that each component has its own repair instance). The basic parameters are: $k = 3$, $\lambda = 1.0$, $\nu = 1.0$, $\mu = 100.0$, $t = 1000.0$. The degree of coupling between the state subset describing the NCD-structure of the model, can be weakened by increasing the coverage probability c. The stiffness index qt will be regulated by μ, the largest model rate. Our experiments yield the following results:

coverage c	0.50		0.80		0.90		0.95		0.99	
randomization	42.6	306454	42.6	306454	42.6	306454	42.6	306454	42.6	306454
random.-stat.	4.9	25541	11.8	61101	21.5	120541	32.8	220621	42.6	306454

Table 3: computational effort of randomization vs. coverage factor.

The experimental results exhibit the predicted behavior. With increasing coverage factor the coupling is weakened, and the larger the coverage factor is, the more vector-matrix products are required until the DTMC reaches stationarity. Thus for our modified randomization algorithm we can characterize computational intractability (or stiffness) by the subsequent definition:

Definition 1 Let Q be the generator matrix of a CTMC. For a given solution interval [0,t], starting with the initial state probability vector P_0, we define t* as follows:

$t* = t_s$, if the Markov chain reaches its equilibrium at time instant $t_s < t$, otherwise $t* = t$. Then qt* is a valid stiffness index for the modified randomization, i.e. stiffness is given, if

$$| q \, t* | \gg 1 . \tag{9}$$

Remarks:

We have to note that the above definition has its pitfalls, too. In contrast to the definitions proposed in the literature, it cannot be decided by the model parameters alone, whether a model has NCD-structure or not. Note, that additionally qt* depends on the initial state probability vector.

Our experience from various experiments is, that, especially for complex models, it is very difficult to predict when stationarity is reached. This is a very important aspect, because otherwise the user could switch to stationary analysis techniques a priori. Thus it will often be the application of our modified randomization which serves to indicate stiffness.

Finally we see that the improvement of our new randomization algorithm is restricted to a special type of models, NCD-models still remain computationally intractable. In the next section we study implicit ODE-methods, which are designed to cope with stiff differential equations, when other numerical methods fail.

4 Implicit ODE-methods

If conventional *explicit ODE-methods* are applied for stiff differential equations, computational problems will arise, see e.g. [13]. The limitation of the global truncation error can only be assured by restrictive conditions on the step size h, i.e. explicit numerical techniques converge to the correct solution as the step size h→0. More precisely, in the context of Markov chains the maximal step size h depends on the largest diagonal element of the generator matrix: i.e. h is proportional to 1/q. The exact condition depends on the appropriate numerical method, for details see [14]. Note that for a given solution interval [0,t] the number of required integration steps is of order qt, i.e. the effort is comparable to that of standard randomization.

To overcome these difficulties a number of *implicit ODE-methods* have been developed [13], [2]. There are methods which are *absolutely stable*, i.e. the global truncation error converges to 0 for all step sizes h. Thus the step length is only restricted by *local truncation errors* (LTE). The LTE is estimated after each integration step, and it determines an appropriate step length for the next step, see [13]. An (nearly) optimal step size control is a crucial feature of any efficient implementation of ODE-solvers. The step size h is adapted to the actual variation of the state probability vector, and for stiff problems it can range several orders of magnitude. In particular, when a system is in equilibrium, h can be arbitrary large.

The application of well-known implicit ODE-methods on Markov chains has been proposed recently. In [3], and [19] a linear multistep method, the TR-BDF-2 method, is proposed, where each integration step consists of two substeps: first the trapezoid rule (TR) is applied, then a BDF-2 step (second order backward difference formula [4]) follows. More precisely, the i-th TR-step integrates from time instant t_i to $t_{i+\gamma}$ h= t_i + γ h, where the constant $\gamma \cong$ 0.5 determines the length of the intermediate step. If P_i and $P_{i+\gamma}$ denote the corresponding state probability vectors, a new vector $P_{i+\gamma}$ is given by solving the linear algebraic system

$$P_{i+\gamma} (2\, I - \gamma\, h_i\, Q) = 2\, P_i + P_i\, \gamma\, h_i\, Q \qquad (10)$$

The subsequent BDF-2 step integrates from $t_{i+\gamma}$ h to $t_{i+1} = t_i + h$, and is determined by

$$P_{i+1} [(2-\gamma)I - (1-\gamma)\, h_i\, Q] = \gamma^1\, P_{i+\gamma} - \gamma^1(1-\gamma)^2\, P_i \qquad (11)$$

The composite method is of order p=2. Such low-order methods require relative small step sizes to provide high accuracy (e.g. smaller than 10^{-6}). Unfortunately, in reliability models we are often interested in very small probabilities, which require high accuracy. In this case higher-order methods are attractive, which allow larger step lengths. In [15] a third-order implicit Runge-Kutta method [13], [14] is used for stiff Markov chains. The algebraic system, which describes an integration step of this implicit Runge-Kutta method (RK), is given by

$$P_{i+1} (I - \frac{2}{3}\, h_i\, Q + \frac{1}{6}\, h_i^2\, Q^2) = P_i (I + \frac{1}{3}\, h_i\, Q) . \qquad (12)$$

The main computational effort of these implicit numerical methods is caused by the solution of the algebraic systems, which are of the form $x \cdot Q^* = r$, where Q^* is the iteration matrix according to (10)-(12). Direct and iterative methods can be applied to solve such systems [20].

- Direct methods, like the LU-factorization, produce additional matrix entries (fill-in), and cause a computational effort of order $O(n^3)$, if the generator matrix Q has the dimension n. Thus it is very expensive to factorize large generator matrices. If only the right-hand side r of the algebraic system is modified, we just need to perform a back-substitution for its solution. But unfortunately, h usually changes for nearly every step, so that a new LU-factorization is required for each step.

- Iterative methods, like Gauss-Seidel or SOR-solver, produce no fill-in. Usually iterative methods are preferred to direct solvers, if the state number exceeds a machine-dependent threshold (e.g. 1000 states). If the bandwidth of the matrix is small, this number can be increased. The computational costs of iterative methods are determined by the number of required iterations. In our special context a very good startvector is known, viz. the state probability vector of the preceeding step. Thus in most applications only a few iterations are required. In [19] and [15] a Gauss-Seidel solver is used to solve the algebraic systems. It is asserted that usually less than 10 iterations are needed.

Implementation aspects:

The TR-BDF-2 method and the Runge-Kutta approach has been implemented using a direct method (LU-factorization (LU)) and an iterative method (Gauss-Seidel (GS)) to solve the generated algebraic systems. Our implementation of the Gauss-Seidel algorithm considers the remarks made in [20]. A correct solution of the ODE-solver is very sensitive on an appropriate convergence test. Especially for NCD-models convergence speed is very slow, thus we suggest a relative convergence test as described in [20]. The error tolerance depends on the desired accuracy.

The step size control for the TR-BDF-2 method is based on the formulas given in [19]. For the implicit Runge-Kutta method the estimation of the LTE according to [15] is not satisfactory, because the proposed direct estimation of the fourth derivative is expensive and leads to an overestimation of the LTE for large h. Therefore the step length cannot exceed a threshold, even if the model has reached its equilibrium, and too many integration steps are required, so that the method becomes inefficient. To cope with this problems, we use the method of Milne's device for moderate h and Richardson extrapolation for larger h [13].

In [15] it is shown that all these methods perform efficiently for some stiff models which are not of NCD-type. This coincides with our experiences. But now we will investigate, how TR-BDF-2 and implicit Runge-Kutta methods work for Markov chains satisfying the new stiffness characterization (9), i.e. those models where the modified randomization algorithm fails. Again we consider our extended machine-repairmen model (see Table 4). The first entry in the Table denotes the CPU time on a SPARC station, the second one the number of integration steps for the direct methods, and the number of Gauss-Seidel steps (or iterations) for the iterative methods, respectively.

coverage c	0.50		0.80		0.90		0.99	
TR-BDF-LU	1.9	208	1.9	207	1.9	205	1.8	203
implicit RK-LU	0.8	82	0.8	85	0.8	85	0.7	72
TR-BDF-GS	1.6	2942	1.8	3443	1.9	4196	3.8	12556
implicit RK-GS	2.7	6846	18.7	59960	61.9	205287	1052	3589266

Table 4: computational effort of implicit ODE-methods vs. coverage factor.

We can observe that the implicit ODE-methods work very efficiently if a direct solver is used - even in those cases with hard NCD-structure, where the modified randomization algorithm fails. For both methods only a few integration steps are necessary. The Runge-Kutta method needs less steps ($\cong$85) than TR-BDF-2 ($\cong$ 200) because of its higher order. Each step requires a matrix factorization, but the state space is so small (18 states), that the computational effort is negligible. Problems arise if we use iterative solvers and, unfortunately, we are dependent on such methods if models (matrices) are large.

If the degree of coupling between state sets is weakened by increasing the coverage factor, the number of required Gauss-Seidel steps grows significantly. What are the reasons for this behavior? After the NCD-model has reached short-term-equilibrium [5], step size h gets relative large, i.e. it corresponds to the rates between the state sets. In this case only the small matrix entries are responsible for variations of the state probability vector, and it requires a lot of Gauss-Seidel steps to compute the next state probability vector. This effect is well-known for the stationary analysis of NCD-models, see e.g. [20].

We can observe different behavior for the TR-BDF-2 and the Runge-Kutta method: TR-BDF-2-GS causes a moderate increase of the computational effort for large coverage factors. But the implicit Runge-Kutta method becomes infeasible, if the degree of coupling is weakened. The solution of the implicit equations (12) is considerably harder to achieve, because the term with the square matrix $\frac{1}{6} h^2 \cdot Q^2$ causes an ill-conditioning of the algebraic system.

A result of this experiments is that both implicit methods cause a very large number of Gauss-Seidel steps, if NCD-models are given. For low accuracies the TR-BDF-2 methods can be recommended. If very high accuracy is desired, the transient analysis gets very expensive, and three alternatives remain:

- *Implicit Runge-Kutta method using direct solvers*

  Applying this method we usually need as many LU-factorizations as integration steps are required. Thus this approach can be recommended only for relative small matrices.

- *Implicit Runge-Kutta method using iterative solvers*

  As just showed, the corresponding algebraic systems are so ill-conditioned that iterative solvers become computationally expensive.

- *TR-BDF-2 with Gauss-Seidel solver*

  The algebraic systems (10) and (11) are not as hard to solve as in the case of Runge-Kutta, but due to the low order of this method the allowed step size is relative small to achieve high accuracies. Hence for TR-BDF-2 the computational effort is very large, too.

Note that the method of choice depends on the concrete model features, viz. the number of states, the degree of coupling between the state sets, and the user-desired accuracy. In the next section we propose some ideas to reduce the number of Gauss-Seidel steps, when using implicit ODE methods.

5 Adaptation of implicit ODE-methods to stiff Markov chains

In this section we try to improve the behavior of implicit ODE-methods, which employ iterative methods to solve the generated algebraic systems. The goal is to reduce the total number of Gauss-Seidel steps.

One first idea is to use successive overrelaxation (SOR) instead of Gauss-Seidel-iteration. One can choose the relaxation factor according to [20], but this unfortunately does not yield a significant improvement in our experiments. This result is in agreement with the behavior of SOR applying on the stationary analysis of NCD-models. In the following subsections two other ideas are presented in more detail.

- modification of step size control to prevent large step sizes,

- application of iterative aggregation/disaggregation techniques to improve the solver performance.

5.1 Modifications of step size control

Iterative solvers need a lot of Gauss-Seidel steps, if the step size h is large. Thus it is an obvious idea, to improve the solver performance by the prevention of large step sizes. A simple way to prevent large h without any modification of the algorithm is to increase the user-desired accuracy. The next Table shows the results for different accuracies, applying the ODE-methods on our extended machine-repairmen model with coverage c = 0.90.

For all numerical methods the number of necessary integration steps increases with the accuracy requirement, i.e. step sizes decrease. Thus, as predicted, the computational effort of the implicit Runge-Kutta with Gauss-Seidel iterations decreases, if higher accuracy is desired. For all other methods an increased number of steps causes also a higher computational effort. This is also true for the TR-BDF using Gauss-Seidel iterations, because its convergence speed is sufficient for our example model. Note, that also for the implicit RK-GS method the CPU-time increases, if the accuracy gets smaller than 10^{-11}. Additionally, each step size modification causes an significant overhead to generate the new iteration matrix corresponding to h.

accuracy δ	10^{-6}		10^{-8}		10^{-10}		10^{-11}	
TR-BDF-LU	1.9	205	8.3	907	38.2	4158	81.1	8941
implicit RK-LU	0.8	85	2.2	235	6.6	712	12.2	1275
TR-BDF-GS	1.9	4196	7.7	15020	33.4	62156	71.4	125866
implicit RK-GS	61.9	205287	50.9	163484	40.8	119839	43.1	116078

Table 5: computational effort of implicit ODE-methods vs. accuracy.

Obviously, the number of required iterations is hard to control by increasing the accuracy, because a priori its effect on computational behavior is not predictable. Alternatively we can modify the step size control algorithm. Usually, the optimal step size of the next (i+1)-th integration step is controlled by estimating the local truncation error (LTE_i) of the last step. If δ denotes the user-desired accuracy, p the order of the method, and h_i the last step size, then the length of the next step can be estimated by

$$h^*_{i+1} = h_i \sqrt[p+1]{\frac{\delta}{LTE_i}} \tag{13}$$

If a decrease of step length is necessary (i.e. $LTE_i > \delta$), the last integration step is discarded, and must be repeated with the new step size. In the same way we can control the step length by the number of required Gauss-Seidel steps n_i of the i-th integration step. Let N denote a (user-defined) threshold for the number of Gauss-Seidel steps, then

$$h^*_{i+1} = h_i \sqrt[p+1]{\frac{N}{n_i}} \tag{14}$$

yields a heuristic condition analogous to (13), which takes the number of Gauss-Seidel steps into account. A step size control considering (13) and (14) is realized by $h_{i+1} = \min(h*_i, h**_i)$. The choice of an appropriate threshold N depends on the state space dimension of the model and the convergence speed of the generated matrices. The next Table shows the results for our machine repairmen model with coverage factor c = 0.90, and accuracy 10^{-6}.

N	10		25		50		100		1000	
TR-BDF-GS	2.0	4675	1.9	4270	1.9	4313	1.9	4225	1.9	4225
implicit RK-GS	59.5	52762	16.1	25315	10.1	23878	18.7	35466	26.8	85518

Table 6: computational effort of implicit ODE-methods vs. N.

We can observe that the modified step size control has nearly no influence on the TR-BDF-2 method, because it has good convergence speed in our example, i.e. it needs less than 10 Gauss-Seidel iterations for each step. But for the Runge-Kutta method there is an optimal threshold (about 50) for the number of Gauss-Seidel steps, which reduces the computational effort significantly. Note that for small N the algebraic systems can be solved only for small step sizes. Thus the number of integration steps and hence the computational effort increases, if N gets too small.

Note, that this modified step size control yields a significant improvement of the RK-GS method (compare also Table 6 with Table 4), though the problem remains to determine a good threshold of N a priori.

To cope with this problem in a third approach we relate the number of required Gauss-Seidel steps n_i of the i-th step to the corresponding step length h_i. Let $K_i = n_i / h_i$ the number of Gauss-Seidel iterations per unit step size, then the step size h_{i+1} of the next step can be obtained by:

$$h*_{i+1} = h_i \sqrt[p+1]{\frac{K_{i-1}}{K_i}} \qquad (15)$$

The results of this new step size control strategy are given in Table 7, where the first lines repeat the results of the standard step size control according to Table 4.

We can ovbserve that the modified step size control yields large improvements especially for the implicit Runge-Kutta method. It works even better than the step size control according to (14) with a good threshold N. Furthermore an advantage of this approach is, that neither an appropriate threshold of the accuracy nor the maximum number of Gauss-Seidel steps has to be specified. Thus this step size control is self-contained, i.e. it works without the help of an experienced user, and can be easily used.

coverage c	0.50		0.80		0.90		0.99	
TR-BDF-GS	1.6	2942	1.8	3443	1.9	4196	3.8	12556
TR-BDF-GS*	1.4	2489	1.8	3300	1.8	3871	3.5	9056
implicit RK-GS	2.7	6846	18.7	59460	61.9	205287	1052	3589266
implicit RK-GS*	2.7	6846	5.7	7886	9.7	15719	48.2	70991

Table 7: computational effort of implicit ODE-methods for the modified step size control.

5.2 Iterative aggregation-disaggregation techniques

Another interesting approach to reduce the computational effort, is the improvement of the solver of the algebraic systems. It is well-known that the stationary analysis of NCD-Markov models can be improved by *iterative aggregation-disaggregation techniques* [6]. These techniques cannot be directly applied, because the iteration matrices in our implicit ODE-methods are not conservative, and the right side is different from the zero-vector. But, recently, iterative aggregation-disaggregation (A/D) techniques have been extended to solve general linear equations [21], so that we can use them for our integration methods. Now we outline the basic idea of iterative aggregation-disaggregation techniques (for details see [21]) to solve a non singular system of linear equations:

$$x \cdot Q = r.\qquad(16)$$

Suppose that there is a partition of the overall state space $\Omega = S_1 \cup S_2 \cup ... \cup S_m$. Then $M = \{1,..,m\}$ denotes the so-called macro state space, and $m(i)$, $i \in \Omega$ is the macro state corresponding to micro state i. Iterative aggregation-disaggregation algorithms alternate between solving the original algebraic system (16) and an aggregate version of the system of the form:

$$x^* Q^* = r^* \text{ with } x^* = (x_1^*,...,x_m^*), \ r^* = (r_1^*,...,r_m^*). \quad(17)$$

This aggregate system is related to the interactions between the different partitions. Let I,J be macro states, then the aggregate matrix $Q^* = (q_{IJ}^*)$ is constructed by

$$q_{IJ}^* = \sum_{i \in S_I} \left(\sum_{j \in S_J} \frac{x_i \, q_{ij}}{\sum_{i \in S_I} x_i} \right).\qquad(18)$$

The right-hand side in (17) is given by $r_I^* = \sum_{i \in S_I} r_i$ for $I = 1,..,m$. The solution of (17) yields the aggregate probabilties $x_I^* = \sum_{i \in S_I} x_i$. Now, iterative A/D techniques proceed in the following manner: Gauss-Seidel steps of the overall system (16) yield an iteration vector $x = (x_1,...,x_n)$, which is used to construct the aggregate system (17), which is solved for x^*. Then a disaggregation step recalculates the vector x

$$x_i = x_i \, \frac{x_{m(i)}^*}{\sum_{i \in S_{m(i)}} x_i},\qquad(19)$$

and the algorithm continues with some Gauss-Seidel steps. Aggregation and disaggregation steps are iterated until convergence is achieved for the overall system (16). Note that there is no additional error caused by iterative A/D techniques. The same convergence tests [20] as for SOR are used, i.e. the same accuracy is obtained.

Implementation aspects:

The acceleration of convergence depends on the state space partition. In the case of NCD-models we have to determine state space partitions in such a way, that states of different subsets are only connected by small rates, i.e. they are loosely coupled. To cope with this problem we investigate the connectivity structure of the state space: we determine the maximal strong components of a state graph, whose edges correspond to state transitions with "large" rates [1]. The corresponding algorithm is very fast even for very large matrices, viz. of the order $O(\eta(n))$, if $\eta(n)$ denotes the number of non-zero entries in Q. Note that this approach allows it to determine the NCD-structure of a Marov chain. For details see [8] where we also study the problem of finding thresholds separating large from small rates.

The aggregate system is usually small, so it can be solved by direct methods. An aggregation step is only inserted if the convergence speed of the Gauss-Seidel iteration is slow. The following Table shows the results of the ODE-methods using the iterative A/D techniques for our machine-repairmen example. To consider the effects of iterative A/D techniques solely, we first use the unmodified step size control. (The results of Table 4 for the conventional Gauss-Seidel solver are repeated.)

coverage c	0.50		0.80		0.90		0.99	
TR-BDF-GS	1.6	2942	1.8	3443	1.9	4225	3.8	12556
TR-BDF-AD	1.6	2942	1.8	3443	1.9	4186	2.9	7397
implicit RK-GS	2.7	6846	18.7	59960	51.9	169616	1052	3589266
implicit RK-AD	2.5	4324	12.2	20119	27.2	47659	99.1	227675

Table 8: computational effort of implicit ODE-methods using Gauss-Seidel and iterative A/D solver.

If the degree of coupling between state sets is weakened by an increasing coverage factor, the improvement of iterative A/D-technique is large for the Runge-Kutta method. TR-BDF-GS is quite fast, so that slight improvements are given only for hard NCD-structures. Comparing the results with those of the modified step size control given in Table 7, we see that iterative A/D-solvers yield smaller computational gains. Since the implementation effort is much larger for the iterative aggregation/disaggregation approach, it can be recommended to use the modified step size control in any case.

Finally we combine the modified step size control according to (15) with iterative A/D solver. We obtain the following results:

coverage c	0.50		0.80		0.90		0.99	
TR-BDF-GS	1.6	2489	1.8	3300	1.7	3838	3.3	8849
implicit RK-AD	2.1	2670	5.5	7711	9.6	15219	44.5	61004

Table 9: computational effort using modified step size control and iterative A/D solver.

It can be seen that the composite application of modified step size control and iterative A/D solvers yields the best results: The solver performance of our modified step size control can be slightly improved, if additionally iterative A/D solvers are used. Especially for hard NCD-models using high order implicit RK-methods significant computational gains can be obtained, and in these cases the application of iterative A/D solver seems to be helpful.

6 Conclusion

We have considered various numerical methods for the transient solution of stiff Markov chains. As recommended in the literature, randomization is the method of choice for benign problems. But we have seen that also large stiffness indices do not always cause computational intractability of randomization. If we test the embedded DTMC for stationarity, it is possible to solve models efficiently which are extremely stiff according to formal stiffness definitions. Computational intractability remains, if models reach their equilibrium "very late". This experience leads to a modified characterization (or concept) of stiff Markov chains. A well-known model class satisfying this characterization is given by the NCD-models. For those models randomization remains infeasible.

Implicit ODE-solvers have been designed to cope with those models, where other numerical methods fail. Thus we have investigated their performance on stiff NCD-models. Experiments show that implicit ODE-methods are fast if the generated algebraic systems are solved by direct methods, which is only possible for small models. If iterative solvers are used, the computational effort increases if the coupling of the NCD-model is weakened. This problem arises because NCD-models lead to ill-conditioned iteration matrices. These difficulties are rendered, if implicit Runge-Kutta methods are used, because the square of the generator matrix is involved in the algebraic systems. For very loosely coupled NCD-models implicit Runge-Kutta methods using Gauss-Seidel iteration become computational intractable.

To cope with this problem we propose a modified step size control, which takes the number of required Gauss-Seidel steps into account, and the usage of iterative aggregation/disaggregation techniques. The composite application of both techniques yields large computational gains, especially for high order implicit Runge-Kutta methods.

As a final result we recommend the TR-BDF-2 method if moderate accuracy is desired. The proposed step size control and iterative A/D techniques may improve the TR-BDF-2 method significant. However, these techniques are very helpful, if implicit Runge-Kutta method are required to yield high accuracy.

Unfortunately, in any case it is an expensive task to employ implicit ODE-solvers. Thus there remains enough room for further extensions, e.g. the application of model partitioning techniques.

References

[1] A. V. Aho, J. E. Hopcraft, J. D. Ullman. *The Design and Analysis of Computer Algorithms*. Addison-Wesley, Reading, MA, 1974.

[2] R. C. Aiken (ed.). *Stiff Computation*. Oxford University Press, Oxford, 1985.

[3] R. Bank, W. Courghan, W. Fichtner, E. Grosse, D. Rose, R. Smith. Transient simulation of silicon devices and circuits. *IEEE Transactions on CAD*, 1985, pp. 436-451.

[4] R. K. Brayton, F. G. Gustavson, G. D. Hachtel. A new efficient algorithm for solving differential-algebraic systems using implicit backward differentiation formulas. *Proceedings of the IEEE*, Vol. 60, 1972, pp. 98-108.

[5] P. J. Courtois. Decomposability: Queueing and computer applications. *Academic Press*, 1977.

[6] W. L. Cao, W. J. Stewart. Iterative aggregation/disaggregation techniques for nearly uncoupled markov chains. *Journal of the ACM*, 1985, pp. 702-719.

[7] J. Dunkel. On the modeling of workload dependent memory faults. *Proc. of the 20th International Symposium of Fault Tolerant Computing (FTCS--20)*, 1990, pp. 348-355.

[8] J. Dunkel, H. Stahl. A heuristical approach to determine state space partitions. *Internal report*, University of Dortmund, Informatik IV, (in German), 1991.

[9] B. L. Fox, P. W. Glynn. Computing poisson probabilities. *Communications of the ACM*, 1988, pp. 440-445.

[10] G. H. Golub, C. F. van Loan. Nineteen dubious ways to compute the exponential of a matrix. *SIAM review*, 1978, pp. 801-835.

[11] D. Gross, D. Miller. Randomization technique as a modeling tool and solution procedure for transient markov processes. *Operations Research*, 1984, pp. 343-361.

[12] W. K. Grassmann. Finding transient solutions in markovian event systems through randomization. *1st Intern. Workshop on the Numerical Solution of Markovian Chains*, Raleigh, 1990, pp. 375-95.

[13] J. Lambert. *Computational methods in ordinary differential equations*. Wiley, London, 1973.

[14] W. L. Miranker. *Numerical methods for stiff equations and singular perturbation problems*. Reidel, Dordrecht, 1981.

[15] M. Malhotra, K. S. Trivedi. High-order methods for transient analysis of stiff markov chains. *Intern. Conf. on the Performance of Distributed Systems and Integrated Communication Networks*, Kyoto, 1991.

[16] J. K. Muppala. Performance and dependability modeling using stochastic reward nets. *Ph.D. thesis*, Duke University, Durham, NC, 1991.

[17] J. K. Muppala, K. S. Trivedi. Numerical transient solution of finite markovian queueing systems. To appear in: *Queueing and Related Models*, U. Bhat ed., Oxford University Press, 1992.

[18] B. Melamed, M. Yadin. Randomization procedure in the computation of cumulative-time distributions over discrete state markov processes. *Operations Research*, 1984, pp. 926-944.

[19] A. L. Reibman, K. S. Trivedi. Numerical transient analysis of markov models. *Computers and Operations Research*, 1988, pp. 19-36.

[20] W. J. Steward, A. Goyal. Matrix methods for large dependability methods. *IBM Research Report RC 11485*, 1985.

[21] P. J. Schweitzer, K. W. Kindle. An iterative aggregation--disaggregation algorithm for solving linear equations. *Applied Mathematics and Computation*, Vol. 18, 1986, pp. 313-353.

APPLICATION OF FORMAL METHODS

FORMAL TECHNIQUES

FOR SYNCHRONIZED

FAULT-TOLERANT

SYSTEMS

Ben L. DI VITO[1], *Ricky W. BUTLER*[2]
[1]ViGYAN, Inc.
30 Research Drive, Hampton, Virginia 23666-1325, USA
[2]NASA Langley Research Center
Hampton, Virginia 23681-0001, USA

Abstract

We present the formal verification of synchronizing aspects of the Reliable Computing Platform (RCP), a fault-tolerant computing system for digital flight control applications. The RCP uses NMR-style redundancy to mask faults and internal majority voting to purge the effects of transient faults. The system design has been formally specified and verified using the EHDM verification system. Our formalization is based on an extended state machine model incorporating snapshots of local processors' clocks.

1 Introduction

NASA is engaged in a major research effort towards the development of a practical validation and verification methodology for digital fly-by-wire control systems. Researchers at NASA Langley Research Center (LaRC) are exploring formal verification as a candidate technology for the elimination of design errors in such systems. In previous reports [5], [4], [2], we put forward a high level architecture for a *reliable computing platform* (RCP) based on fault-tolerant computing principles. Central to this work is the use of formal methods for the verification of a fault-tolerant operating system that schedules and executes the application tasks of a digital flight control system. Phase 1 of this effort established results about the high level design of RCP. This paper discusses our Phase 2 results, which carry the design, specification, and verification of RCP to lower levels of abstraction. Complete details of the Phase 2 work are available in technical report form [3].

The major goal of this work is to produce a verified real-time computing platform, both hardware and operating system software, useful for a wide variety of control-system applications. Toward this goal, the operating system provides a user interface that "hides" the implementation details of the system such as the redundant processors, voting, clock synchronization, etc. We adopt a very abstract model of real-time computation, introduce three levels of decomposition of the model towards a physical realization, and rigorously prove that the decomposition correctly implements the model. Specifications and proofs have been mechanized using the EHDM verification system [22].

A major objective of the RCP design is to enable the system to recover from the effects of transient faults. More than their analog predecessors, digital flight control systems are vulnerable to external phenomena that can temporarily affect the system without permanently damaging the physical hardware. External phenomena such as electromagnetic interference (EMI) can flip the bits in a processor's memory or temporarily affect an ALU. EMI can come from many sources such as cosmic radiation, lightning or High Intensity Radiated Fields (HIRF).

RCP is designed to automatically purge the effects of transients periodically, provided the transient is not massive, that is, simultaneously affecting a majority of the redundant processors in the system. Of course, there is no hope of recovery if the system designed to overcome transient faults contains a design flaw. Consequently, emphasis has been placed on techniques that mathematically show when the desired recovery properties are obtained.

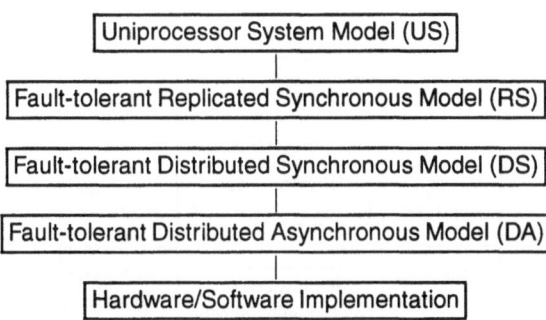

Figure 1: Hierarchical specification of RCP.

1.1 Design of RCP

We propose a well-defined operating system that provides the applications software developer a reliable mechanism for dispatching periodic tasks on a fault-tolerant computing base that *appears* to him as a single ultra-reliable processor. A four-level hierarchical decomposition of the reliable computing platform is shown in Figure 1.

The top level of the hierarchy describes the operating system as a function that sequentially invokes application tasks. This view of the operating system will be referred to as the *uniprocessor model*, which forms the top-level requirement for the RCP.

Fault tolerance is achieved by voting the results computed by the replicated processors operating on identical inputs. Interactive consistency checks on sensor inputs and voting of actuator outputs requires synchronization of the replicated processors. The second level in the hierarchy describes the operating system as a synchronous system where each replicated processor executes the same application tasks. The existence of a global time base, an interactive consistency mechanism and a reliable voting mechanism are assumed at this level.

Although not anticipated during the Phase 1 effort, another layer of refinement was inserted before the introduction of asynchrony. Level 3 of the hierarchy breaks a frame into four sequential phases. This allows a more explicit modeling of interprocessor communication and the time phasing of computation, communication, and voting. The use of this intermediate model avoids introducing these

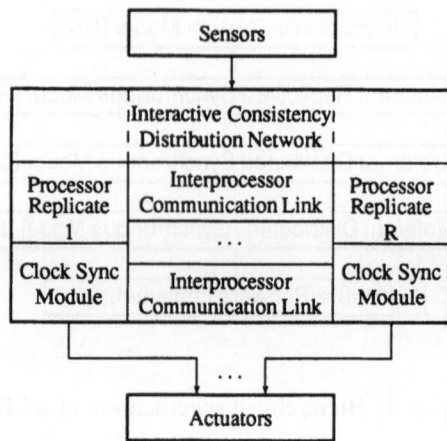

Figure 2: Generic hardware architecture.

issues along with those of real time, thus preventing an overload of details in the proof process.

At the fourth level, the assumptions of the synchronous model must be discharged. Rushby and von Henke [17] report on the formal verification of Lamport and Melliar-Smith's [10] interactive-convergence clock synchronization algorithm. This algorithm can serve as a foundation for the implementation of the replicated system as a collection of asynchronously operating processors. Dedicated hardware implementations of the clock synchronization function are a long-term goal.

Figure 2 depicts the generic hardware architecture assumed for implementing the replicated system. Single-source sensor inputs are distributed by special purpose hardware executing a Byzantine agreement algorithm. Replicated actuator outputs are all delivered in parallel to the actuators, where force-sum voting occurs. Interprocessor communication links allow replicated processors to exchange and vote on the results of task computations. As previously suggested, clock synchronization hardware will be added to the architecture as well.

1.2   Previous efforts

Many techniques for implementing fault-tolerance through redundancy have been developed over the past decade, e.g. SIFT [6], FTMP [7], FTP [9], MAFT [23], and MARS [8]. An often overlooked but significant factor in the development process is the approach to system verification. In SIFT and MAFT, serious consideration was given to the need to mathematically reason about the system. In FTMP and FTP, the verification concept was almost exclusively testing.

Among previous efforts, only the SIFT project attempted to use formal methods [13]. Although the SIFT operating system was never completely verified [14], the concept of Byzantine Generals algorithms was developed [11] as was the first fault-tolerant clock synchronization algorithm with a mathematical performance proof [10]. Other theoretical investigations have also addressed the problems of replicated systems [12].

Some recent work has focused on problems related to the style of fault-tolerant computing adopted by RCP. Rushby has studied a fault masking and transient recovery model and created a formalization of it using EHDM [15], [16]. Rushby's model is more general than ours, but assumes a tighter degree of synchronization where voting takes place after every task execution. In addition, Shankar has undertaken the formalization of a general scheme for modeling fault-tolerant clock synchronization algorithms [19], [20]. Several efforts in hardware verification are likewise relevant. Bevier and Young have verified a circuit design for performing interactive consistency [1], while Srivas and Bickford have carried out a similar activity [21]. Schubert and Levitt have verified the design of processor support circuitry, namely a memory management unit [18].

2 Modeling approach

The specification of the Reliable Computing Platform (RCP) is based on state machine concepts. A system state models the memory contents of all processors as well as *auxiliary variables* such as the fault status of each processor. This latter type of information may not be observable by a running system, but provides a way to express precise specifications. System behavior is described by specifying an initial state and the allowable transitions from one state to another. A transition specification must determine (or constrain) the allowable destination states in terms of the current state and current inputs. The intended interpretation is that each component of the state models the local state of one processor and its associated hardware.

RCP specifications are given in relational form. This enables one to leave unspecified the behavior of a faulty component. Consider the example below.

$$R_{tran} : \text{function[State, State} \rightarrow \text{bool]} =$$
$$(\lambda s, t : \text{nonfaulty}(s(i)) \supset t(i) = f(s(i)))$$

In the relation R_{tran}, if component i of state s is nonfaulty, then component i of the next state t is constrained to equal $f(s(i))$. For other values of i, that is, when $s(i)$ is faulty, the next state value $t(i)$ is unspecified. Any behavior of the faulty component is acceptable in the specification defined by R_{tran}.

It is important to note that the modeling of component hardware faults is for specification purposes *only* and reflects no self-cognizance on the part of the running system. We assume a nonreconfigurable architecture that is capable of masking the effects of faults, but makes no attempt to detect or diagnose those faults. Transient fault recovery is the result of an automatic, continuous voting process; no explicit invocation is involved.

2.1   RCP state machines

The RCP specification consists of four separate models of the system: Uniprocessor System (US), Replicated Synchronous (RS), Distributed Synchronous (DS), Distributed Asynchronous (DA). Each of these specifications is in some sense complete; however, they are written at different levels of abstraction and describe the behavior of the system with different degrees of detail.

1. **Uniprocessor System layer (US).** This constitutes the top-level specification of the functional system behavior defined in terms of an idealized, fault-free computation mechanism. This specification is the correctness criterion to be met by all lower level designs.

2. **Replicated Synchronous layer (RS).** Processors are replicated and the state machine makes global transitions as if all processors were perfectly synchronized. Interprocessor communication is implicit at this layer. Fault tolerance is achieved using exact-match voting on the results computed by the replicated processors operating on identical inputs.

3. **Distributed Synchronous layer (DS).** Next, the interprocessor communication mechanism is modeled and transitions for the RS layer machine are broken into a series of subtransitions. Activity on the separate processors

is still assumed to occur synchronously. Interprocessor communication is accomplished using a simple mailbox scheme.

4. **Distributed Asynchronous layer (DA).** Finally, the lowest layer relaxes the assumption of synchrony and allows each processor to run on its own independent clock. Clock time and real time are introduced into the modeling formalism. The DA machine requires an underlying clock synchronization mechanism.

Most of this paper will concentrate on the DA layer specification and its proof.

The basic design strategy is to use a fault-tolerant clock synchronization algorithm as the foundation for the operating system, providing a global time base for the system. Although the synchronization is not perfect, it is possible to develop a reliable communications scheme where the system clock skew is strictly bounded. For all working clocks p and q, the synchronization algorithm provides a bounded clock skew δ between p and q, assuming that the number of faulty clocks, say m, does not exceed $(\mathsf{nrep}-1)/3$, where nrep is the number of replicated processors. This property enables a simple communications protocol to be established whereby the receiver waits until $\mathsf{maxb} + \delta$ after a pre-determined broadcast time before reading a message (maxb is the maximum communication delay).

Each processor in the system executes the same set of application tasks during every cycle of a continuously repeating task schedule. A schedule comprises a fixed number of frames, each $\mathsf{frame\_time}$ units of time long. A frame is further decomposed into four phases: $\mathsf{compute}$, $\mathsf{broadcast}$, vote and sync. During the $\mathsf{compute}$ phase, all of the applications tasks scheduled for this frame are executed.[1] The results of all tasks that are to be voted this frame are then loaded into the outgoing mailbox, initiating a broadcast send operation. During the next phase, the $\mathsf{broadcast}$ phase, the system merely waits a sufficient amount of time ($\mathsf{maxb} + \delta$) to allow all of the messages to be delivered. During the vote phase, each processor retrieves all of the replicated data from each processor and performs a voting operation. Typically, majority voting is used for each of the selected state elements. The processor then replaces its local memory with the voted values. Finally, the clock synchronization algorithm is executed during the sync phase. Although conceptually this can be performed in either software or hardware, we intend to use a hardware implementation.

---

[1]Multi-rate scheduling is accomplished in RCP by having a task execute every n frames, where n may be chosen differently for each task.

2.2   Extended state machine model

Formalizing the behavior of the Distributed Asynchronous layer requires a means of incorporating time. We accomplish this by formulating an extended state machine model that includes a notion of local clock time for each processor. It also recognizes several types of transitions or operations that can be invoked by each processor. The type of operation dictates which special constraints are imposed on state transitions for certain components.

The time-extended state machine model allows for autonomous local clocks on each processor to be modeled using snapshots of clock time coinciding with state transitions. Clock values within a state represent the time at which the last transition occurred (time current state was entered). If a state was entered by processor p at time T and is occupied for a duration D, the next transition occurs for p at time $T + D$ and this clock value is recorded for p in the next state. A function $c_p(T)$ is assumed to map local clock values for processor p into real time. Notationally, $s(i)$.lclock refers to the (logical) clock-time snapshot of processor i's clock in state s.

Clocks may become skewed in real time. Consequently, the occurrence of corresponding events on different processors may be skewed in real time. A state transition for the DA state machine corresponds to an aggregate transition in which each processor experiences the same event, such as completing one phase of a frame and beginning the next. Each processor may experience the event at different real times and even different clock times if duration values are not identical.

Four classes of operations are distinguished:

1. **L:** Purely local processing that involves no broadcast communication or mailbox access.

2. **B:** Broadcast communication where a send is initiated when the state is entered and must be completed before the next transition.

3. **R:** Local processing that involves no send operations, but does include reading of mailbox values.

4. **C:** Clock synchronization operations that may cause the local clock to be adjusted and appear to be discontinuous.

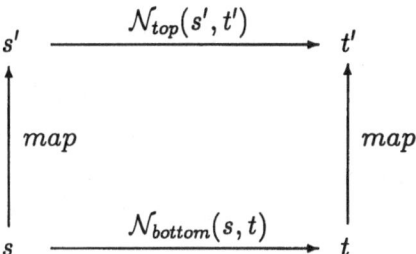

Figure 3: States, transitions, and mappings.

We make the simplifying assumption that the duration spent in each state, except those of type C, is nominally a fixed amount of clock time. Allowances need to be made, however, for small variations in the actual clock time used by real processors. Thus if ν is the maximum rate of variation and D_I, D_A are the intended and actual durations, then $|D_A - D_I| \le \nu D_I$ must hold.

2.3 The proof method

The proof method is a variation of the classical algebraic technique of showing that a homomorphism exists. Such a proof can be visualized as showing that a diagram "commutes" (Figure 3). Consider two adjacent levels of abstraction, called the top and bottom levels for convenience. At the top level we have a current state, s', a destination state, t', and a transition that relates the two. The properties of the transition are given as a mathematical relation, $\mathcal{N}_{top}(s', t')$. Similarly, the bottom level consists of states, s and t, and a transition that relates the two, $\mathcal{N}_{bottom}(s, t)$. The state values at the bottom level are related to the state values at the top level by way of a mapping function, map. To establish that the bottom level implements the top level one must show that the diagram commutes (in a sense meant for relations instead of functions):

$$\mathcal{N}_{bottom}(s, t) \supset \mathcal{N}_{top}(map(s), map(t))$$

where $map(s) = s'$ and $map(t) = t'$ in the diagram. One must also show that initial states map up:

$$\mathcal{I}_{bottom}(s) \supset \mathcal{I}_{top}(map(s))$$

An additional consideration in constructing such proofs is that only states reachable from an initial state are relevant. Thus, it suffices to prove a conditional form of commutativity that assumes transitions always begin from reachable states. A weaker form of the theorem is then called for:

$$\mathcal{R}(s) \wedge \mathcal{N}_{bottom}(s, t) \supset \mathcal{N}_{top}(map(s), map(t))$$

where $\mathcal{R}$ is a reachability predicate. This form enables proofs that proceed by first establishing state invariants. Each invariant is shown to hold for all reachable states using a specialized induction schema and then invoked as a lemma in the main proof.

By carrying out such proofs for each adjacent pair of specification layers in Figure 1, we construct a transitive argument that the lowest layer correctly implements the top-most layer. This is equivalent to a direct proof from bottom to top using the functional composition of all the mappings. Such a large proof is difficult to accomplish in practice; hence the use of a layered approach.

2.4 EHDM language and verification system

Design verification in RCP has been carried out using EHDM. The EHDM verification system [22] is a mature tool, which has been under development by SRI International since 1983 and followed their earlier work on HDM. It comprises a highly integrated environment for formal system development. The specification language is based on a higher-order logic with features supporting module structure and parameterization. An operational subset of the language can be automatically translated to Ada.

EHDM contains an automated theorem prover to support proving in the higher-order logic. Decision procedures for several arithmetic domains are embedded in the system. Users invoke the prover by writing a proof directive in the specification language, stating explicit premises and any necessary substitutions.

3   Clock time and real time

In this section we discuss the synchronization theory upon which the DA specification depends. Although the RCP architecture does not depend on any particular clock synchronization algorithm, we have used the specification for the interactive consistency algorithm (ICA) [10] since EHDM specifications for ICA already exist [17].

The formal definition of a clock is fundamental. A clock can be modeled as a function from real time t to clock time T: $C(t) = T$ or as a function from clock time to real time: $c(T) = t.^2$ Since the ICA theory was expressed in terms of the latter, we will also be modeling clocks as functions from clock time to real time. We must be careful to distinguish between an uncorrected clock and a clock being resynchronized periodically. We use the notation $c(T)$ for an uncorrected clock and $rt^{(i)}(T)$ to represent a synchronized clock during its ith frame.[3]

3.1 Fault model for clocks

In addition to requirements conditioned on having a nonfaulty processor, the DA specifications are concerned with having a nonfaulty clock as well. It is assumed that the clock is an independent piece of hardware whose faults can be isolated from those of the corresponding processor. Although some implementations of a fault-tolerant architecture such as RCP could execute part of the clock synchronization function in software, thereby making clock faults and processor faults mutually dependent, we assume that RCP implementations will have a dedicated hardware clock synchronization function. This means that a clock can continue to function properly during a transient fault period on its adjoining processor. The converse is not true, however. Since the software executing on a processor depends on the clock to properly schedule events, a nonfaulty processor having a faulty clock may produce errors. Therefore, a one-way fault dependency exists.

Good clocks have different drift rates with respect to perfect time. Nevertheless, this drift rate can be bounded. Thus, we define a good clock as one whose drift rate is strictly bounded by $\rho/2$. A clock is "good", i.e., a predicate $\mathsf{good\_clock}(T_0, T_n)$ is true, between clock times T_0 and T_n iff:

$$\forall T_1, T_2 : T_0 \leq T_1 \leq T_n \wedge T_0 \leq T_2 \leq T_n$$
$$\supset |c_p(T_1) - c_p(T_2) - (T_1 - T_2)| \leq \tfrac{\rho}{2} * |T_1 - T_2|$$

The synchronization algorithm is executed once every frame of duration $\mathsf{frame\_time}$. The notation $T^{(i)}$ is used to represent the start of the ith frame at time $T^0 + i * \mathsf{frame\_time}$. The notation $T \in R^{(i)}$ means that T falls in the ith frame, that is,

$$\exists \Pi : 0 \leq \Pi \leq \mathsf{frame\_time} \wedge T = T^{(i)} + \Pi$$

During the ith frame the synchronized clock on processor p, rt_p, is defined by

---

[2] We will use the now standard convention of representing clock time with capital letters and real time with lower case letters.

[3] This differs from the notation, $c^{(i)}(T)$, used in [17].

$rt_p(i,T) = c_p(T + \text{Corr}_p^{(i)})$, where Corr is the cumulative sum of the corrections that have been made to the (logical) clock.

Note that in order for a clock to be nonfaulty in the current frame it is necessary that it has been working continuously from time zero[4]:

$$\text{goodclock}(p, T^{(0)} + \text{Corr}_p^{(0)}, T^{(i+1)} + \text{Corr}_p^{(i)})$$

From these definitions we state the condition of having enough good clocks to maintain synchronization:

```
enough_clocks: function[period → bool] =
    ( λ i : 3 * num_good_clocks(i, nrep) > 2 * nrep)
```

3.2   Clock synchronization

Clock synchronization theory provides two important properties about the clock synchronization algorithm, namely that the skew between good clocks is bounded and that the correction to a good clock is always bounded. The maximum skew is denoted by δ and the maximum correction is denoted by Σ. More formally, for all nonfaulty clocks p and q, two conditions obtain:

S1: $\forall T \in R^{(i)} : |rt_p^{(i)}(T) - rt_q^{(i)}(T)| < \delta$

S2: $|\text{Corr}_p^{(i+1)} - \text{Corr}_p^{(i)}| < \Sigma$

The value of δ is determined by several key parameters of the synchronization system: $\rho, \epsilon, \delta_0, m,$ nrep. The parameter ϵ is a bound on the error in reading another processor's clock. δ_0 is an upper bound on the initial clock skew and m is the maximum number of faulty clocks.

The main synchronization theorem is:

```
sync_thm:  Theorem enough_clocks(i) ⊃
    ( ∀ p, q :  ( ∀T : T ∈ R^(i) ∧
      nonfaulty_clock(p, i) ∧ nonfaulty_clock(q, i)
      ⊃ |rt_p^(i)(T) − rt_q^(i)(T)| ≤ δ))
```

The proof that DA implements DS depends crucially upon this theorem.

---

[4]This is a limitation not of RCP, but of existing, mechanically verified fault-tolerant clock synchronization theory. Future work will concentrate on how to make clock synchronization robust in the presence of transient faults.

3.3 Implementation restrictions

Recall that the DA extended state machine model recognized four different classes of state transition: L, B, R, C. Although each is used for a different phase of the frame, the transition types were introduced because operation restrictions must be imposed on implementations to correctly realize the DA specifications. Failure to satisfy these restrictions can render an implementation at odds with the underlying execution model, where shared data objects are subject to the problems of concurrency. The set of constraints on the DA model's implementation concerns possible concurrent accesses to the mailboxes.

While a broadcast send operation is in progress, the receivers' mailbox values are undefined. If the operation is allowed sufficient time to complete, the mailbox values will match the original values sent. If insufficient time is allowed, or a broadcast operation is begun immediately following the current one, the final mailbox value cannot be assured. Furthermore, we make the additional restriction that all other uses of the mailbox be limited to read-only accesses. This provides a simple sufficient condition for noninterfering use of the mailboxes, thereby avoiding more complex mutual exclusion restrictions.

> **Operation Restrictions.** Let s and t be successive DA states, i be the processor with the earliest value of $c_i(s(i).\text{lclock})$, and j be the processor with the latest value of $c_j(t(j).\text{lclock})$. If s corresponds to a broadcast (B) operation, all processors must have completed the previous operation of type R by time $c_i(s(i).\text{lclock})$, and the next operation of type B can begin no earlier than time $c_j(t(j).\text{lclock})$. No processor may write to its mailbox during an operation of type B or R.

By introducing a prescribed discipline on the use of mailboxes, we ensure that the axiom describing broadcast communication can be legitimately used in the DA proof. Although the restrictions are expressed in terms of real time inequalities over all processors' clocks, it is possible to derive sufficient conditions that satisfy the restrictions and can be established from local processor specifications only, assuming a clock synchronization mechanism is in place.

4   Design specifications

The RCP specifications are expressed in terms of some common types and constants, declared in EHDM as follows:

```
Pstate: Type   (* computation state *)
inputs: Type   (* sensor inputs *)
outputs: Type  (* actuator outputs *)
nrep: nat      (* number of processors *)
```

Mailboxes and their unit of information exchange are provided with types:

```
MB : Type       (* mailbox entry *)
MBvec: Type = array [processors] of MB
```

This scheme provides one slot in the mailbox array for each replicated processor.

In the following, we present a sketch of the specifications for the US and DA layers. To keep the presentation brief, we omit the RS and DS specifications. Details can be found in [3].

4.1   US specification

The US specification is very simple:

$$\mathcal{N}_{us}: \text{function}[\text{Pstate}, \text{Pstate}, \text{inputs} \rightarrow \text{bool}] = (\lambda s, t, u : t = f_c(u, s))$$

The function $\mathcal{N}_{us}$ defines the transition relation between the current state and the next state. We require that the computation performed by the uniprocessor system be deterministic and can be modeled by a function $f_c : \text{inputs} \times \text{Pstate} \rightarrow \text{Pstate}$. To fit the relational, nondeterministic state machine model we simply equate $\mathcal{N}_{us}(s, t, u)$ to the predicate $t = f_c(u, s)$.

External system outputs are selected from the values computed by f_c. The function $f_a : \text{Pstate} \rightarrow \text{outputs}$ denotes the selection of state variable values to be sent to the actuators. The type outputs represents a composite of actuator output types.

While there is no explicit mention of time in the US model, it is intended that a transition correspond to one frame of the execution schedule.

The constant initial_proc_state represents the initial Pstate value when computation begins.

initial_us: function[Pstate → bool] = (λ s : s = initial_proc_state)

Although the initial state value is unique, initial_us is expressed in predicate form for consistency with the overall relational method of specification.

4.2 DA specification

The DA specification permits each processor to run asynchronously. Every processor in the system has its own clock and task executions on one processor take place at different times than on other processors. Nevertheless, the model at this level explicitly takes advantage of the fact that the clocks of the system are synchronized to within a bounded skew δ.

```
da_proc_state: Type =
    Record healthy : nat,
           proc_state : Pstate,
           mailbox : MBvec,
           lclock : logical_clocktime,
           cum_delta : number
    end record


da_proc_array: Type =
    array [processors] of da_proc_state


DAstate: Type =
    Record phase : phases,
           sync_period : nat,
           proc : da_proc_array
    end record
```

The phase field of a DAstate indicates whether the current phase of the state machine is compute, broadcast, vote, or sync. The sync_period field holds the current (unbounded) frame number.

The state for a single processor is given by a record named da_proc_state. The first field of the record is healthy, which is 0 when a processor is faulty. Otherwise, it indicates the (unbounded) number of state transitions since the last transient fault. A permanently faulty processor would have zero in this field for all subsequent frames. A processor that is *recovering* from a transient fault is indicated by a value of healthy less than the constant recovery_period. A processor is said to be *working* whenever healthy $\geq$ recovery_period. The

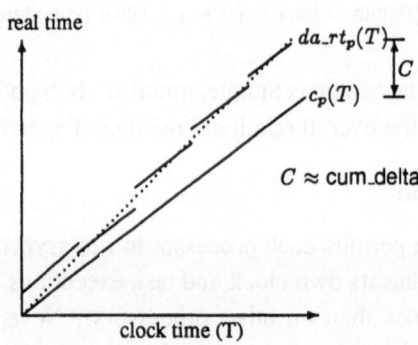

Figure 4: Relationship between c_p and da_rt.

proc_state field of the record is the computation state of the processor. The mailbox field of the record denotes the incoming mailbox mechanism on each processor.

The lclock field of a DAstate stores the current value of the processor's local clock. The real-time corresponding to this clock time can be found through use of the auxiliary function da_rt.

da_rt: function[DAstate, processors, logical_clocktime → realtime] =
   (λ da, p, T : c_p(T + da.proc(p).cum_delta))

This function corresponds to the rt function of the clock synchronization theory. Thus, da_rt(s, p, T) yields the real time corresponding to processor p's synchronized clock. Given a clock time T in the current frame (s.sync_period), da_rt returns the real-time at which processor p's clock reads T. The current value of the cumulative correction is stored in the field cum_delta.

Every frame the clock synchronization algorithm is executed, and an adjustment given by the Corr function of the clock synchronization theory is added to cum_delta. Figure 4 illustrates the relationship among c_p, da_rt, and cum_delta.

The specification of time-critical behavior in the DA model is accomplished using the da_rt function.

$\mathcal{N}_{da}$: function[DAstate, DAstate, inputs $\rightarrow$ bool] =
    ($\lambda\, s, t, u$: enough_hardware(t)
       $\wedge\, t$.phase = next_phase(s.phase)
       $\wedge\, (\forall i :$ **if** s.phase = sync
               then $\mathcal{N}_{da}^s(s, t, i)$
               else t.proc(i).healthy = s.proc(i).healthy
                    $\wedge\, t$.proc(i).cum_delta = s.proc(i).cum_delta
                    $\wedge\, t$.sync_period = s.sync_period
                    $\wedge\,$ (nonfaulty_clock(i, s.sync_period) $\supset$
                         clock_advanced(s.proc(i).lclock,
                                t.proc(i).lclock, duration(s.phase))
                          $\wedge\, (s$.phase = compute $\supset \mathcal{N}_{da}^c(s, t, u, i))$
                          $\wedge\, (s$.phase = broadcast $\supset \mathcal{N}_{da}^b(s, t, i))$
                          $\wedge\, (s$.phase = vote $\supset \mathcal{N}_{da}^v(s, t, i)))$
       end if))

Figure 5: DA transition relation.

For example, the broadcast_received function is expressed in terms of da_rt:

broadcast_received:
function[DAstate, DAstate, processors $\rightarrow$ bool] =
    ($\lambda\, s, t, q : (\forall p : (s$.proc($p$).healthy > 0
       $\wedge$ da_rt(s, p, s.proc(p).lclock) + max_comm_delay
       $\leq$ da_rt(t, q, t.proc(q).lclock))
      $\supset t$.proc(q).mailbox(p) = s.proc(p).mailbox(p)))

Thus, the data in the incoming bin p on processor q is defined to be equal to the value broadcast by p, s.proc(p).mailbox(p), only when the real time on the receiving end, da_rt(t, q, t.proc(q).lclock), is greater than the real time at which the send was initiated, da_rt(s, p, s.proc(p).lclock), plus max_comm_delay. This specification anticipates the design of a communications system that can deliver a message within max_comm_delay units of time.

In the DA level there is no single transition that covers the entire frame. There is only a phase-based state transition relation, $\mathcal{N}_{da}$, shown in Figure 5. Note that the transition to a new state is only valid when enough_hardware holds in the next state:

```
enough_hardware:
    function[DAstate → bool] =
        ( λ t : maj_working(t) ∧ enough_clocks(t.sync_period))
```

The transition relation $\mathcal{N}_{da}$ is defined in terms of four subrelations (not shown): $\mathcal{N}_{da}^c$, $\mathcal{N}_{da}^b$, $\mathcal{N}_{da}^v$ and $\mathcal{N}_{da}^s$, each of which applies to a particular phase type.

As defined by the compute phase relation $\mathcal{N}_{da}^c$, the proc_state field is updated with the results of task computation, $f_c(u, s.\text{proc}(i).\text{proc\_state})$, and the mailbox is loaded with the subset of these results to be broadcast. Note that each nonfaulty replicated processor is required to behave deterministically with respect to task computation; in particular, f_c is the same computation function as specified in the US layer. Moreover, the local clock time is changed in the new state. This is accomplished by the predicate clock_advanced, which is not based on a simple incrementation operation because the number of clock cycles consumed by an instruction stream will exhibit a small amount of variation on real processors. The function clock_advanced accounts for this variability, meaning the start of the next phase is not wholly determined by the start time of the current phase.

```
clock_advanced:
    function[logical_clocktime, logical_clocktime, number → bool] =
        ( λ X,Y,D : X + D * (1 − ν) ≤ Y ∧ Y ≤ X + D * (1 + ν))
```

$ν$ represents the maximum rate at which one processor's execution time over a phase can vary from the *nominal* amount given by the duration function. $ν$ is intended to be a nonnegative fractional value, $0 \leq ν < 1$. The nominal amount of time spent in each phase is specified by a function named duration:

```
duration: function[phases → logical_clocktime]
```

The predicate initial_da puts forth the conditions for a valid initial state. The initial phase is set to compute and the initial sync period is set to zero. Each element of the DA state array has its healthy field equal to recovery_period and its proc_state field equal to initial_proc_state.

```
initial_da: function[DAstate → bool] =
    ( λ s : s.phase = compute ∧ s.sync_period = 0 ∧
        ( ∀ i : s.proc(i).healthy = recovery_period ∧
            s.proc(i).proc_state = initial_proc_state ∧
            s.proc(i).cum_delta = 0 ∧ s.proc(i).lclock = 0 ∧
            nonfaulty_clock(i, 0)))
```

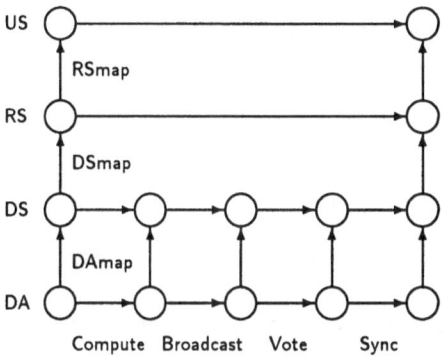

Figure 6: RCP state machine and proof hierarchy.

By initializing the **healthy** fields to the constant **recovery_period** we are starting the system with all processors *working*. Note that the mailbox fields are *not* initialized; any mailbox values can appear in a valid initial **DAstate**.

5 Summary of system proof

Figure 6 shows the complete state machine hierarchy and the relationships of transitions within the aggregate model. By performing three layer-to-layer state machine implementation proofs, the states of **DA**, the lowest layer, are shown to correctly map to those of **US**, the highest layer. This means that any implementation satisfying the **DA** specification will likewise satisfy **US** under our chosen interpretation, which is given by a functional composition:

$$DAmap \circ DSmap \circ RSmap$$

5.1 Overall hierarchy

The two theorems required to establish that **RS** implements **US** are the following.

RS_frame_commutes: **Theorem**
$$reachable(s) \land \mathcal{N}_{rs}(s,t,u) \supset \mathcal{N}_{us}(RSmap(s), RSmap(t), u)$$

RS_initial_maps: **Theorem** $initial\_rs(s) \supset initial\_us(RSmap(s))$

The theorem **RS_frame_commutes** shows that a successive pair of reachable **RS** states can be mapped by **RSmap** into a successive pair of **US** states (upper tier of Figure 6 commutes). The theorem RS_initial_maps shows that an initial RS state can be mapped into an initial US state.

To establish that DS implements RS, the following formulas must be proved.

> DS_frame_commutes: **Theorem**
> s.phase = compute $\wedge$ frame_N_ds(s, t, u) $\supset$
> $\mathcal{N}_{rs}($DSmap$(s),$ DSmap$(t), u)$

> DS_initial_maps: **Theorem** initial_ds(s) $\supset$ initial_rs(DSmap(s))

Note that DS transitions have finer granularity than RS transitions: one per phase (four per frame). Therefore, to follow the proof paradigm, we must consider only DS states found at the beginning of each frame, namely those whose phase is compute. frame_N_ds is a predicate that composes four sequential phase transitions using $\mathcal{N}_{ds}$.

> frame_N_ds: function[DSstate, DSstate, inputs $\rightarrow$ bool] =
> ($\lambda s, t, u : (\exists x, y, z :$
> $\mathcal{N}_{ds}(s, x, u) \wedge \mathcal{N}_{ds}(x, y, u) \wedge \mathcal{N}_{ds}(y, z, u) \wedge \mathcal{N}_{ds}(z, t, u)))$

Using this device, we can show that the second tier of Figure 6 commutes.

Finally, to establish that DA implements DS, the following formulas must be proved:

> phase_commutes: **Theorem**
> reachable$(s) \wedge \mathcal{N}_{da}(s, t, u) \supset \mathcal{N}_{ds}($DAmap$(s),$ DAmap$(t), u)$
> DA_initial_maps: **Theorem** initial_da(s) $\supset$ initial_ds(DAmap(s))

Since DA and DS transitions are both one per phase, the proof is completed by showing that each of the four lower cells of Figure 6 commutes.

5.2   DA layer proof

We provide a brief sketch of the key parts of the DA to DS proof. First, note that the two specifications are very similar in structure. The primary difference is that the DS specification lacks all features related to clock time and real time. A DSstate structure is similar to a DAstate, lacking only the lclock, cum_delta, and sync_period fields. Thus, in the DA to DS mapping function, these fields are not mapped (i.e., are abstracted away) and all of the other fields are mapped identically. Additionally, the DS transition relation is very similar to $\mathcal{N}_{da}$:

$\mathcal{N}_{ds}$: function[DSstate, DSstate, inputs → bool] =
$\quad(\lambda\, s, t, u : \mathsf{maj\_working}(t)$
$\qquad\wedge\, t.\mathsf{phase} = \mathsf{next\_phase}(s.\mathsf{phase})$
$\qquad\wedge\, (\forall\, i: \mathbf{if}\ s.\mathsf{phase} = \mathsf{sync}$
$\qquad\qquad\mathbf{then}\ \mathcal{N}_{ds}^s(s, t, i)$
$\qquad\qquad\mathbf{else}\ t.\mathsf{proc}(i).\mathsf{healthy} = s.\mathsf{proc}(i).\mathsf{healthy}$
$\qquad\qquad\quad\wedge\, (s.\mathsf{phase} = \mathsf{compute} \supset \mathcal{N}_{ds}^c(s, t, u, i))$
$\qquad\qquad\quad\wedge\, (s.\mathsf{phase} = \mathsf{broadcast} \supset \mathcal{N}_{ds}^b(s, t, i))$
$\qquad\qquad\quad\wedge\, (s.\mathsf{phase} = \mathsf{vote} \supset \mathcal{N}_{ds}^v(s, t, i))$
$\qquad\qquad\mathbf{end\ if}))$

The phase_commutes theorem must be shown to hold for all four phases. Thus, the proof is decomposed into four separate cases, each of which is handled by a lemma of the form:

phase_com_$\mathcal{X}$: **Lemma**
$\quad s.\mathsf{phase} = \mathcal{X} \wedge \mathcal{N}_{da}(s, t, u) \supset \mathcal{N}_{ds}(\mathsf{DAmap}(s), \mathsf{DAmap}(t), u)$

where $\mathcal{X}$ is any one of {compute, broadcast, vote, sync}. The proof of this theorem requires the expansion of the $\mathcal{N}_{da}$ relation and showing that the resulting formula logically implies $\mathcal{N}_{ds}(\mathsf{DAmap}(s), \mathsf{DAmap}(t), u)$.

The proof of each lemma phase_com_$\mathcal{X}$ is facilitated by using a common, general scheme for each phase that further decomposes the proof by means of four subordinate lemmas. The general form of these lemmas is as follows:

Lemma 1: $s.\mathsf{phase} = \mathcal{X} \wedge \mathcal{N}_{da}(s, t, u) \supset (\forall\, i: \mathcal{N}_{da}^{\mathcal{X}}(s, t, i))$

Lemma 2: $s.\mathsf{phase} = \mathcal{X} \wedge \mathcal{N}_{da}^{\mathcal{X}}(s, t, i) \supset \mathcal{N}_{ds}^{\mathcal{X}}(\mathsf{DAmap}(s), \mathsf{DAmap}(t), i)$

Lemma 3: $ss.\mathsf{phase} = \mathcal{X} \wedge \mathsf{DS.maj\_working}(tt) \wedge (\forall\, i: \mathcal{N}_{ds}^{\mathcal{X}}(ss, tt, i))$
$\qquad\quad \supset \mathcal{N}_{ds}(ss, tt, u)$

Lemma 4: $s.\mathsf{phase} = \mathcal{X} \wedge \mathcal{N}_{da}(s, t, u) \supset \mathsf{DS.maj\_working}(\mathsf{DAmap}(t))$

A few differences exist among the lemmas for the four phases, but they adhere to this scheme fairly closely. The phase_com_$\mathcal{X}$ lemma follows by chaining the four lemmas together:

$\mathcal{N}_{da}(s, t, u) \supset (\forall\, i: \mathcal{N}_{da}^{\mathcal{X}}(s, t, i)) \supset$
$(\forall\, i: \mathcal{N}_{ds}^{\mathcal{X}}(\mathsf{DAmap}(s), \mathsf{DAmap}(t), i)) \supset$
$\mathcal{N}_{ds}(\mathsf{DAmap}(s), \mathsf{DAmap}(t), u)$

In three of the four cases above, proofs for the lemmas are elementary. The proof of Lemma 1 follows directly from the definition of $\mathcal{N}_{da}$. Lemma 3 follows directly from the definition of $\mathcal{N}_{ds}$. Lemma 4 follows from the definition of $\mathcal{N}_{da}$, enough_hardware, and the basic mapping lemmas.

Furthermore, for three of the four phases, the proof of Lemma 2 is straightforward. For all but the broadcast phase, Lemma 2 follows from the definition of $\mathcal{N}_{ds}^{\mathcal{X}}$, $\mathcal{N}_{da}^{\mathcal{X}}$, and the basic mapping lemmas.

However, in the broadcast phase, Lemma 2 from the scheme above, which is named com_broadcast_2, is a much deeper theorem. The broadcast phase is where the effects of asynchrony are felt: we must show that interprocessor communications are properly received in the presence of asynchronously operating processors. Without clock synchronization we would be unable to assert that broadcast data is received. Hence the need to invoke clock synchronization theory and its attendant reasoning over inequalities of time.

The lemma com_broadcast_2 deals with the main difference between the DA level and the DS level—the timing constraint in the function broadcast_received. The timing constraint

$$\text{da\_rt}(s, p, s.\text{proc}(p).\text{lclock}) + \text{max\_comm\_delay} \leq \text{da\_rt}(t, q, t.\text{proc}(q).\text{lclock})$$

must be satisfied to show that the DS level analog of broadcast_received holds. A key lemma relating real times on two processors is instrumental for this purpose:

ELT: **Lemma**
$\quad T_2 \geq T_1 + \text{bb} \wedge (T_1 \geq T^0)$
$\quad \wedge\ (\text{bb} \geq T^0) \wedge T_2 \in R^{(\text{sp})} \wedge T_1 \in R^{(\text{sp})}$
$\quad \wedge\ \text{nonfaulty\_clock}(p, \text{sp}) \wedge \text{nonfaulty\_clock}(q, \text{sp})$
$\quad \wedge\ \text{enough\_clocks}(\text{sp})$
$\quad \supset rt_p^{(\text{sp})}(T_2) \geq rt_q^{(\text{sp})}(T_1) + (1 - \frac{\rho}{2}) * |\text{bb}| - \delta$

This lemma establishes an important property of timed events in the presence of a fault-tolerant clock synchronization algorithm. Suppose that on processor q an event occurs at $T1$ according to its own clock and another event occurs on processor p at time $T2$ according to its own clock. Then, assuming that the clock times fall within the current frame and enough clocks are nonfaulty, then the following is true about the real times of the events:

$$rt_p^{(\text{sp})}(T_2) \geq rt_q^{(\text{sp})}(T_1) + (1 - \frac{\rho}{2}) * |\text{bb}| - \delta$$

where $\mathsf{bb} = T_2 - T_1$, $T_1 = s.\mathsf{proc}(p).\mathsf{lclock}$, and $T_2 = t.\mathsf{proc}(q).\mathsf{lclock}$.

If we apply this lemma to the broadcast phase, letting T1 be the time that the sender loads his outgoing mailbox bin and T2 be the earliest time that the receivers can read their mailboxes (i.e., at the start of the vote phase), we know that these events are separated in time by more than $\left(1 - \frac{\rho}{2}\right) * |\mathsf{bb}| - \delta$. By choosing the value $\mathsf{bb} = \mathsf{duration(broadcast)}$ in such a way that this real time quantity exceeds max_comm_delay, accounting for ν variation as well, we can prove that all broadcast messages are properly received.

5.3 Proof mechanization

All proofs sketched above as well as the other RCP proofs have been carried out with the assistance of EHDM [3]. Although the first phase of this work was accomplished without the use of an automated theorem prover [5], we found the use of EHDM beneficial to this second phase of work for several reasons.

- Increasingly detailed specifications emerge in the lower level models.

- The strictness of the EHDM language forced us to elaborate the design more carefully.

- Most proofs are not very deep but contain substantial detail. Without a mechanical proof checker, it would be far too easy to overlook a flaw in the proofs.

- The proof support environment of EHDM assures us that our proof chains are complete and we have not overlooked some unproved lemmas.

- The decision procedures for linear arithmetic and propositional calculus relieved us of the need to reduce many formulas to primitive axioms of arithmetic. Especially useful was EHDM's reasoning ability for inequalities.

6 Conclusion

We have described a formalization of the synchronizing aspects of a reliable computing platform (RCP). The top level specification is extremely general and should serve as a model for many fault-tolerant system designs. The successive refinements in the lower levels of abstraction introduce, first, processor replication and voting, second, interprocess communication by use of dedicated mailboxes, and finally, the asynchrony due to separate clocks in the system.

Key features of the overall RCP work completed during Phase 2 and improvements over the results of Phase 1 include the following.

- Specification of redundancy management and transient fault recovery are based on a very general model of fault-tolerant computing similar to one proposed by Rushby [15], [16], but using a frame-based rather than task-based granularity of synchronization.

- Specification of the asynchronous layer design uses modeling techniques based on a time-extended state machine approach. This method allows us to build on previous work that formalized clock synchronization mechanisms and their properties.

- Formulation of the RCP specifications is based on a straightforward fault model, providing a clean interface to the realm of probabilistic reliability models. It is only necessary to determine the probability of having a majority of working processors and a two-thirds majority of nonfaulty clocks.

- A four-layer tier of specifications has been completely proved to the standards of rigor of the EHDM mechanical proof system. The full set of proofs can be run on a Sun SPARCstation in less than one hour.

- Important constraints on lower level design and implementation constructs have been identified and investigated.

Based on the results obtained thus far, work will continue to a Phase 3 effort, which will concentrate on completing design formalizations and develop the techniques needed to produce verified implementations of RCP architectures.

Acknowledgments

The authors would like to acknowledge the many helpful suggestions given by Dr. John Rushby of SRI International. His suggestions during the early phases of model formulation and decomposition lead to a significantly more manageable proof activity. We are also grateful to John and Sam Owre for the timely assistance given in the use of the EHDM system. We are likewise grateful to Paul Miner of NASA Langley for his careful review of our work. This research was supported (in part) by the National Aeronautics and Space Administration under Contract No. NAS1-19341.

References

[1] W. R. Bevier, W. D. Young. The proof of correctness of a fault-tolerant circuit design. *Second IFIP Conference on Dependable Computing For Critical Applications*, Tucson, Arizona, February 1991, pp. 107–114.

[2] R. W. Butler, J. L. Caldwell, B. L. Di Vito. Design strategy for a formally verified reliable computing platform. *6th Annual Conference on Computer Assurance (COMPASS 91)*, Gaithersburg, MD, June 1991.

[3] R. W. Butler, B. L. Di Vito. Formal design and verification of a reliable computing platform for real-time control (phase 2 results). *NASA Technical Memorandum 104196*, January 1992.

[4] B. L. Di Vito, R. W. Butler, J. L. Caldwell. High level design proof of a reliable computing platform. *Dependable Computing for Critical Applications 2*, Dependable Computing and Fault-Tolerant Systems, Springer Verlag, Wien New York, 1992, pp. 279–306. Also presented at *2nd IFIP Working Conference on Dependable Computing for Critical Applications*, Tucson, AZ, Feb. 18–20, 1991, pp. 124–136.

[5] B. L. Di Vito, R. W. Butler, J. L. Caldwell, II. Formal design and verification of a reliable computing platform for real-time control (phase 1 results). *NASA Technical Memorandum 102716*, October 1990.

[6] J. Goldberg et al. Development and analysis of the software implemented fault-tolerance (SIFT) computer. *NASA Contractor Report 172146*, 1984.

[7] A. L. Hopkins, Jr., T. Basil Smith, III, J. H. Lala. FTMP — A highly reliable fault-tolerant multiprocessor for aircraft. *Proc. IEEE*, 66(10), October 1978, pp. 1221–1239.

[8] H. Kopetz, A. Damm, C. Koza, M. Mulazzani, W. Schwabl, C. Senft, R. Zainlinger. Distributed fault-tolerant real-time systems: The Mars approach. *IEEE Micro*, 9, February 1989, pp. 25–40.

[9] J. H. Lala, L. S. Alger, R. J. Gauthier, M. J. Dzwonczyk. A Fault-Tolerant Processor to meet rigorous failure requirements. *Technical Report CSDL-P-2705*, Charles Stark Draper Lab., Inc., July 1986.

[10] L. Lamport, P. M. Melliar-Smith. Synchronizing clocks in the presence of faults. *Journal of the ACM*, 32(1), January 1985, pp. 52–78.

[11] L. Lamport, R. Shostak, M. Pease. The Byzantine Generals problem. *ACM Transactions on Programming Languages and Systems*, 4(3), July 1982, pp. 382–401.

[12] L. V. Mancini, G. Pappalardo. Towards a theory of replicated processing. *Lecture Notes in Computer Science*, Vol. 331, Springer Verlag, 1988, pp. 175–192.

[13] L. Moser, M. Melliar-Smith, R. Schwartz. Design verification of SIFT. *NASA Contractor Report 4097*, September 1987.

[14] NASA. Peer review of a formal verification/design proof methodology. *NASA Conference Publication 2377*, July 1983.

[15] J. Rushby. Formal specification and verification of a fault-masking and transient-recovery model for digital flight-control systems. *NASA Contractor Report 4384*, July 1991.

[16] J. Rushby. Formal specification and verification of a fault-masking and transient-recovery model for digital flight-control systems. *Second International Symposium on Formal Techniques in Real Time and Fault Tolerant Systems*, Nijmegen, The Netherlands, January 1992, pp. 237–258, Vol. 571, *Lecture Notes in Computer Science*, Springer Verlag.

[17] J. Rushby, F. von Henke. Formal verification of a fault-tolerant clock synchronization algorithm. *NASA Contractor Report 4239*, June 1989.

[18] T. Schubert, K. Levitt. Verification of memory management units. *Second IFIP Conference on Dependable Computing For Critical Applications*, Tucson, Arizona, February 1991, pp. 115–123.

[19] N. Shankar. Mechanical verification of a schematic Byzantine clock synchronization algorithm. *NASA Contractor Report* 4386, July 1991.

[20] N. Shankar. Mechanical verification of a generalized protocol for byzantine fault-tolerant clock synchronization. *Second International Symposium on Formal Techniques in Real Time and Fault Tolerant Systems*, Nijmegen, The Netherlands, January 1992, pp. 217–236, Vol. 571 of *Lecture Notes in Computer Science*, Springer Verlag.

[21] M. Srivas, M. Bickford. Verification of the FtCayuga fault-tolerant microprocessor system (Volume 1: A case study in theorem prover-based verification). *NASA Contractor Report* 4381, July 1991.

[22] F. W. von Henke, J. S. Crow, R. Lee, J. M. Rushby, R. A. Whitehurst. EHDM verification environment: an overview. *11th National Computer Security Conference*, Baltimore, Maryland, 1988.

[23] C. J. Walter, R. M. Kieckhafer, A. M. Finn. MAFT: a multicomputer architecture for fault-tolerance in real-time control systems. *IEEE Real-Time Systems Symposium*, December 1985.

COMPILER CORRECTNESS AND

INPUT/OUTPUT

Paul CURZON
University of Cambridge, Computer Laboratory,
New Museums Site, Pembroke Street, Cambridge CB2 3QG, U.K.

Abstract

We describe the formal machine-checked verification of a compiler from a subset of the Vista structured assembly language to the flat assembly language Visa. In particular, we describe the problems associated with input and output commands. We present an oracle based model of I/O. We show how this model can be incorporated into both the relational semantics of Vista and the interpreter semantics of Visa. We illustrate how the compiler correctness theorem proved is sufficient to deduce correctness properties of compiled code from properties of the original program.

1 Introduction

An important consideration when developing dependable systems is ensuring that the compiler preserves the meaning of the program. In this paper, we describe the formal verification of a compiler for a structured assembly language. In particular, we describe the problems associated with input and output commands. Our motivation for verifying the compiler is to allow us to infer properties about the code executed from properties proven about source programs. This problem has been previously considered for programs without I/O commands [5]. Here we extend those ideas to include I/O behaviour. The source and target languages of our compiler are Vista and Visa, respectively. Vista is a structured assembly language designed for use with the VIPER microprocessor [4] at the Defence Research Agency (Electronics division), formerly RSRE. The statements of Vista are VIPER machine instructions together with structural commands. Visa is a flat VIPER assembly language. VIPER is a 32-bit computer, designed for safety critical applications. Some aspects of it have been formally verified [2, 3].

The compiler and the semantics of the source and target languages are defined in higher-order logic. It is an extension of first-order logic in which functions may take other functions as arguments or return them as results. A basic understanding of typed first-order logic should be sufficient to follow the main points of this paper. We use the standard logical connectives: negation ($\neg$), conjunction ($\wedge$), disjunction ($\vee$), implication ($\supset$), equality ($=$), universal quantification ($\forall$), existential quantification ($\exists$) and conditional ($..\Rightarrow...|...$). All terms must have a well defined type. It may be an atomic type such as a natural number (num) or a string (string). Compound types such as pairs (#) and lists (list) may be constructed. Also, type definitions (defining syntax, states, etc.) similar to the datatypes of programming languages may be given. Constructors in type definitions are separated by "|". Type variables may be used for arbitrary types. Theorems and definitions are implicitly universally quantified over their free variables. All the proofs described have been mechanically checked by the HOL system [10].

2 Formal semantics

The compiler correctness problem is often illustrated by a diagram such as that given in Figure 1. A prerequisite is a formal semantics of both source and target language. A relation, compare, comparing the meanings of programs obtained is also required. Informally, the compiler will be correct if the meaning of every Vista program is related to the meaning of the Visa code resulting from compiling it.

The formal semantics of a language can be given in many styles. Much of the early compiler correctness work used denotational semantics for the source language and operational (interpreter) semantics for the target language, as established by McCarthy and Painter [18]. An alternative is to use a denotational semantics for both source and target language [22, 24]. Plotkin's work on structural operational semantics [21] has inspired recent work to be based completely on operational semantics. For example, Despeyroux's proof of correctness of a translator from MiniML to code for the Categorical Abstract Machine used a structural operational semantics for both source and target language [8]. Martin and Toal [17] and Simpson [23] also took this approach. It has the advantage that inner computation inductions can be avoided. The compiler correctness work performed at Computational Logic Inc. used an interpreter semantics defined in the Boyer-Moore logic for both source and target languages [20, 25]. As interpreters were also used to describe the semantics of the lower levels of the target microprocessor, this had the advantage that the proofs of different levels could easily be combined to give a verified stack of system components [1]. Joyce split his compiler for the Tamarack microprocessor into two phases, compiling first into an intermediate language [13]. The semantics of the intermediate language and target language were given as interpreters. The semantics of the source language was described using a relational semantics in higher-order logic. That is, the semantics was given as a relation between initial and final states. We take a similar approach to Joyce but modified to include a model of I/O.

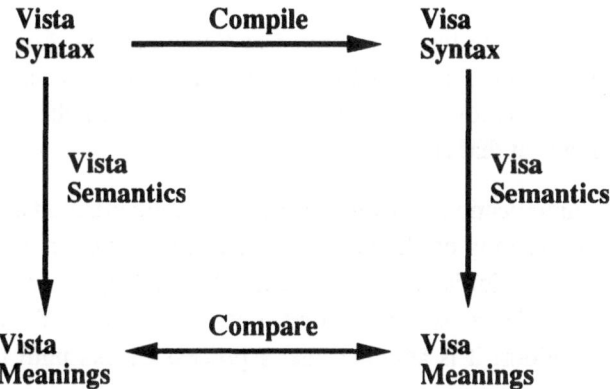

Figure 1: The compiler correctness problem.

3 Input/Output

To give the semantics of languages with input and output commands a model of I/O is needed. In particular, a mechanism for recording the events which have occurred or will occur is required. The details can vary greatly. For example, the data-structures may store the events that have happened (a *trace*), predict events that will happen (an *oracle*), or give all events; past, present and future (a *history*. We consider here just a few examples from the literature.

A simple approach is to use a file based model [11]. Input and output take the form of reading and writing to files. A file is a sequential data-structure and is included as part of the state. Input files are initially full of values and are gradually emptied by the program (an oracle). Output files start empty and are filled (a trace). Only the ordering of events within individual files is recorded. There is no ordering between different files, and in particular the causality between input and output events is not recorded. Both Polak [22] and Stepney *et al* [24] have used this model in their compiler correctness work.

In the hardware and microprocessor verification work performed with higher-order logic a model with an explicit notion of time is used. State variables are represented by history functions from time to data values; the unit of time depending on the level of abstraction considered [19].

In the system verification work performed at Computational Logic Inc. [20], an oracle represented by a list of tuples is used for input. On each cycle of the processor, a tuple is removed from the oracle. It represents the values on the asynchronous inputs at that clock cycle.

In CSP [12], communications between processes are recorded on a trace which gives the relative ordering of all events. There is no distinction between input and output. Events are initiated simultaneously by the "input" process and the "output" process. There is no explicit notion of time. To ensure the synchronous behaviour, it is assumed that a process blocks until the other is available before the communication occurs.

For the semantics of Vista, we require a model of I/O which does not include a notion of time, since Vista does not have a real time semantics. We do need to determine the relative ordering of all events, however, since Vista programs can interact with external devices. This suggests that we should use one data-structure for all events, input or output and whichever channel they are related to. We would also like to reason about non-terminating behaviour. Vista cannot perform events in parallel, so simultaneous events are not an issue. We use an oracle model based on a suggestion by Joyce [15]. The future behaviour of a process is given by a single oracle represented as a list of events. This gives the relative orderings of all the events which occur whilst a process is active. The oracle is not part of the state. This has the advantage that the behaviour of the process is given by a single oracle rather than by the difference between the oracles in the initial and final states. Events contain information such as the data passed, the channel they are associated with and whether they are input or output events with respect to the program. The details may vary, depending on the language being described. For Vista we use the three classifications: input, output and done events; the latter corresponding to the program terminating. A simple model of time is added for the semantics of the flat assembly language Visa by using special non-events whenever no I/O event occurs.

4 Generic definitions and proofs

The use of formal methods can be both time consuming and expensive. Therefore, an important consideration is that the definitions and proofs be useful for more than one application. Rather than describe the semantics of the source and target language directly, we have defined language *schemas*. The precise details of the ALU operations available are not specified. The type variables monop, aluop and compop are used to represent the monadic ALU operations, binary ALU operations and comparison operations, respectively. They can be instantiated to give the Vista and Visa instructions which correspond to the VIPER microprocessor or alternatively to other machines with a similar architecture. We give a generic compiler which compiles source schema programs to target ones. This is possible because the Vista ALU commands are mapped to Visa instructions with the same semantics. The compiler correctness proof need be performed only once for the language schemas. A proof for a particular language can then be obtained by instantiation [5]. The types used to represent machine words are also left as variables—word20 and word32. The names are suggestive of the word sizes for the VIPER microprocessor, though they could be instantiated to word types with any suitable word length.

A representation tuple, rep, provides functions which give the semantics of the operators and conversion functions between the word types. The tuple is used as an argument to all definitions which need this information. Part of the instantiation process is providing this tuple. This technique was first suggested by Joyce [14]. Theorems about the generic definitions, may depend on properties of the functions in the representation tuple. The representation tuple must be instantiated with functions for which such properties hold. A proof obligation of the instantiation process is to prove these theorems. They will be explicit assumptions in the generic theorems proved. The assumption (W32To20ConversionCorrect rep) in the compiler correctness theorem given in Section 8 is one example.

5 The source language

The source language we have considered is a language schema based on the structured assembly language Vista [16]. Since Vista is an assembly language, the general purpose registers are accessible and the basic commands correspond directly to those available on the underlying machine; here VIPER.

```
PROGRAM
    DATA arg1, arg2, t, ans;
    INPUT X arg1;
    INPUT Y arg2;
    t := Y;
    A := 0;
    WHILE X~=0 DO
        A := A + t;
        X := X - 1
    OD;
    OUTPUT A ans;
    STOP
FINISH
```

Figure 2: A Vista multiplication program.

Vista also has some properties of a high level language. Variables may be declared and used, and structuring commands such as while loops are provided. We consider a subset of the language. For example, the only structuring commands considered are sequencing and a while loop. Simple commands include generic binary and monadic machine functions, variable assignment, input and output. Expressions are limited to literal words and variables. Conditions which test the B condition code and compare a register with an expression are provided.

An example program, written in an instantiated version of the subset, which performs multiplication by repeated addition is given in Figure 2. Rather than use the concrete syntax of the example, we use the abstract syntax given in Figure 3. Each syntactic domain is defined as a new type. In fact the subset we have considered also includes procedure declaration and call, though we do not describe it here due to space considerations.

```
Variable =      string
Register =      A | X | Y
Expression =    WORD word32 | NAME Variable
Condition =     B |
                COMPARE compop Register Expression
Command =       SKIP |
                STOP |
                SEQ Command Command |
                MCHFUNC aluop Register Register Expression |
                MONADIC monop Register Expression |
                ASSIGN Variable Register |
                INPUT Variable Register |
                OUTPUT Variable Register |
                WHILE Condition Command
Declarations =  DSKIP |
                DATA Variable |
                DSEQ Declarations Declarations
Program =       PROGRAM Declarations Command
```

Figure 3: The abstract syntax of the Vista schema subset.

The semantics of a Vista program is given by a relation between an oracle, an initial state and a final state. The relation is true if from the initial state the program could have I/O behaviour as predicted by the oracle and end in the final state. As a simple example, consider the Skip command. Informally its behaviour is that it does nothing. Formally, it could be defined by the relation SemSkip.

```
SemSkip oracle q₁ q₂ =
     (q₂ = q₁) ∧ (oracle = [])
```

This relation will be true if the initial and final states are identical and the oracle is empty; that is, no input or output occurs.

A Vista state may be one of three kinds. An *Error* state indicates that the program is invalid (such as containing undeclared variables). A *Halt* state indicates that the program has terminated. A *Run* state indicates that the program is executing normally. The latter two contain a memory store and register store, giving the values of memory and registers, respectively. Throughout this paper we use the variables q, q_1 and q_2 to range over Vista states. We use three classifications of I/O events for Vista : input, output and done events; the latter indicating that the program has terminated. Input and output events have associated with them information about the address written to and the data transmitted (a generic word). We use the variables oracle, $oracle_1$ and $oracle_2$ to range over Vista oracles.

```
VistaState = ERROR | HALT Ms Rs | RUN Ms Rs
```

```
Event = DONE | IN address word32 | OUT address word32
```

The semantic relation SemCom defines the semantics of commands. We will concentrate here on the semantics of the I/O commands and the ramifications of using oracles. The semantic relations for the basic commands are similar to each other. If the initial state is not a run state, then it must be identical to the final state and the oracle must be empty (as a halted or erroneous program will perform no further I/O). This is described by the relation CheckIsRun. The predicate IsRun tests if a state is a run state. If so the semantics is given by the argument sem. Otherwise the states should be identical and the oracle empty.

```
CheckIsRun sem oracle q1 q2 =
    (IsRun q1) ⇒
        (sem oracle q1 q2) |
        ((q2 = q1) ∧ (oracle = []))
```

The semantics of the Stop command is then given by:

```
SemStop oracle q1 q2 =
    (q2 = HALT (RsOf q1) (MsOf q1)) ∧
    (oracle = [DONE])
```

```
SemCom STOP rep env = CheckIsRun SemStop
```

From a run state, it halts the processor, i.e., the final state is a halt state with register store and memory store as the initial state (extracted by the functions RsOf and MsOf). The oracle holds a single done event. SemCom has a representation tuple argument and an environment argument. The latter is built by the declarations and maps variable names to their corresponding locations. It is not used here since no variables are referred to.

Commands which access variables require that the variable has previously been declared. For each declared variable the environment will have an entry mapping it to a location. If there is no entry, the meaning of the variable should be an error. Consequently, the final state after a command containing an undeclared variable has executed will be an error and no I/O will be performed. The relation CheckError is useful when giving the semantics of such a command.

```
CheckError test sem oracle q1 q2 =
    test ⇒
        ((q2 = ERROR) ∧ (oracle = [])) |
        sem oracle q1 q2
```

An example of its use is in the semantics of Output which writes the contents of a source register to a peripheral variable whilst leaving the state unchanged.

```
SemOut 1 src oracle q1 q2 =
    (oracle = [OUT (LocOf 1)(RegisterOf src (RsOf q1))]) ∧
    (q2 = q1)
```

```
SemOutput 1 src = CheckError (IsUnbound 1) (SemOut 1 src)

SemCom (OUTPUT v src) rep env =
        CheckIsRun (SemOutput (SemVexp v env) src)
```

The semantics of the variable is first determined (using **SemVexp**). If it was not declared (as tested by **IsUnbound**) the oracle and final state are given by **CheckError**. Otherwise, the oracle must be a single output event to the variable location of the value in the source register. **LocOf** extracts the value from a variable result. **RegisterOf**, given a register and a register store, returns the value of the register.

The Input command loads a destination register with a value input from a peripheral variable.

```
SemIn 1 dest oracle q1 q2 =
    (LENGTH oracle = 1) ∧
    IsIn (HD oracle) ∧
    (EventAddr (HD oracle) = LocOf 1) ∧
    (q2 = RUN
            (UpdateRegstore dest (RsOf q1)
                    (EventValue (HD oracle)))
            (MsOf q1))
SemInput 1 dest = CheckError (IsUnbound 1) (SemIn 1 dest)

SemCom (INPUT v dest) rep env =
    CheckIsRun (SemInput (SemVexp v env) dest)
```

Provided the initial state is a run state and the target location has been declared, the oracle should consist of a single input event from the indicated peripheral address. The destination register in the final state should hold the value indicated by the oracle. **LENGTH** returns the length of an oracle. **HD** returns the first event on an oracle. **IsIn** tests whether an event is an input event. **EventAddr** and **EventValue** return the address and value, respectively, associated with an input or output event. **UpdateRegstore**, updates a given register in a register store with a new value.

The above commands perform at most one event. If commands are sequenced, then the I/O behaviour of the sub-commands must be combined. If a sequence of two commands uses up an oracle to move from an initial state to a final state, the first command should use some initial part of the oracle and move to an intermediate state. The second command should then use the remainder of the oracle ending in the final state. The oracle for the whole sequence is found by appending together the oracles of the subcommands.

```
SemSeq csem₁ csem₂ oracle q₁ q₂ =
    ∃q oracle₁ oracle₂.
        (csem₁ oracle₁ q₁ q) ∧
        (csem₂ oracle₂ q q₂) ∧
        (oracle = APPEND oracle₁ oracle₂)

SemCom (SEQ c₁ c₂) rep env =
    SemSeq (SemCom c₁ rep env) (SemCom c₂ rep env)
```

6 The target language

The target language of our compiler, Visa, is a flat assembly language schema for VIPER-like microprocessors. It was designed to be an intermediate language for the verified Vista compiler. It does not support all the features available on VIPER; only those required to implement the Vista subset. The instruction set is given in Figure 4.

```
Operand =       LITERAL  word20 |
                CONTENTS bool address

Instruction =   STP |
                JMP address |
                JNB address |
                INP Register address |
                OUT Register address |
                ALU aluop Register Register Operand |
                CMP compop Register Operand |
                MON monop Register Operand |
                STR Register address
```

Figure 4: The instruction set of Visa.

It consists of stop, jump, conditional jump, binary ALU, monadic ALU, comparison, store, input and output commands. The ALU and comparison commands act on an operand. This can either be a literal value, or an address where the value can be found. At this level the address space consists of named regions of infinite size. A flag in the operand indicates whether an address refers to the constant store or data store.

A Visa state is represented as a tuple, (data, p, (a, x, y), b, stop). It consists of a data store, a program counter, a register store holding the A, X and Y registers, the B condition code and a stop flag. Throughout this paper we use variables s, s_1 and s_2 to range over Visa states. The program and constants are stored in separate read-only stores.

We use four classifications of Visa events: those of Vista with an additional no-action event. Each event in the oracle corresponds to a single Visa machine cycle. We use the variables toracle and tevent to range over Visa oracles and events, respectively.We use a relational-based interpreter semantics for Visa. The semantics of each instruction is given by a relation between a single event, an initial state and a final state. For example, the Stop instruction performs a DONE event, setting the stop flag. Its semantics is defined by VisaSemStp.

```
VisaSemStp tevent (data, p, r, b, stop) s₂ =
    (tevent = DONE) ∧
    (s₂ = (data, p, r, b, T))
```

The Output instruction causes an output event to occur. The program counter is incremented using AddAddress. The value output is the contents of a source register, determined using REG. The address written to is given in the instruction.

```
VisaSemOut src addr tevent (data, p, r, b, stop) s₂ =
    (tevent = OUT addr (REG src r)) ∧
    (s₂ = (data, AddAddress p 1, r, b, F))
```

The Input instruction causes an input event to occur from the address indicated in the instruction. The value input is determined by the oracle. It is accessed from the event with TEventValue. A destination register is updated with the value given in the event using UpdateReg.

```
VisaSemInp dest addr tevent (data, p, r, b, stop) s2 =
    (tevent = IN addr (TEventValue tevent)) ∧
    (s2 = (data, AddAddress p 1,
              UpdateReg dest r (TEventValue tevent), b, F))
```

The fetch—decode—execute cycle of the machine is defined by the relation visa in terms of the semantics for the instructions. It first checks that the processor has not stopped. As for Vista, it relates an oracle, initial state and final state. Additionally, the constant store, code store, representation tuple and an end address giving the bound of the fragment of code under consideration must be provided. The relation will be true if the code can be executed from the initial state using the oracle to reach the final state. In addition, the program should either halt the processor or jump beyond the end address for the first time on the last cycle.

```
 0:  INP X 0
 1:  INP Y 1
 2:  STR Y 2
 3:  MON := A (LITERAL 0)
 4:  CMP ~= X (LITERAL 0)
 5:  JNB 9
 6:  ALU + A A (CONTENTS F 2)
 7:  ALU - X X (LITERAL 1)
 8:  JMP 4
 9:  OUT A 3
10:  STP
```

Figure 5: The compiled multiplication program.

7 The compiler

The compiler is defined in higher-order logic by a function CompileProgram which translates Vista programs to Visa assembly code. It returns either the compiled code or a compile-time error. Compiled code is represented as a list of generic instructions and a list of the constants used by the program. The latter must be loaded into the constant store before the program is executed.

Figure 5 shows the instructions of the compiled version of the multiplication program given in Figure 2.

The compiler first creates a symbol table from the declarations. This maps variable names to locations in Visa memory where the variable's value will be stored. The symbol table is passed to the command translator **TransCom**. When translating structural commands, sub-commands are first translated and the results combined. The simple non-structural commands translate to single Visa commands. For example, the translation of the Input command is given by the following definitions.

```
TransInput v d =
    (v = ERROR) ⇒
        COMPILE_ERROR |
        (LoadInst (INP d (ErrValOf v)) [])

TransCom (INPUT v dest) rep symbtab base cstbase =
    TransInput (TransVar v symbtab) dest
```

The memory location corresponding to the variable is looked up in the symbol table using **TransVar**. If it is an error, a compile-time error results. Otherwise, the result is extracted using **ErrValOf**, and a Visa **INP** command formed. This is turned into compiled code with an empty constant list using the function **LoadInst**. The base address of the code and constant stores, **base** and **cstbase**, are not used here.

8 The compiler correctness statement

The compiler correctness theorem needed, and thus the relation between the source and target semantics needed, will depend on the purpose for which it is to be used. Here we use it to map properties about Vista programs to equivalent properties about the compiled code. For example, if we have proved the total correctness of a Vista program, we would like the compiled program to also be totally correct. In general, the relation between semantics required for mapping partial correctness properties of programs is different to that required for total correctness. However, as has been previously noted [8, 5], if the target language is deterministic, then only a single relation is required: partial correctness will follow from total correctness.

The relation required has the following form, where **vista** is the semantic relation of a Vista program or command and **visa** is that for the Visa code:

```
∀q1 q2 oracle s1.
    vista oracle q1 q2 ∧
    CompareStates q1 s1 ⊃
        ∃s2 toracle.
            visa toracle s1 s2 ∧
            CompareStates s2 q2 ∧
            (oracle = AbstractOracle toracle)
```

This states that if we assume

- the Vista program takes some initial state q_1 to final state q_2 using oracle **oracle**, and

- q_1 corresponds to initial Visa state s_1 as specified by the relation **CompareStates**

then we can deduce that there is a Visa state s_2 and Visa oracle **toracle** such that

- the Visa semantics from an initial state s_1 will have I/O behaviour as specified by **toracle** and end in state s_2,

- the final Visa state will correspond to the final Vista state, and

- the Vista oracle and Visa oracle will correspond, as specified by the function **AbstractOracle**. It removes the non-events.

The actual relation used is more complex because extra assumptions must be made that the compiled code and constants are loaded into their respective memory stores, and that the program counter is initially loaded with the start address of the code. Error states must also be considered. **ComCorrect** formalises the relation. It relates the semantics of Vista with the *syntax* of a Visa program rather than with its semantics. The semantic relation **visa** is built in. The correctness diagram really has the form shown in Figure 6. A compiler correctness statement based on this relation, is sufficient to prove total correctness properties of compiled code from total correctness properties of Vista code.

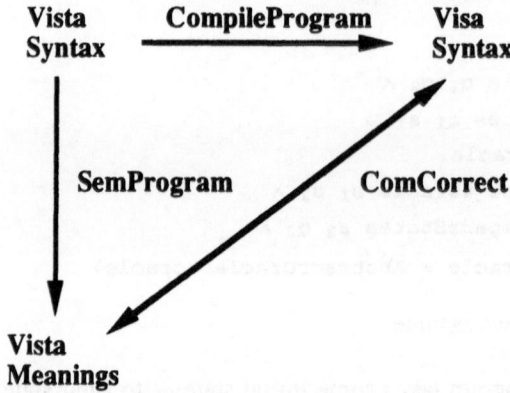

Figure 6: The compiler correctness problem restated.

The compiler will be correct if the semantics of every Vista program is related to the Visa code resulting from compiling it. The correctness theorem is:

```
(Compiles (CompileProgram p rep)) ∧
(W32To20ConversionCorrect rep) ⊃

    ComCorrect rep (SemProgram p rep) (CompileProgram p rep)
```

This statement has two assumptions. The first is that the program actually compiles; nothing is proved about invalid programs. The second assumption is a theorem about the representation tuple. When the tuple is instantiated the functions for converting between 32-bit words and 20-bit words must have the following property: if a 32-bit word is converted to a 20-bit and back again, the original 32-bit word is obtained, provided it fits into 20 bits. This is specified by W32To20ConversionCorrect. The correctness statement tells us nothing if the instantiation of the languages does not fulfil this condition. SemProgram gives the semantics of a program.

The main correctness statement is proved by proving appropriate correctness statements for each of the Vista syntactic domains: declarations, commands, etc. These are each proved by structural induction on the syntax concerned.

The correctness statement for declarations states that they preserve an appropriate correspondence between the symbol table and the environment. The translation of a program uses an empty initial symbol table and the semantics uses an empty environment. Since these correspond, we can use the declaration correctness theorem to deduce that the correspondence holds between the symbol table used when translating commands and the environment used in their semantics. The correctness statement for commands is proved by structural induction. For each command, we consider three cases, depending on whether the initial Vista state is a run, halt or error state. We then compare the results of executing the Vista command with those for executing the Visa instruction, showing that they correspond.

9 The correctness of application programs

The compiler correctness statement we have proved is sufficient to deduce total correctness specifications of Visa programs from total correctness specifications of Vista programs. The definition of total correctness must take account of the use of oracles. Conditions are extended to take an oracle argument. The precondition takes as an argument the oracle which the program uses up. The postcondition takes an empty oracle, since after the execution of a program there should be no outstanding I/O. The definition of total correctness states that if some oracle and initial state satisfy the precondition then there exists some final state such that the semantics with the oracle take the initial state to the final state and the postcondition holds of that final state with an empty oracle.

```
Total P semantics Q =
    ∀q₁ oracle. P oracle q₁ ⊃
        ∃q₂. semantics oracle q₁ q₂ ∧ Q NIL q₂
```

Using the compiler correctness theorem we have derived an inference rule which, from the total correctness of a program which compiles, deduces the total correctness of the compiled code. We have also derived a programming logic for the Vista subset [6] based on the above definition of total correctness using the methods of Gordon [9]. We can thus prove properties of Vista programs using the programming logic and formally deduce that corresponding properties hold of the compiled code. Further details of this work can be found elsewhere [7].

10 Conclusions and further work

We have described an oracle model of I/O behaviour and have shown how it can be incorporated into both the relational semantics of Vista and interpreter semantics of Visa. We have proved a compiler correctness statement based on these semantics for a translator from a subset of Vista to Visa. We have extended the definition of total correctness for programs to include oracles. The compiler correctness theorem can thus be combined with a programming logic based on this definition.

The work could not have been completed with any degree of assurance without some form of mechanisation. We have found the HOL system to be a versatile tool, giving the user control over the proof whilst also providing some automation of mundane tasks. It is useful for the user to direct the proof, as a greater understanding of why the program is correct is obtained. This is one of the advantages of performing formal proofs. It gives more faith that the theorems proved are useful, and makes it easier to correct mistakes when theorems cannot be proved.

The only axioms we used were those upon which the HOL system is built. This approach is very time consuming since all theorems that are not pre-proved in the system have to be proved. However, it increases our confidence in the correctness of the work. Proving theorems which are true is often straightforward. The effort is expended in coming up with the suitable definitions and lemmas in the first place. We spent much time working with definitions and goals that were later modified when lemmas could not be proved. With an axiomatic approach, such mistakes could be missed. Using an expressive specification language such as higher order logic helps avoid mistakes. Whilst perhaps not essential, it is useful in that ideas can be expressed more naturally than in a first order logic and so mistakes are less likely to be made. For example, when defining the semantics we frequently pass semantic relations as arguments. Other techniques such as animation of definitions prior to formal proof would have helped to avoid some of the problems. Our use of generic definitions also saved much proof work. Not only can the work be retargeted to different microprocessors, but also, for example, a single correctness proof suffices for all binary machine instructions.

Our use of oracles for I/O in conjunction with a relational style semantics appears to be very versatile. We have used it to describe the semantics of both the Vista subset and Visa . It could be adapted for other languages. However, a drawback of the relational style semantics as described is that it cannot be used to directly reason about the properties of non-terminating programs. This could be overcome by breaking a program into separate terminating parts, possibly by a process similar to verification condition generation from an annotated program. Further research is required.

Ultimately, we would like to compile (generic) Vista code into (generic) VIPER machine code, proving appropriate correctness properties. The research described here only addresses the first stage of that process. We must prove correct an assembler from Visa to VIPER code. Further work is also required to extend the subset of Vista considered. The languages and correctness theorems described are generic with respect to the ALU and comparison operations available. We would also like to make the model generic in other ways; for example, with respect to parts of the state such as the register set.

Acknowledgement

I am grateful to Gavin Bierman, Mike Gordon, John Kershaw, Jeff Joyce, Clive Pygott and the members of the Hardware Verification Group at Cambridge for their help and advice. The anonymous referees also made useful comments. This work has been funded by MoD research agreement AT2029/205.

References

[1] W. R. Bevier, Jr., W. A. Hunt, W. D. Young. Towards verified execution environments. *Proc. of the 1987 IEEE Symp. on Security and Privacy*, 1987.

[2] A. Cohn. A proof of correctness of the Viper microprocessor: The first level. *VLSI Specification, Verification and Synthesis*, G. Birtwistle, P. A. Subrahmanyam Eds., Kluwer Academic Publishers, 1988, pp. 1-91.

[3] A. Cohn. Correctness properties of the Viper block model: The second level. *Current Trends in Hardware Verification and Automated Theorem Proving*, G. Birtwistle, P. A. Subrahmanyam Eds., Springer-Verlag, 1989, pp. 27-72.

[4] W. J. Cullyer. Implementing safety critical systems: The Viper Microprocessor. *VLSI Specification, Verification and Synthesis*, G. Birtwistle, P. A. Subrahmanyam Eds., Kluwer Academic Publishers, 1988, pp. 1-25.

[5] P. Curzon. A verified compiler for a structured assembly language. *Proc. of the 1991 Int. Workshop on the HOL Theorem Proving System and its Applications*, IEEE Computer Society Press, 1992, pp. 253-262.

[6] P. Curzon. A programming logic for a verified structured assembly language. *Logic Programming and Automated Reasoning*, A. Voronkov Ed., Lecture Notes in Artificial Intelligence Vol. 624, Springer-Verlag, 1992, pp. 403-408.

[7] P. Curzon. Deriving correctness properties of compiled code. *Proc. of the 1992 Int. Workshop on Higher Order Logic Theorem Proving and its Applications*, L. Claesen, M. Gordon Eds., 1992.

[8] J. Despeyroux. Proof of translation in natural semantics. *IEEE Symp. on Logic in Computer Science*, 1986, pp. 193-205.

[9] M. J. C. Gordon. Mechanizing programming logics in higher order logic. *Current Trends in Hardware Verification and Automated Theorem Proving*, G. Birtwistle, P. A. Subrahmanyam Eds., Springer-Verlag, 1989, pp. 387-439.

[10] M. J. C. Gordon, T. F. Melham Eds. *Introduction to HOL: A Theorem Proving Environment for Higher Order Logic*. Cambridge University Press, 1992.

[11] C. A. R. Hoare, N. Wirth. An axiomatic definition of the programming language PASCAL. *Acta Informatica*, Vol. 2, 1973, pp. 335-355.

[12] C. A. R. Hoare. *Communicating Sequential Processes*. Prentice Hall Int., 1985.

[13] J. J. Joyce. A verified compiler for a verified microprocessor. *Technical Report 167*, University of Cambridge, Computer Laboratory, March 1989.

[14] J. J. Joyce. Generic specification of digital hardware. *Technical Report 90-27*, The University of British Columbia, Department of Computer Science, September 1990.

[15] J. J. Joyce. Private communication. 1990.

[16] J. Kershaw. Vista user's guide. *Technical Report 401-86*, RSRE, 1986.

[17] D. Martin, R. Toal. Case studies in compiler correctness using HOL. *Proc. of the 1991 Int. Workshop on the HOL Theorem Proving System and its Applications*, IEEE Computer Society Press, 1992, pp 242-252.

[18] J. McCarthy, J. Painter. Correctness of a compiler for arithmetic expressions. *Proc. of Symp. in Applied Mathematics*, J. T. Schwartz Ed., Vol. XIX, 1966, pp. 33-41.

[19] T. F. Melham. Abstraction mechanisms for hardware verification. *VLSI Specification, Verification and Synthesis*, G. Birtwistle and P. A. Subrahmanyam Eds., Kluwer Academic Publishers, 1988, pp. 267-292.

[20] J. S. Moore. A mechanically verified language implementation. *Journal of Automated Reasoning*, Vol. 5, 1989, pp. 461-492.

[21] G. D. Plotkin. A structural approach to operational semantics. *Technical report*, University of Aarhus, Denmark, Computer Science Department, 1981.

[22] W. Polak. Compiler Specification and Verification. *Lecture Notes in Computer Science Vol. 124*, Springer-Verlag, 1981.

[23] T. G. Simpson. Design and verification of IFL: a wide-spectrum intermediate functional language. *Technical Report 91/440/24*, The University of Calgary, Department of Computer Science, July 1991.

[24] S. Stepney, D. Whitley, D. Cooper, C. Grant. A demonstrably correct compiler. *Formal Aspects of Computing*, Vol. 3, 1991, pp. 58-101.

[25] W. D. Young. A mechanically verified code generator. *Journal of Automated Reasoning*, Vol. 5, 1989, pp. 493-519.

ON LINE ERROR DETECTION

CONTROL FLOW CHECKING

IN OBJECT-BASED

DISTRIBUTED SYSTEMS

Nasser A. KANAWATI, Ghani A. KANAWATI, Jacob A. ABRAHAM
Computer Engineering Research Center, The University of Texas at Austin
Austin, Texas 78758, USA

Abstract

Object-based distributed systems are becoming increasingly popular since objects provide a secure and easy means of using the abstraction of shared memory. In this paper we develop a new object-based control flow checking technique called (COTM) to detect errors due to hardware faults in such systems. The proposed technique monitors objects and thread flow across objects in two stages. The first stage applies control flow checking for every object invocation. In the second stage, the legality of a terminating thread is examined. Results of fault injection experiments on several applications written in C++ and modified to incorporate the object-based checks show that the proposed technique achieves high fault coverage with low performance overhead.

This research was supported in part by the SIDO Innovative Science and Technology Office and managed by the Office of Naval Research (ONR) under Contract N00014-89-K-0089.

1   Introduction

Distributed systems have recently emerged as a cost-effective solution to many complex computational problems and for process control and related applications. When used in situations where the cost of failures is very high, these systems have to meet requirements of high reliability in addition to high performance. Highly dependable operation can be achieved by designing the systems to be fault-tolerant. Such systems have to detect errors (due to faults in the system) concurrently with the normal operation in order to preserve the integrity of the results. Several fault tolerance techniques have been designed to increase the reliability of systems [1]. In order to reduce the overhead of achieving reliable operation, recent fault tolerance techniques have used information about the computation being performed, referred to later as the behavioral approach, and have been implemented at the system architecture level [3], [5], [4].

Concurrent control flow checking is the technique whereby the sequence of addresses of the executing program is monitored at run time and is compared to that of allowable sequences defined by the system model. The control flow error detection mechanisms are typically embedded in the user application code. Several control flow checking techniques have appeared in the literature and have been implemented in numerous systems [3], [10], [11], [14], [17], [16], [19]. In [3], a program is divided into loop-free intervals and code is inserted for performing concurrent checking. Recent work in control flow checking has concentrated on signature monitoring [16], [10], [12], [17], [19]. In signature monitoring, an application program is partitioned, at assembly time, into several segments. A segment is a sequence of instructions with one entry point and one exit point. A signature is generated, commonly off-line by the compiler, as a function of the sequence of instructions within the segment. The signature is either embedded in the original code of the application program [17], [13], [19], or is stored in the local memory of a watchdog processor [20]. The watchdog monitor compares run-time generated signatures with these precomputed reference signatures.

Several techniques to improve the dependability properties of signature control flow monitoring have been presented [17], [13] and [16]. These improvements included 1) increasing error coverage, 2) reducing the cost of storage, 3) minimizing performance overhead by reducing the number of signatures fetched from memory, as well as 4) decreasing the latency of error detection. In [16], signatures were placed at locations that minimize memory overhead while the detectability of control flow errors is increased. The study in [16] has found that undetected control flow errors can be due to correlation of intermediate signatures. In [22], run-time program behavior was utilized to reduce the number of signatures as well as to optimize the placement of signatures. In [23], no reference signatures were required to be stored, instead selected memory locations were tagged in the block identification phase. A signature at a tagged memory location should form an $m - out - of - n$ code. At run-time, these previously tagged locations initiate checking.

An asynchronous control flow approach was proposed in [20]. In this approach, the signatures of several tasks executing on a number of processors, on behalf of an application, are stored in a linked list data structure forming a signature graph. Signatures received by the watchdog processors from each participating processor are checked according to the current position in the respective signature graph. The approach in [20] is similar to that presented in [25] where a roving emulator was used to monitor one or more functional units. In the roving emulator, a time slice is allocated to each monitored functional unit. The roving emulator performs the same operations as the monitored unit and compares the two results. A special architecture for the approach presented in [20] was proposed in [21]. An error detection technique in multiprocessor systems similar to that in [25] and using on-line signature analysis and signature verifications was presented in [24]. In this technique, generation of a reference signature is performed during the normal execution of the application program in a phase called the learning phase during which the application is submitted to a large number of test vectors. Experiments have shown that control flow checking detects a large percentage of errors in computer systems due to hardware and software faults [7].

Object based systems are becoming increasingly popular, since objects provide a secure and easy method to use the abstraction of shared memory. This concept is considered by many to be attractive for programming fault-tolerant distributed systems [8]. The proposed fault tolerance techniques, however, have dealt primarily with the problem of recovering the system to a working state using several schemes of replication, and assuming a fail-safe model [8], [18], [27]. In the fail-safe model, a computer node is considered to be producing correct output, or is considered to have failed and to have informed other sites in the network of its failure. These fault tolerance techniques do not consider control flow monitoring in object-based distributed systems.

In this paper we present a new approach for concurrent error detection by monitoring object invocations and thread control flow called **(COTM)** in object-based distributed systems. Hardware implementation of control flow checking is not feasible in a distributed environment such as a network of several nodes (computers) of possibly different architectures, since in these systems, addresses and signatures become irrelevant across nodes. The proposed technique is designed to be insensitive to different architectures since its checking mechanism is applied at the system level and does not deal with sequence of addresses of the executing process. In addition, the COTM technique utilizes the inherent nature of object-based systems regarding encapsulation of data in objects. Little redundancy is added to improve the system reliability since the COTM technique exploits several features embedded in the object-oriented paradigm. This results in low performance overhead (less than 20%) while still maintaining a high fault coverage (over 98%). In addition, all the implementations details are kept hidden from the user. It should be mentioned that the proposed technique augments the other lower-level system error detection techniques. Section 2 provides an overview of the control flow checking technique. Section 3 presents the approach adopted in COTM. Section 4 discusses the implementation of COTM. The dependability properties are evaluated analytically in Section 5. The results of experiments to determine the overhead and to validate the fault coverage of the technique are summarized in Section 6.

2   Controi flow checking for object-based systems

This section provides a summary of the model used as the basis of the proposed control flow checking scheme.

An *object* is a named address space which contains code and data. An object encapsulates a set of data and defines a set of operations that act on the set of data. These operations are implemented as *methods* similar to subroutines in conventional programming. Objects that encapsulate identical sets of data and share a set of operations are said to belong to a *class* of objects. A method in an object may invoke other methods of the same object, of different objects of the same class, or even methods defined for different classes. A *thread* is the only form of activity in this model [8], [9]. It is comparable to a process in conventional programming. In this model, data is restrained from moving in or out of an object, but can only be transferred while passing parameters to methods.

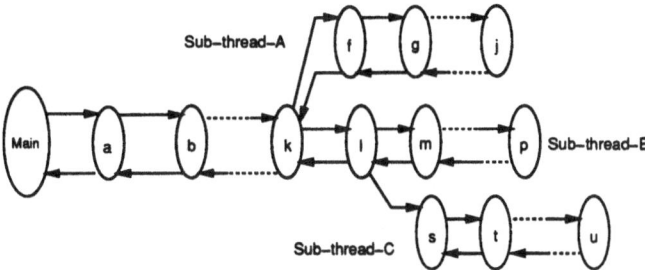

Figure 1: Object-thread model.

A thread, while executing, invokes methods of local objects or methods of remote objects scattered across the distributed system. The traversal of a thread proceeds in two directions: 1) forward, and 2) backward. The forward traversal is the course taken when methods are invoked, while the backward traversal involves the case when execution of the invoked methods is completed and control is transferred back to the calling objects.

Figure 1 shows a thread starting at object (*main*) which diverges into several subthreads at objects (*k*) and (*l*). In this figure, object (*k*) initiates two synchronous invocations. In a synchronous invocation, the invoking object initiates a thread, waits until the thread completes, and resumes when control returns to it. Subthreads that span objects (*f,g,...,j*) and objects (*l,m,...,p*) are two synchronous invocations. On the other hand, object (*l*) starts an asynchronous thread (*s,t,...,u*). In this case, object (*l*) is not requesting a returned value from the invoked object (*s*) and hence, *sub-thread-B* can be executed in parallel to *sub-thread-C*. Note that the calling object (*k*) in Figure 1 remains blocked during the execution of *sub-thread-A* and *sub-thread-B*. On the other hand, execution inside object (*l*) can be modified so that *sub-thread-C* executes in parallel to *sub-thread-B*. In Unix systems, using TCP/IP protocols, parallel thread execution is accomplished by *forking* a process (creating a new process called the *child* process which is a copy of the *parent* calling process) before invoking the method in object (*s*). The child process starts a remote procedure call (RPC), invokes the method in object (*s*), and waits until the RPC returns. Meanwhile, the parent process, *sub-thread-B*, will be executing concurrently with the child process *sub-thread-C*. In this scheme, the checking objects would commence its error detection operation on a different node while execution proceeds in the invoked object.

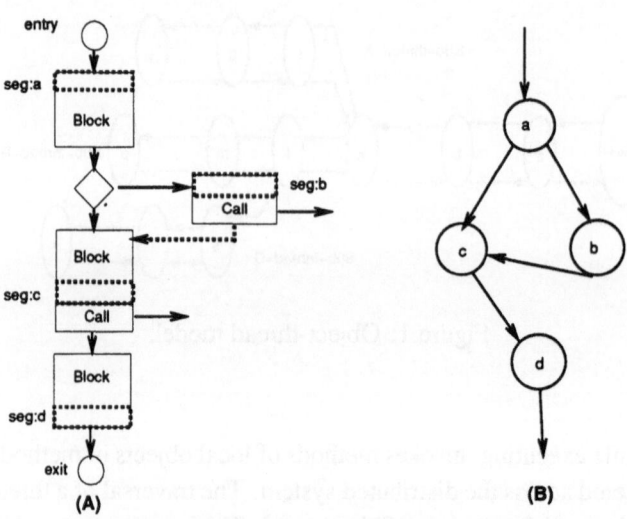

Figure 2: Example program flow graph.

Control flow information in the object-thread model is derived from the application program at two levels. The first level represent the control flow transfer which takes place while the thread is traversing objects. This model is represented by a directed flow graph G1= {V1,E1} where the set of nodes, V1, represents the objects across which the thread traverses in its course of execution. The set of edges, E1, that connect these nodes represents the method invocations and control transfer between objects. Figure 1 shows an example of this type of graph. Traversal of a thread of execution in presence of control flow errors may deviate from the normal traversal of this thread with no errors. The second level deals with the execution of code inside a method. This is similar to a procedure in conventional programming. This model is represented by the *program graph* G2 = {V2,E2} [3], where each node in this graph is a block of the program, which we call a *segment*. Figure 2a shows the flow chart of part of the code within an object and 2b shows the corresponding program graph G2.

3   Concurrent object-thread monitoring (COTM)

The COTM technique uses the control flow information in the G1 and G2 graphs in order to partition the application program into unique segments. This involves selecting objects from G1 and assigning new segment identifiers (*ids*) at the beginning of each node in the G2 graph. Assignment of segments inside a method call is similar to that adopted in [3]. Note that assignment of these segments is performed at the high language level and not at the assembly language level. The unique segment *ids* are inserted at the entry point of a method, before conditional branching, before any procedure call or method invocation, and before exit conditions, as shown in Figure 2a.

In the COTM technique, a method m of class c can be invoked only from a limited number of valid segments in the application program which can be identified from the source code. In addition, the sequence of segments traversed in a thread is unique. In order to obtain high error coverage and low error detection latency with a low performance overhead, control flow checking is achieved in two stages. In order to reduce the overhead, the checking of valid sequences is activated only at the end of the execution of a thread, rather than checking the sequence of the traversed segments for every method invocation. Such a checking procedure would, however, result in high error detection latency since a control error will only be detected towards the end of the execution of a thread. An unacceptable error detection latency can be reduced by examining the validity of a received segment for every invoked method.

The application program is modified to record all segment *ids* it has traversed during its course of execution. At run time, the COTM technique monitors the recorded segments to check the control flow in the two stages. Control flow checking in the first stage is applied during forward as well as backward traversal of objects. During forward traversal of objects, the segment *id* where an invocation was initiated will be transmitted to the invoked object as part of the parameter list. The invoked method of an object checks whether this control transfer is valid. This procedure is similar to the relay runner scheme presented in [2]. On the other hand, while a thread is traversing backward, the segment *id* of a returning method is validated. Control flow checking in the second stage includes encoding the sequence of the segment *ids* during run time by producing a checksum signature. After the traversed objects in a thread returns control to the first object in that thread, the produced signature is compared to the set of allowable signatures defined for the application program. The advantage of this approach is that it decreases the overhead incurred when searching the flow graph in order to validate a sequence of segments. Combining the two stages in the COTM technique provides higher coverage and lower error detection latency with a lower performance overhead.

It is important to distinguish between valid and correct execution sequences. A correct sequence is a valid sequence but the opposite is not true [6]. Although the above technique for the detection of control faults can identify valid sequences, it cannot detect incorrect sequences. Valid sequences that are not correct are the result of faults which produce control errors that affect the decision logic. Previous work has assumed that any control error will produce a deviation from the correct execution sequence [3]. Our experimental results (described in Section 6) show that faults will, in most cases, cause the thread to traverse an incorrect sequence of object invocations.

4   Implementation

Several approaches were considered in order to reduce the overhead incurred in applying the proposed COTM technique as well as to automate the process for incorporating the COTM software in an application program. A preprocessor was constructed to insert segment *ids* and code to initiate control flow checking inside the methods of the classes. The inserted code includes the calls to the checking routines at three locations inside every method. The locations are: 1) at the beginning of every method, 2) after every invocation of other methods and, 3) before exiting the method. The code inserted at the end of a method m is to check a flag, *pass-check*, set previously when method m execution started. If the *pass-check* flag was not set, then an illegal jump may have transferred control to method m, and the error is detected. We obtained the flow information from G1 and G2 by monitoring the program behavior during normal execution. In this approach segment *ids* are recorded when the application program is submitted to a number of test vectors. This approach is the same as the one presented in [24]. We chose to generate the control flow information at run time because it allowed us to concentrate more on the effectiveness of the COTM technique without the effort needed to code and debug a parser.

Two approaches were studied to monitor valid control transfer among objects for stage one. These were: 1) segment verification approach, and 2) code verification approach. Selection of either approach for stage one checking is based on a tradeoff between memory overhead and performance overhead. This tradeoff will be discussed later in the section on performance evaluation. Note, however, that

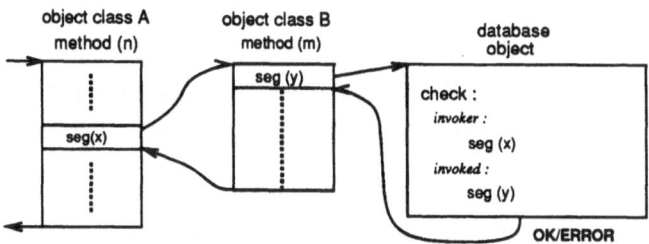

Figure 3: COTM using the database construction approach.

both approaches degrade the system performance and require additional memory.

4.1 Segment verification approach (SV)

A data structure that contains all the allowable control transfer information among objects is constructed. Each entry in the data structure has two fields. The first field is the first segment *id* of a method *m*. The second field is a list of all segment *ids* where an invocation to method *m* is initiated. At run time, the invoked method compares the received segment *id* (sent from the invoking method, along with the rest of its parameters) with the list of segment *ids* stored in the second field of the data structure for method *m*, as shown in Figure 3. Generation of the control flow information for this approach is similar to the one adopted in stage two. In essence, valid segments for stage one and valid sequences of segments for stage two are generated in one step.

4.2 Code verification approach (CV)

At compile time, a preprocessor generates an encoded word referred to as a *signature* for every segment (defined before) for all methods. Examples of code words are Berger codes, m-out-of-n codes and prime numbers. At run time, the invoking object appends the signature to the argument list passed to the invoked method. Signature verification in this approach simply checks whether the received signature is a code word. The advantage of this approach is that it does not require storage of any signatures, thus reducing memory overhead. The drawback of this approach, however, is the time overhead incurred while the checking procedure is verifying the signature. In addition, the error detection coverage is reduced due to aliasing of signatures. This problem will be considered in the next section.

An implementation problem in the COTM technique involves checking the library and system function calls, collectively referred to as library function calls. The presented technique performs the checking at the invoked object. As a result, library function calls will not be validated, since modifying their code to include the COTM technique routine is not feasible in many systems. One solution is to incorporate an object of class $class_{lib}$ in the user code. All library functions called by the user application are then defined as methods in class $class_{lib}$, and every library function call is a method invocation to an object of this class. This object will then initiate the checking operation.

5 Evaluation of the dependability properties of COTM

Evaluation of the dependability properties for the COTM technique includes: 1) error coverage, 2) performance degradation and memory overhead. These properties were found to be related to the implementation of the checking procedure, the number of objects called during thread traversal, and the complexity of the computation in each method.

5.1 Error coverage

The proposed technique detects control bit errors as well as control sequence errors that result in illegal execution sequences. Error coverage is increased as the frequency of initiating the control flow checking technique is increased. This frequency depends on a number of parameters of the application program among which are the frequency of the object invocation, and the size of the invoked object. Let n denote the number of methods in the application program, m denote the number of high level executable program statements, c denote the average number of object invocations inside a method where $1 \leq c \leq m$, and l denote the average time a method invocation is within a loop. The frequency of performing control flow checking is $(2ncl/m)$. This includes checking at the entry of the invoked method and after control flow returns to the invoking object.

Undetected control flow errors across objects in stage one are due to (1) aliasing in the segments/signature, and (2) multiple compensating control flow errors. In the following subsections we evaluate the percentages of these undetected errors. The analysis assumes the following: 1) an error may occur in any program location and transfer control to any other location, 2) an error corresponds to a single bit error, and 3) w is the word length of the processor internal data bus.

5.1.1 Aliasing in segments/signatures

Aliasing occurs when an erroneous segment/signature (s_{err}) sent by the invoking object correlates with another valid segment/signature in the set S, i.e $s_{err} \in S$. Aliasing probabilities for the two proposed approaches, SV and CV are presented in the following subsections.

Aliasing in the SV approach

Assume that every method in an object of class $class_z$ will be invoked from S segments of other methods. Consequently, every method maintains S signatures. Signatures in the database mechanism, as mentioned earlier, are generated as follows: A preprocessor utilizes a random number generator to produce unique segment *ids* of size w for all methods in the application program. The probability of undetected errors, Un, due to aliasing is then: $Un = S/(2^w - 1)$. For $S = 1000$ and $w = 32$ bits, then $Un = 2.328\text{e-}7$.

Aliasing in the CV approach

An encoding technique is utilized to generate the signature. Two codes were considered in our study and their aliasing effects are presented.

(i) *m-of-n* code: an *m-of-n* code correlates with another *m-of-n* node in

$$\frac{n!}{m!(n-m)!} - 1 \quad ways.$$

If n is selected as the word length w, then the probability of undetected errors Un, due to aliasing is:

$$Un = \frac{\frac{w!}{m!(w-m)!} - 1}{2^w - 1}$$

For $m = 4$ and $w = 32$, $Un = 8.372\text{e-}6$.

(ii) *Berger code* [26]: A Berger code detects all multiple unidirectional errors in addition to detecting all single errors. Let i denote the number of information bits in a w bit word, and k denote the number of check bits. The check bits k represent the count of the number of zeroes in i and k is equal to $\lceil log_2(i+1) \rceil$. There are 2^{w-k} Berger code words in 2^w words. Hence, the probability of undetected errors is given by

$$Un = \frac{2^{w-k} - 1}{2^w - 1}$$

When w is large, the probability of undetected errors due to aliasing becomes 2^{-k}. For $w = 32$, $i = 27$, $k = 5$, $Un = 0.031$.

5.1.2 Multiple compensating errors

Compensating errors arise when an error transfers control from method x to a segment of another method y. Later, another error may transfer control back to method x. Here two (or more) errors caused error checking to be bypassed at two checking points, (1) at entry to method x, and (2) at exit from method y. The contribution of compensating control flow errors to the overall undetected errors is likely to be very low and we do not analyze this any further.

5.2 Performance and memory overhead tradeoffs

In the implementations of the SV and the CV approaches, memory as well as performance overhead is incurred. Memory overhead in both implementations is due to the additional memory space for the checking routines as well as the call statements to initiate these routines. The SV approach, however, incurs higher memory overhead than the CV approach since it has to store all valid segments discussed in Section 3. Segment verification is performed similar to the approach in [24] whereby a segment is compared to a set of valid segments and the drawback of storing control flow information, extracted from the flow graph model, is thus averted. On the other hand, the CV approach does not store any signature. Its drawback, however, is that it incurs higher performance (time) overhead since it has to execute a routine to verify whether the received signature is one of the valid code words. Consequently, selection of the codes for the CV approach has to consider the complexity involved in decoding. For example, an algorithm to find whether a number is an m-out-of-n code or a Berger code is simply to count the number of 1's in a word. In the prototype implementation, we implemented the SV approach.

6 Experimental evaluation of proposed technique

In order to experimentally evaluate the dependability properties of COTM and to obtain realistic performance overheads, we employed fault injection experiments. These experiments were conducted on a SUN4 workstation. FERRARI (Fault and ERRor Automatic Real-time Injector), a software implemented fault injector, was used [28]. FERRARI emulates hardware faults and errors by modifying the executable program modules.

In our fault and error injection experiments, we were interested in intermittent faults of a short duration since they are hard to detect. Permanent faults, on the other hand, eventually cause detectable errors, as has been observed in previous studies [15]. FERRARI supports the injection of permanent and intermittent faults and transient errors. The mechanisms for fault and error injection are identical, and the only difference is the duration of the injected fault and error. For injection of intermittent faults, the duration is defined to be one instruction cycle. On the other hand, the duration of permanent faults may be several instruction cycles, or may span the entire execution interval of the application. The selected intermittent fault models are listed in Table 1. In this table, inserting a data line fault, for example, models several actual faults in the processor hardware. These include faults in the external and internal data lines, faults in the PC and its internal registers, faults in address calculation circuitry, and faults in memory.

Fault Model	Description
1	address line fault resulting in executing a different instruction
2	address line fault resulting in executing two instructions
3	address line fault when a data operand is fetched
4	address line fault when a data operand is stored
5	data line fault when an opcode is fetched
6	data line fault when an operand is loaded
7	data line fault when an operand is stored
8	faults in the task memory image (code and data)

Table 1: Fault models used in the experiment.

Tables 2 presents the characteristics for selected benchmark applications used in the experimental evaluation. *cycle* is a small program which consists of six classes of 12 objects, and a thread which cycles among all classes of objects and their methods. *gtroff* is a text formatter from the Free Software Foundation, and *geqn* is an equation formatting program which can be linked with *gtroff*. The objective was to observe the behavior of the targeted system against several tasks of different computational complexity and I/O requests, and the inputs were chosen to exercise the example programs in this fashion.

Appl program	no. of classes	no. of methods	no. of method invocations
cycle	6	12	12
gtroff	117	810	1528
geqn	42	212	337

Table 2: Characteristics of selected application programs.

Over 150,000 fault injection experiments were conducted to measure the effectiveness of the error detection capabilities of the COTM technique. Figure 4 shows the distribution of errors, detected errors, and timeouts for the *cycle* program when stage one checking was applied using the SV approach. In this figure, latent errors were excluded from coverage calculations.

Over 60% of the detected errors were trapped by the system error detection mechanisms. These mechanisms include traps for segmentation faults, bus errors, illegal instructions, system aborts, etc. The coverage obtained for all fault models was higher than 90%. In Figure 4, "forward" refers to the coverage obtained while traversing the threads of the application program in the forward direction and "backward" indicates the coverage resulting when method invocations were returning; "prog exit" is a combination of several programming robustness features, including checking status of I/O operations when opening and closing files. As shown in the figure, the COTM technique provides, on average, an additional coverage exceeding 10% for most of the applied intermittent fault models. For fault model (7), the coverage increase was over 20%.

Table 3 shows the contribution of the system detection mechanisms, the COTM technique, and program exit conditions for two application programs. In these experiments faults were injected in the memory image (text and data segments). Only stage two checking was applied in this experiment. The overall coverage for both applications was over 98%, with the COTM technique providing an additional 30% coverage of the errors produced by the faults.

In Figure 5 we present the behavior of the targeted system for different combinations of error detection mechanisms when faults are injected in data lines while opcodes are fetched. The application program chosen for this experiment is the

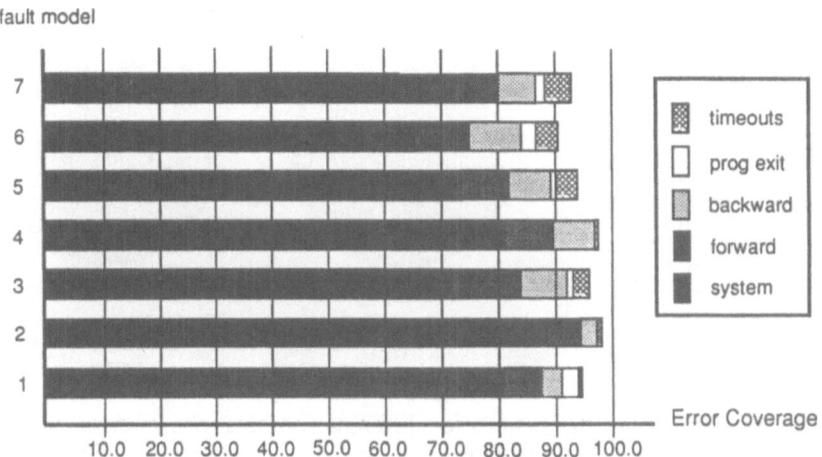

Figure 4: Accumulated error detection coverage for the cycle program using COTM stage one.

gtroff text formatting package. The error coverage in Figure 5(a) is due to the system error detection mechanisms and program robustness features. In Figure 5(b), stage two of the COTM technique is applied. In this figure, error coverage has increased to 98.9%. The COTM technique was able to detect many of those errors which were not detected by the system as well as those errors previously detected by the system. This result was also obtained when both stages of the COTM technique were applied, shown in Figure 5(c). Note that the coverages in Figure 5(b) and Figure 5(c) are equal. The reason, as explained earlier in Section 2 is that in both experiments, the validation of sequences of segments was done at the end of execution of the gtroff application.

appl. program	system detection	COTM	prog exit	overall coverage
gtroff	68.2	29.6	0.3	98.1
geqn	66.3	30.9	0.8	98.0

Table 3: Coverage for two benchmarks when faults were injected in the memory image (text and data).

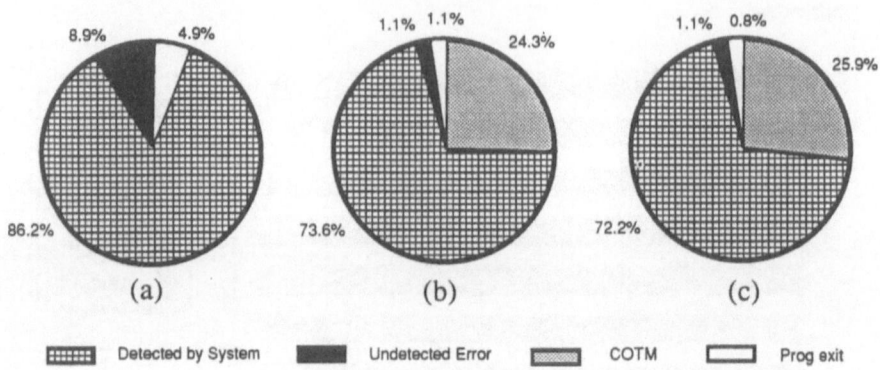

(a)   (b)   (c)

▦ Detected by System   ■ Undetected Error   ▧ COTM   □ Prog exit

Figure 5: Behavior of the system for different combinations of error detection mechanisms for gtroff. (a) No COTM. (b) Stage two of COTM. (c) Both stages of COTM.

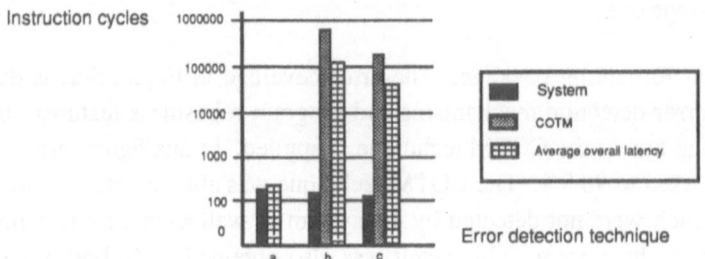

Figure 6: Average latency for different combinations of error detection techniques for the gtroff application program. (a) No COTM. (b) Stage two of COTM. (c) Both stages of COTM.

Figure 6 presents the average error detection latency for the COTM technique and the system error detection mechanisms for the experiments presented in figure 5. As shown in the figure, error detection latency when both stages of the COTM technique were applied is lower than the latency measured when only stage one is applied. Note that the overall latency for case (a) in this figure (COTM is not applied), is smaller than case (b) and (c) since most errors detected by the system are detected within a few instruction cycles, and the undetected errors are not included in the application.

Table 4 shows the error coverage results for the gtroff application program. In this table, stages one and two in the COTM technique are applied for every run. Stage one, combined with the system error detection techniques, is able to detect over 95% of the errors produced by all the of the injected fault types. The overhead when applying stage one was between 8% and 20% for several test input vectors. When both stages of the COTM technique were applied, including the system error detection techniques and the application program exit conditions, coverage was over 98%. The performance overhead incurred, however, increased an additional 12% to 15%. As argued previously, although stage one provides lower coverage than stage two, error detection latency is reduced since the COTM technique is activated at every method invocation.

In Table 5 we present error coverages for the *gtroff* application program as a function of the size of selected formatted manual pages. In these experiments, faults were injected in the memory images (text and data). Results obtained from these experiments as well as other experiments using other types of fault models presented in Table 1 show that coverage due to the different error detection mechanisms is independent on the size of the manual pages.

fault model	system	COTM stage 1	COTM stage 2	prog exit	overall coverage
1	84.5	8.1	6.6	0	99.2
2	92.4	5.3	2.0	0	99.7
3	92.7	4.9	2.2	0	99.8
4	91.8	6.4	1.6	0	99.8
5	71.2	20.8	6.5	0.5	99.0
6	66.2	26.6	5.9	0	98.7
8	68.2	26.1	3.5	0.3	98.1

Table 4: Percentage coverage contributions of several error detection techniques for the gtroff application program using the SV approach when several types of faults are injected in the system.

manual page	number of pages	system	COTM stage 1	COTM stage 2	prog exit	overall coverage
issecure.3	0.5	68.2	26.1	3.5	0.3	98.1
cp.1	2	70.9	24.4	2.5	0.4	98.2
scanf.3v	5	65.8	28.2	4.0	0.3	98.3
ftp.1c	11	67.1	27.2	3.4	0.6	98.3
streamio.4	15	67.0	28.5	2.6	0.3	98.4
termio.4	25	64.3	27.4	4.5	1.4	97.6

Table 5: Percentage coverage contributions of several error detection techniques for the gtroff application program formatting manual pages of different sizes.

7 Conclusion

In this paper, we have presented the concepts and implementation of a software based approach for on-line control flow monitoring in object-based distributed systems.

The approach is suitable for distributed systems, where addresses and signature calculations used for control flow checking in uniprocessors become irrelevant across processor nodes. Experiments show that this technique requires minimal memory overhead (less than 10% for the checking code with some additional memory for storing the valid segments), low performance overhead, and no additional hardware. Error detection coverages measured for selected applications were over 98%. Experiments have shown that the technique is capable of detecting errors due to several types of faults, and we believe that the higher level control flow checking is applicable to a wide variety of systems.

References

[1] A. Avizienis. Fault-Tolerance: the survival attribute of digital systems. *Proc. IEEE*, Vol-66, October 1978, pp. 1109-1125.

[2] S. S. Yau, R. C. Cheung. Design of self-checking software. *Proc. Int. Conf. Reliable Software*, April 1975, pp. 450-457.

[3] S. S. Yau, F.-C. Chen. An approach to concurrent control flow checking. *IEEE Trans. soft. Eng.*, Vol. SE-6, No. 2, March 1980, pp. 126-137.

[4] K. Hua, J. Abraham. Design of systems with concurrent error detection using software redundancy. *Proc. ACM/IEEE Fall Joint Computer Conference*, Dallas, Texas, November 1986, pp. 826-834.

[5] K. Huang, J. A. Abraham. Algorithm-based fault tolerance for matrix operations. *IEEE Trans. on Computers*, Vol. C-33, No. 6, June 1984, pp. 518-528.

[6] J. R. Kane, S.S. Yau. Concurrent software fault detection. *IEEE Trans. Soft. Eng.*, Vol. SE-1, No. 1, March 1975.

[7] M. Schmid, R. Trapp, A. Davidoff, G. Masson. Upset exposure by means of abstraction verification. *Dig FTCS-12*, June 1982, pp. 237-244.

[8] M. Ahamad, *et al*. Fault tolerant computing in object based distributed operating systems. *IEEE symp. on Reliability in distributed soft. and database systems*, 1987, pp. 115-124.

[9] P. Dasgupta, *et al*. The clouds distributed operating system. *IEEE symp. on Reliability in distributed soft. and database systems*, 1988, pp. 2-9.

[10] M. Schuette, J. Shen. Processor control flow monitoring using signatured instruction streams. *IEEE Trans. Comput.*, Vol. C-36, March 1987, pp. 264-276.

[11] A. Mahmood, E. J. McCluskey. Watchdog processors: error coverage and overhead. *Dig FTCS-15* June 1985, pp. 214-219.

[12] A. Mahmood, E. J. McCluskey. Concurrent error detection using watchdog processors - a survey. *IEEE Trans. Comput.*, Vol. 37, February 1988, pp. 160-174.

[13] J. P. Shen, M. Schuette. On-line self-monitoring using signatured instruction streams. *Int. Test Conf.*, 1983, pp. 275-282.

[14] K. Wilken, J. P. Shen. Continuous signature monitoring: efficient concurrent detection of processor control errors. *Int. Test Conf.* 1988, pp. 914-925.

[15] P. Banerjee, J.T. Rahmeh, C. Stunkel, V. S. Nair, K. Roy, V. Balasubramanian, J. A. Abraham. Algorithmic-based fault tolerance on hypercube multiprocessor. *IEEE Trans. on Computers*, Vol. 39, No. 9, September 1990, pp. 1132-1146.

[16] K. Wilken, J.P. Shen. Continuous signature monitoring: low-cost concurrent detection of processor control errors. *IEEE Trans. Computer-Aided Design*, Vol. 9, No. 6, June 1990, pp. 629-641.

[17] M. Namjoo. Techniques for concurrent testing of VLSI processor operation. *Proc., 12th IEEE ITC*, 1982, pp. 461-468.

[18] L. Lin, M. Ahamad. Checkpointing and rollback recovery in distributed object based system. *Dig FTCS-20*, June 1990, pp. 97-103.

[19] J. Sosnowski. Detection of control flow errors using signature and checking instructions. *Proc. 18th IEEE ITC*, 1982, pp. 81-88.

[20] J. B. Eifert, J. P. Shen. Processor monitoring using asynchronous signatured instruction streams. *Dig FTCS-14*, June 1984, pp. 394-399.

[21] J. P. Shen, S. P. Thomas. A roving monitoring processor for detection of control flow errors in multiple processor systems. *Microprocessors and Microprogramming*, 20, 1987, pp. 249-269.

[22] N. J. Warter, W. W. Hwu. A software based approach to achieving optimal performance for signature control flow checking. 1990, pp. 442-449.

[23] S. J. Upadhyaya, B. Ramamurthy. A new efficient signature technique for process monitoring in critical systems. *2nd Int. Working Conf. DCCA-2*, Tucson, Arizona, February 1991, pp. 178-185.

[24] H. Madeira, J. G. Silva. On-Line Signature Learning and Checking. *2nd Int. Working Conf. DCCA-2*, Tucson, Arizona, February 1991, pp. 170-177.

[25] M. A. Breuer, A. A. Ismaeel. Roving emulator as a fault detection mechanism. *Dig FTCS-13*, June 1983, pp. 206-215.

[26] J. M. Berger. A note on error detection codes for asymmetric channels. *Information and Control*, Vol. 4, March 1973, pp. 68-73.

[27] S. K. Shrivastava, S. M. Wheater. Implementing fault-tolerant distributed applications using objects and multi-colored actions. *IEEE Trans. Soft. Eng.*, 1990, pp. 347-356.

[28] G. A. Kanawati, N. A. Kanawati, J. A. Abraham. FERRARI- A Fault and ERRor Automatic Real-time Injector. *Dig FTCS-22*, Boston, 1992, pp. 336-344.

SAFETY-CRITICAL INDUSTRIAL SYSTEMS

A "STRONGLY-FAIL-SAFE MAJORITY VOTED OUTPUT" CIRCUIT USED FOR DESIGNING DEPENDABLE COMPUTER SYSTEMS

Serge NORAZ, Michel PRUNIER
MERLIN GERIN
Safety and Electronic Systems (SES) Department
23, Rue du Vieux-Chêne, 38050 Grenoble Cedex, France

Abstract

As part of critical application used in railways transportation, space, chemical and nuclear industries, the processing part which controls the actuators of the electromechanical part is realized with fail-safe circuits. But these give us to the following problems:

- Complexity of design in case of redundant solutions, done till now, which require conventional fail-safe circuits.

- Necessity of off-line test sequences in case of non-redundant solutions involving "strongly-fail-safe" circuits.

This paper aims to provide a practical solution using "strongly-fail-safe" circuit for designing dependable computer systems aimed at critical processes. The goal is to make use of such systems easier, to avoid the drawbacks generated by off-line test phase and to obtain the best "cost-safety" compromise. First of all, we introduce the scheme of a "strongly-fail-safe" basic cell without any off-line test equipments. The advantage is to obtain a "strongly-fail-safe" circuit which requires few components. Then, we suggest an architecture of a "strongly-fail-safe majority voted output" circuit designed from the "strongly-fail-safe" basic cell. The reliability, the availability and the safety of this "majority voted output" mechanism is assessed. For information a triple modular redundancy computer system including this last mechanism is compared with well-known dependable computer systems. The good results obtained show that such a solution may meet a wide range of safe applications because of its low complexity and its fitness to be easily implemented. In conclusion a quantitative and a comparative study is performed. It involves two dependable redundant computer architectures suggested for an industrial refinery project need.

1 Introduction

Most of digital fail-safe systems aimed for control of critical applications have been developed by using conventional fail-safe techniques, well approved but not suited to realize today's complex computer systems.

The MERLIN GERIN's Safety and Electronic Systems Department proposed practical schemes [6] of "strongly-fail-safe" circuits to be easily implemented and used in existing critical processes architectures designs. But these solutions have a major shortcoming because they include off-line test mechanisms requested during defined test phase modes for which a part of the application is shut off.

In this paper we first set out a simple "strongly-fail-safe" circuit without inherent fault detection mechanisms. Then from this basic cell, called "Compact Secure Interface (CSI)", we show how to obtain easily a "Strongly-fail-safe majority voted output" component, a voter as those usually used in computing architectures aimed for critical applications. Next for information only, we compare a triple modular redundancy (TMR) computer system including this "majority voted output" circuit with well known TMR or redundant computer systems. Finally we perform a quantitative and a comparative study which involves two dependable redundant computer architectures suggested for an industrial refinery project need. From these assessments we give some conclusions.

2 Fail-safe circuits

In this paragraph, we recall the main ideas of the generalized theory of fail-safe systems [5].

2.1 Fail-safe circuits: basic definition

Mine and Koga [3] gave the first formal definition of conventional fail-safe circuits. It may be extended to output spaces having any number of states. In this case, the more general definition of fail-safe systems is the restricted Definition 1 usually used when developing practical fail-safe implementation techniques for circuits.

Definition 1 A system G is fail-safe with respect to a fault set F, an input space X and a safe output space Os if :

$$\forall \, a \in X, \forall \, f \in F, G(a,f) = G(a,\emptyset), \text{ or } G(a,f) \in Os,$$

where $G(a,\emptyset)$ denotes the correct function and $G(a,f)$ denotes the function under fault f.

Here Os is the set of the safe output words that never involves a dangerous situation in the system, even if they occur erroneously. All other words compose the non-safe output space On.

2.2 The "strongly-fail-safe" property

In most fail-safe systems the occurrence of the erroneous outputs belonging to the normal operation output space will not involve fault detection. So suppose a system is fail-safe according to the Definition 1 with respect to a fault set F. If a fault f1 in F occurs in the system then it is ensured that it never produces non-safe erroneous output and the safety is ensured.

However if the system is used a long time after the occurrence of the first fault a second fault f2 may occur. The combined fault [f1, f2] may not belong to F and the fail-safe property could be lost. Therefore in a fail-safe system some mechanism is needed to quickly detect the occurrence of a fault avoiding the use of service delivered from the faulty system, for a long time after the fault occurrence.

This requirement, the self-testing property, is ensured during a special mode (off-line testing mode) for systems which have not inherent fault detection mechanisms, while the normal operation mode (on-line mode) may be sufficient for systems with inherent fault detection mechanisms (output coding).

Definition 2 A circuit is totally fail-safe (TFS) if it is fail-safe and self-testing.

In a fail-safe circuit the fact that several erroneous outputs can be produced before fault detection is not a drawback as far as these outputs are safe states. This required goal is called the totally fail-safe goal.

Totally fail-safe circuits will achieve the TFS goal under the following hypothesis.

Hypothesis 1

a) faults occur one by one,

b) between the occurrence of two faults a sufficient time elapses so that the circuit is sufficiently exercised by means of the fault detection modes.

"Strongly-fail-safe" circuits (Definition 3) are the most general class of circuits allowing to achieve the TFS goal, under Hypothesis 1.

Definition 3 A circuit G is "strongly-fail-safe" with respect to a fault set F if for each fault $f \in F$ either :

a) G is totally fail-safe or,

b) G is fail-safe and if a new fault in F occurs, for the obtained multiple fault, case a) or b) is true.

Now we discuss the possibility of designing "strongly-fail-safe" systems without the use of fault detection mechanisms. The following theorem gives the necessary and sufficient condition for the existence of such systems.

Theorem 1 A system G which has not fault detection capabilities is "strongly-fail-safe" with respect to a fault set F if and only if G is fail-safe with respect to F*.

In which for a given fault set F, F* is the set of multiple faults made up of faults belonging to F.

A common technique used for computer systems aimed for critical applications is to use more than one computing element generating results by their combined operation. An odd number such as three is usually chosen, allowing the combined result, called the "majority-voted output", to be generated by the majority vote of individual results.

This method does not make the system completely free of failure points. In particular, separating out the computing elements of the system into several units and combining their result by majority vote still leaves the system with a single point of failure, the majority vote mechanism itself.

In this way, there is a need for majority-voted output systems which are not subject to a single point of failure able to bring the entire system to a halt or to be the source of a catastrophic event.

This is the topic of the following sections in which we propose the implementation of a "strongly-fail-safe" circuit by using coding techniques and aimed to design a dependable "strongly-fail-safe majority voted output" mechanism.

3 Design of a "Strongly Fail-Safe" basic cell

For most critical applications using fail-safe systems where each binary output drives a critical function, as electromechanical engine (Actuators), what is needed is to obtain signals that are individually fail-safe. A convenient way to code such signals is to use frequency coding. That is to say a range of frequencies represents the non-safe state (says "1" state) and any other electrical state represent the safe state (says "0" state). In practice, many existing fail-safe systems use binary signals with frequency coding [2].

What we are looking at is to make a device by using frequency coding technique in such a way that, regarding its inputs and its output, the circuit is "strongly-fail-safe" according to the Definition 3.

3.1 Functional description

To do this we propose in Figure 1 the basic cell whose principle is laid down below.

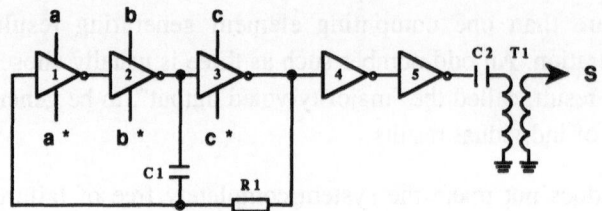

Inverters 1, ..., 4   : NMOS inverters.
Inverter 5   : Power CMOS inverter.

Figure 1: A basic cell designing with coding techniques.

The device, called "Compact Secure Interface" (CSI), is built with an odd number (3) of logic gates (inverters) performing a dynamic circuit and a suitable transformer part. Its functional inputs, generated by example by Totally Self-Checking circuits [4], correspond to the power supply inputs of the first three NMOS inverters. The two others inverters, the NMOS one and the CMOS one, are respectively used to improve the signal to squares and amplify them in order to drive the transformer correctly.

For a voltage coded binary input signals ("1" = 5v, "0" = 0v), a, a*, .. , c, c*, code words are those for which a* = .. = c* = ¬a = .. = ¬c.

That is to say, from these words, the CSI generates a frequency coded output signal on its output S.

In one case, a* = .. = c* = ¬a = .. = ¬c = "0" (defined as "action code word"), the output signal is a frequency signal Fø corresponding to the "1" output word.

In the other case, a* = .. = c* = ¬a = .. = ¬c = "1" (defined as "non-action code word"), due to the fact that all the logic gates are supplied, the output signal is a normal low level voltage corresponding to the "0" output word.

For non-code words, which are those for whom $\exists$ (i, i*) = (0,0) or (1,1), i $\in$ {a, b, c} the output signal of the CSI is in the "0" word.

3.2 "Strongly-fail-safe" property

By construction and by the use of power supply inputs of its logic gates, we can ensure that the above depicted cell is fail-safe according to Definition 1 with respect to a predicted single fault set F likely to occur (Figure 2) in a logic gate, transformer, capacitor, resistor, ... Indeed on a demonstration model, we made a fault injection test sequence including each kind of this predicted single fault set F (Figure 2).

Component	Fault
Switch of any inverter	"Stuck at" one input
	"Stuck at" zero input
	Open circuited input
	Short between gate/drain
	Short between gate/source
	Short between drain/source
	Open circuited source
	Open circuited drain
Transformer	Opened line
Capacitor	Opened and shorted
Resistor	Opened and shorted
Connection	Opened line

Figure 2: Predicted single fault set F

Each time, we checked that the behavior of the circuit under fault (Figure 3) is conformable to the basic Definition 1 (with "1" the set of the non-safe output words On and "0" the set of the safe output words Os).

Injected predicted single fault	Coded binary input signal		
	Action code word	Non action code word	No code words
No fault	⊓⊔⊓ 150KHZ	—	—
Opened Capacitor C1	⊓⊔⊓ 300KHZ	—	—
Shorted resistor R1	⊓⊔⊓ 280KHZ	—	—
Shorted capacitor C2	⊓⊔⊓ 150KHZ	—	—
Open circuited input of the NMOS of CMOS inverter	Uncertain behavior ⊓⊔⊓ or ___	—	—
Open circuited source of the PMOS of CMOS inverter	Uncertain behavior ⊓⊔⊓ or ___	—	—
All other single faults	—	—	—

Figure 3: Results of the single fault injection test sequence.

So the one output basic cell is a "0" fail-safe single output circuit.

In the same way (fault injection test sequence) and for any predicted combined fault f" = f x f' (f and f' belonging to the above predicted single fault set F), the result shows that the fail-safe property is maintained (Figure 4).

Injected predicted combined fault	Coded binary input signal		
	Action code word	Non action code word	No code words
No fault	⊓⊔⊓ 150KHZ	—	—
Short between gate/drain of the switch of the NMOS inverter 4 and Short between gate/source of the NMOS of the CMOS inverter 5	⊓⊔⊓ 150KHZ	—	—
Open capacitor C1 and shorted resistor R1	⊓⊔⊓ 300KHZ	—	—
All other combined faults	—	—	—

Figure 4: Results of the combined fault injection test sequence.

We analysed the behavior of the CSI under single and double faults with

SPICE simulator. We obtained the same results that the fault injection test sequences on the demonstration model.

Therefore, according to theorem 1, the CSI is "Strongly Fail-Safe" with respect to any predicted single fault f in F which may occur.

3.3   Evaluation of the estimated failure rate

First of all we consider the non-safe failure rate λ_{ns} of the CSI as the failure rate of the failure modes which lead its output S to the "1" non-safe state when its functional inputs are "non-action code word" or non-code words. As well its safe failure rate λ_S is the failure rate of the failure modes for which its output S is in the "0" safe state whatever its functional inputs may be.

Due to its "strongly-fail-safe" property, the estimated non-safe failure rate of the CSI, performing the safety is:

$$\lambda_{ns}CSI < 10^{-9} \text{ h}^{-1}.$$

Its estimated safe failure rate, performing the reliability and availability, is obtained in adding up the estimated failure rates (CNET 86 reliability review) of all the components. That is to say, with a pessimistic hypothesis, all the single failure which may occur in the CSI lead its output S to the "0" safe state. With the following estimated reliability data :

$$\lambda_{gate} \quad\quad = \lambda_G = 1{,}15 * 10^{-6} \text{ h}^{-1},$$

$$\lambda_{capacitor} \quad = \lambda_C = 2{,}28 * 10^{-8} \text{ h}^{-1},$$

$$\lambda_{transformer} = \lambda_T = 1{,}00 * 10^{-7} \text{ h}^{-1},$$

$$\lambda_{resistor} \quad\; = \lambda_R = 2{,}26 * 10^{-8} \text{ h}^{-1},$$

the safe failure rate $\lambda_S CSI$ is:

$$\lambda_S CSI \leq 3\lambda_G + 2\lambda_C + \lambda_T + \lambda_R \leq 3{,}62 * 10^{-6} \text{ h}^{-1}.$$

Now, such a cell is used as a basic block in order to design a "Strongly-fail-safe majority voted output" circuit.

4 A strongly fail-safe "majority voted output" circuit

4.1 Functional description

Assuming that we have three identical Totally Self Checking (TSC) [4] processing systems A, B, C doing the same tasks at the same time and delivering the same data. In order to design a TMR computer system, we use "majority voted output Si" circuits as the one given in Figure 5 where each single output Si must be used for the safe drive of a critical function.

The "majority voted output Si" (MVO) circuit depicted down includes three CSI as is done in section 3. For each interface, a signal adaptator device and a diode, required for the vote function, are added.

Regarding from the processing systems, (a1,a1*), .. , (ai,ai*), .. , (an,an*), (CA,CA*), are the data delivered by TSC processor A, (b1,b1*), .. , (bi,bi*), .. , (bn,bn*), (CB,CB*), the data delivered by TSC processor B, and (c1,c1*), .. , (ci,ci*), .. , (cn,cn*), (CC,CC*), the data delivered by TSC processor C with :

$\forall\, i \in \{1,..., n\}$, $ai^* = bi^* = ci^* = \neg ai = \neg bi = \neg ci$, and, $\forall\, I \in \{A, B, C\}$, $CI^* = \neg CI = $ "0".

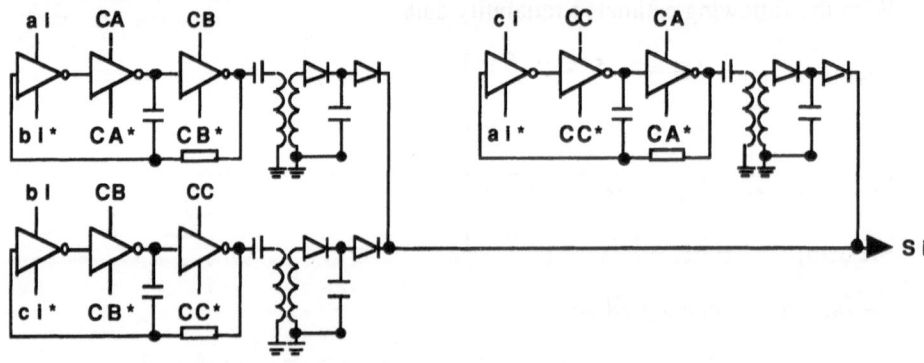

Figure 5: The "majority voted output Si" circuit

The pair (CI,CI*) is the double rail encoded watch dog signal of the TSC processing system I, $\forall\, I \in \{A, B, C\}$. It indicates for each processor I if that processor has failed and its data is no longer valid.

The functional truth table of the "majority voted output Si" circuit, which attempts to form its majority output from (ai,ai*), (bi,bi*), (ci,ci*) valid data only, is shown in Figure 6 with:

$\mathcal{A}$ = ai * CA, $\mathcal{B}$ = bi * CB and $\mathcal{C}$ = ci * CC.

The obstructed squares are impossible cases because of the correlation $\mathcal{K}$ = ki * CK with:

$\mathcal{K} = \mathcal{A}, \mathcal{B}, \mathcal{C}$; k = a, b, c.

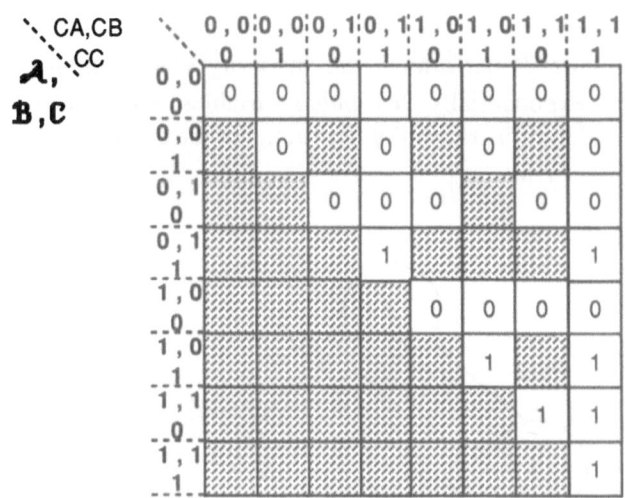

Figure 6: Truth table of the depicted "majority voted output Si" circuit

4.2 "Strongly-fail-safe" property

As shown in Figure 4, the "majority voted output Si" circuit is made from three subcircuits including both "strongly-fail-safe" CSI and some components used to achieve the "vote" function.

We verified, by fault injection test sequence on a demonstrating model, that the voting element is fail-safe according to Definition 1 with respect to any predictable single fault set F (done in section 3.2) likely to appear in any subcircuit (CSI and components required for the vote function).

In the same way, we can ensure that if a new predicted fault $f \in F$ occurs, in the same subcircuit or in another one, for the combined fault the "majority voted output Si" circuit does not lose its fail-safe property. Therefore (Theorem 1), the "majority voted output Si" circuit is "strongly-fail-safe" with respect to any predicted single fault which may occur.

4.3   Evaluation of the estimated failure rate

Due to its "strongly-fail-safe" property, the "majority voted output Si" circuit has an estimated non-safe failure rate $\lambda_{ns}MVO < 10^{-9}\ h^{-1}$ according to the fault hypothesis.

The graph (Figure 7) performing the "majority voted output Si" behavior, with regard to exponentially distributed variables, allows to determine the analytic expression of its reliability: $R_{MVO}(t)$ or the availability $A_{MVO}(t)$.

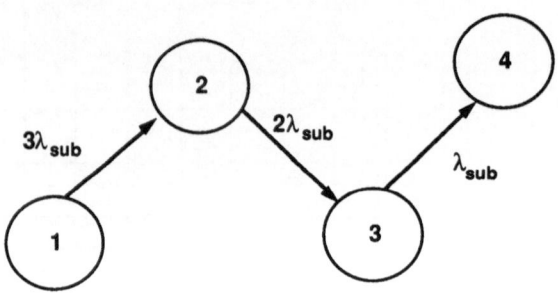

$\lambda_{sub} =$ $\lambda_{CSI} + \lambda_{capacitor} + 2\lambda_{diode}.$

State 1:   The structure operates correctly.

State 2:   1 among 3 subcircuits has failed.

State 3:   2 among 3 subcircuits has failed, the remaining subcircuit ensures the "2/2" function.

State 4:   Failed state, all the subcircuits are out of order.

Figure 7: Behavior graph of the "majority voted output Si".

Then, $R_{MVO}(t)$ $=$ $\exp - (3 \quad \lambda_{sub} * t)$

$- \quad 3 \exp - (2 \quad \lambda_{sub} * t)$

$+ \quad 3 \exp - (\quad \lambda_{sub} * t)$

With :

λ_{sub} $= \lambda_{CSI} + \lambda_{capacitor} + 2\lambda_{diode}$

$= 3{,}86 * 10^{-6} \, h^{-1}$

Note that,

$1 - R_{MVO}(t) \cong 8 * 10^{-4}$ with $t = 18$ months.

So, because of its non-safe failure rate, this mechanism is efficient from the safety point of view. It is not the same from the availability point of view in case of widely available requirements. Therefore, to improve this last measure, the only way consists in making use of the "majority voted output" circuit so that it can be checked on-line. In the next paragraph, we briefly describe this purpose.

4.4 MVO circuit and availability

So that the "majority voted output" circuit can be tested on-line, its output value, corresponding to the standby status of the function which it controls, should be the unsafe state (Si = "1" = 5v). This function may be an undervoltage actuator by example.

In this case, the function downstream to the "majority voted output Si" should operate with voltage loss. So, during the standby status, all the block including a CSI circuit and the components added are checked. Then, if an output of one block changes ("1" --> "0"), following a failure, it is signaled immediately (monitoring each output by a message error "0"). This is the only way to act on the availability of the assembly (Figure 8).

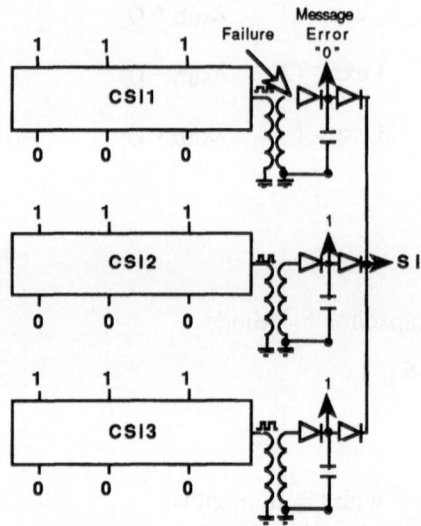

Figure 8: An on-line tested "majority voted output Si" circuit

If the "majority voted output circuit" is used as depicted before, its behavior under fault is pictured in Figure 9

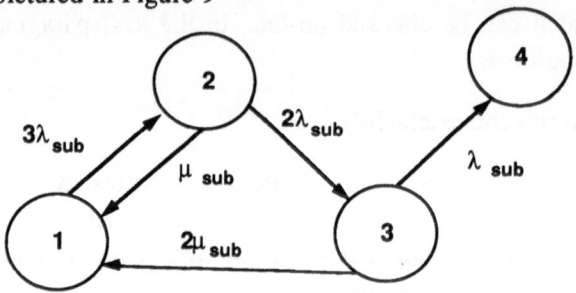

λ_{sub} $\quad$ = λ_{CSI} + $\lambda_{capacitor}$ + $2\lambda_{diode}$.

μ_{sub} $\quad$ = $\frac{1}{T}$ with T = 1 h.

State 1 : The 3 subcircuits operate correctly.

State 2 : 2 among 3 subcircuits operate, the other one is under repair because of a hardware fault.

State 3 : Only one subcircuit operates and ensures the 2/2 "vote", the other ones are under repair because of a hardware fault.

State 4 : Failed state, all the subcircuits are out of order.

Figure 9: Behavior graph of on-line available "majority voted output Si" circuit

The results of the reliability and availability assessments show that, over one year period, the non-safe failure rate of the "majority voted output" circuit (λ_{ns}MVO) and the safe one (λ_sMVO) are :

λ_{ns}MVO $< 10^{-9}$ h^{-1} (Safety) and,

λ_sMVO (t) $< 10^{-9}$ h^{-1} (Availability), t $\leq$ 1 year,

consequently to exponentially distributed variables.

5   Which solution for what problem ?

5.1   A qualitative study

The next scheme (Figure 10) is given for information only. It is used to compare various dependable architectures but not to obtain the features of each one. Moreover, the failure rate of each processing unit is the same in the duplex, triplex or non-redundant architectures.

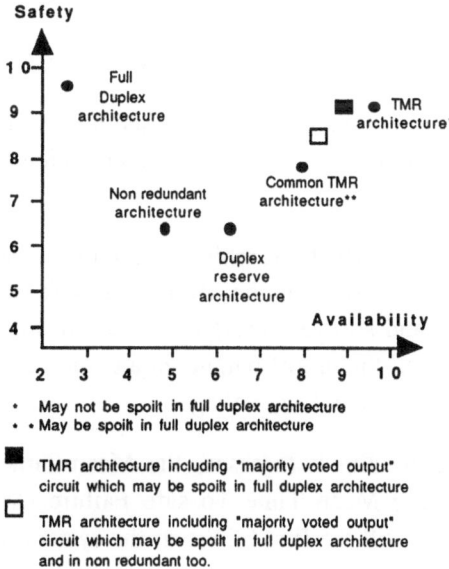

Figure 10: Comparative board for some well-known architectures

We carry over the x-axis and the y-axis respectively $\lambda A(t)$ (A(t) is the availability) and $\lambda S(t)$ (S(t) is the safety) where the first value corresponds to the transition rate to the unavailable state and the second one, the transition rate to the unsafe state. These transition rates are given for a determined time.

It is interesting to note that a TMR architecture including a "strongly-fail-safe majority voted output" circuit which may be degraded in full duplex system may be a very attractive solution. Because it has about the same features with a non-degradable TMR system. On the other hand and from a safety point of view, with such a solution the results are a little worse than a full duplex architecture. Moreover such an architecture is better than common TMR systems which may be degraded in full duplex system.

From an availability point of view, our solution is better than full duplex architecture and in the same order for the common degradable (full duplex) TMR systems.

At last, it is easy to show that such a solution is better than a TMR system using "strongly-fail-safe majority voted output" circuit which may be not only spoilt in full duplex architecture but in non-redundant architecture too.

5.1   A quantitative and comparative study for an industrial project need

For an industrial refinery project need, we assessed the Mean Time To Failure (MTTF) of the two proposed redundant architectures pictured below (Figure 11).

Whatever the architecture may be (architecture 1 or 2), during the normal operation all the binary outputs are in the defined non-safe state "1" (high level voltage). By hardware and software design, a detected failure on a board of the system leads one or all the on-off outputs in an emergency shutdown state defined as the safe state "0" (low level voltage).

For each architecture, the aim is to assess the Mean Time To catastrophic Failure (MTTFc) and the Mean Time To safe Failure (MTTFs). The first measure is the mean time till the first failure in the system in such a way one on-off output at least is stuck at "1" (command masking). The second measure is the mean time till the first failure in the system in such a way one on-off output at least shut down in the safe state (spurious command).

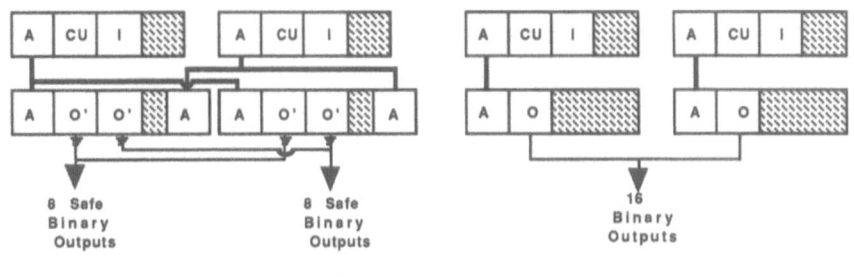

| Architecture 1 | Architecture 2 |

A   : Power supply board
I   : 32 On-off inputs board
CU : Control Unit board
O   : 16 On-off Binary outputs board
O' : 8 Safe on-off outputs board including 8 majority voted output circuits.

Figure 11: Two dependable architectures proposed for an industrial project

With the reliability data performed from estimated failure rates (MIL-HDBK 217 E reliability review) a Failure Modes, Effects and Criticality Analysis (FMECA),

λ_{ud} Power supply board $\qquad = \quad 3 \quad * 10^{-6}$ h,

λ_{ud} 32 On-off inputs board $\qquad = \quad 10,5 * 10^{-6}$ h,

λ_{ud} Control Unit board $\qquad = \quad 4,7 \; * 10^{-6}$ h,

λ_{ud} 16 On-off outputs board $\qquad = \quad 16,4 * 10^{-6}$ h,

λ_{ud} 8 On-off safe binary outputs board $\quad = \quad 47 \quad * 10^{-6}$ h,

λ_{d} Power supply board $\qquad = \quad 23 \quad * 10^{-6}$ h,

λ_{d} 32 On-off inputs board $\qquad = \quad 4 \quad * 10^{-6}$ h,

λ_{d} Control Unit board $\qquad = \quad 19,4 * 10^{-6}$ h,

λ_{d} 16 On-off outputs board $\qquad = \quad 4,4 \; * 10^{-6}$ h,

λ_d 8 On-off safe binary outputs board   =   90 $* 10^{-6}$ h,

λ_{ud} X: Undetected failure rate of X board,

λ_d X: Detected failure rate of X board,

the results of the comparative study is the following (Figure 12).

Architecture	Arch. 1	Arch. 2
MTTFc	$7,4 * 10^7$h	$8,2 * 10^5$h
MTTFs	$6,7 * 10^5$h	$1,2 * 10^5$h

Figure 12: Comparative results of two dependable architectures

6 Conclusions

In order to safely drive electromechanical actuators in critical applications, a well known method consists in using fail-safe techniques. But on the one hand the major shortcoming of conventional fail-safe circuits is that they require redundant solutions which involve complex designs. On the other hand, in the case of "strongly-fail-safe" circuits made till now, off-line test mechanisms may involve too much material which is not "cost-safety" in keeping with a better compromise.

In this paper a practical alternative of these solutions well adapted to industrial needs is given. Firstly by giving the scheme of a "strongly-fail-safe" basic cell without off-line test equipments and made with few components. Secondly by designing an architecture of a "strongly-fail-safe majority voted output" circuit from basic cells. The good results obtained as much for reliability as for availability, the low complexity of its architecture and its fitness to be easily implemented, and to be easily modified for special needs, show that this "vote" hardward function is able to meet a wide range of safe and available applications [1] with a good "cost/dependability" compromise.

References

[1] G. Chaumontet, V. Castro Alves, M. Nicolaidis, B. Courtois. MAPS: a safety microcontroller for railways signalling. *Proc. 13 rd FTSD*, Varna, Bulgaria, June 1990.

[2] K. Futsuhara, N. Sugimoto, M. Mukaido. Fail-safe logic elements having upper and lower thresholds and their application to safety control. *Proc. 18th International Symposium on Fault-Tolerant Computing*, Tokyo, Japan, June 1988.

[3] H. Mine, Y. Koga. Basic properties and a construction method for fail-safe logical systems. *IEEE Transactions on Electronic Computers*, June 1967.

[4] M. Nicolaidis, S. Noraz, B. Courtois. A Unified Built In Self Test Scheme: UBIST. *Proc. 18th International Symposium on Fault-Tolerant Computing*, Tokyo, Japan, June 1988.

[5] M. Nicolaidis, S. Noraz, B. Courtois. A generalized theory of fail-safe systems. *Proc. 19th International Symposium on Fault-Tolerant Computing*, Chicago, USA, June 1989.

[6] S. Noraz, M. Nicolaidis, B. Courtois. VLSI implementation for control of critical systems. *Proc. SAFECOMP'89*, Vienna, Austria, December 1989.

DEPENDABLE COMPUTING FOR

RAILWAY CONTROL SYSTEMS

Giorgio MONGARDI
ANSALDO Trasporti
Via dei Pescatori 35, 16129 Genova, Italy

Abstract

This paper deals with a dependable microprocessor system applied to control equipment and train movements in a railway station. First, application general requirements are outlined and basic principles and adopted techniques for dependability are shown; hardware and software vital architecture are described. Then some details about application special features are given, in order to present a suitable software verification and validation environment and to explain procedures and tools for system design. At last, some hints about first installations and relevant results are given.

1 Introduction

Microprocessors started to be used in the field of railway traffic control (railway signalling) when their high performance allowed to combine the control function required by the process with the diagnostic and redundancy management functions required to ensure "fail-safe" operation. Once the limits of the previous relay technology were overcome, clients asked for increasingly more complex and sophisticated functions as well as for a full integration of signalling and traffic automation systems - to enhance the quality of the service, from the point of view of the users.

Railway station interlocking systems based on microprocessors (Solid State Interlocking, SSI) were developed in all technologically advanced countries and have been used since a few years by those Railway Authorities which wish to have a good cost/benefit ratio.

In Europe and Japan SSIs were used in passenger transportation networks with medium/large stations and heavy-medium range traffic; in these applications complex interlocking systems were designed, including central and remote peripheral units, with vital data transmission between them [2], [8], [9], [11], [13]. In the U.S.A. small SSIs have been produced since the 80's, usually applied to freight transportation lines; the data exchange with remote systems is not vital and mainly related to centralised control functions [1], [5].

This paper deals with the Computer Based Interlocking (CBI), developed by Ansaldo Trasporti under the supervision of the Italian Railways.

2 Requirements

The CBI is a microprocessor real-time control system, applied to a railway line or station; it must be designed and realised according with very high level of safety integrity [6], in order to avoid dangerous situations for passengers and plants. Goals in term of timeliness constraints and safety and availability levels were determined from the statistical examination of relay interlocking behaviour.

Embedded diagnostic functions have to be added to basic interlocking, to detect failures and make maintenance easier. Its structure must be modular, to cope with a wide range of station complexity and also with plants distributed in large areas; last, they must be able to connect CBI to remote Traffic Management Systems.

European Railway Authorities adopted O.R.E. Recommendations as a guideline for realisation of SSIs [10]; they give some suggestions about:

• typical system architectures;

- general requirements for safety related software;

- list of system documents;

- operator interface special features, etc.

This year the Institution of Railway Signal Engineers issued a guideline for validation of safety related systems [7], while at CENELEC some Working Groups of Technical Committee 9XA are preparing devoted standards.

3 System Architecture

3.1 Overview

The design principles and implementation techniques adopted ensure to a reasonable extent avoidance of design errors both in hardware and software:

- the project was organised and planned according to the International Standards applicable;

- a Quality Assurance Plan was drawn up in compliance with the relevant International Standards;

- a specific methodology relevant to design, verification and approval was defined and approved both for hardware and software;

- together with project development plans, a general documentation plan was defined in accordance with International Standards.

After examination of the literature about dependable computer systems for critical applications and careful analysis of the above general requirements, the CBI architecture was defined, including different redundant processing nodes, interconnected by means of serial buses (Figure 1).

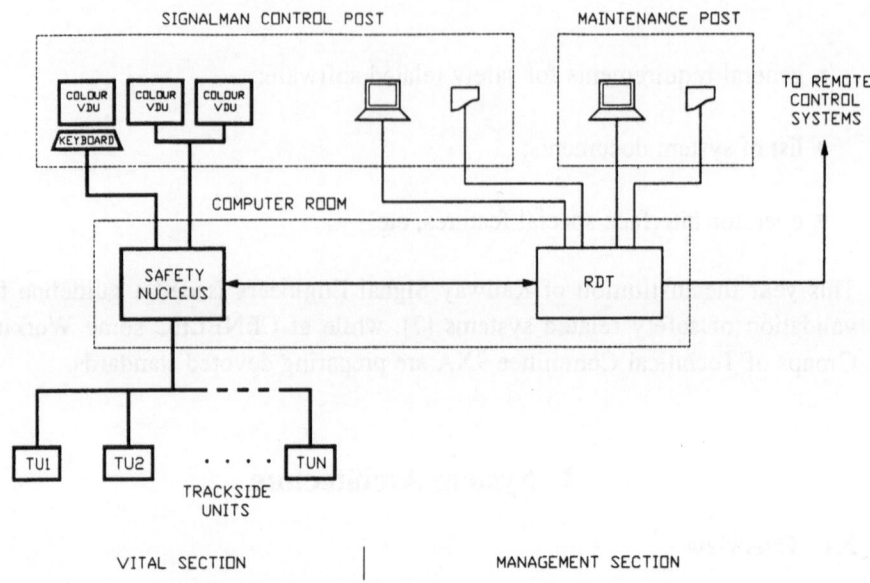

Figure 1: ACC - Computer based interlocking

In more detail, the CBI includes:

- a vital functions subsystem, with a central part - the Safety Nucleus - and a peripheral one, made by as many Trackside Units as station size requires;
- a supervision subsystem, that performs events recording and diagnostic functions, for both the controlled station and the CBI itself.

The Safety Nucleus and the supervision subsystem, with their own operator interfaces, are the CBI Central Post. Trackside Units are designed and manufactured in such a way they can be installed wayside, when this may be suitable.

Safety Nucleus and Trackside Units are linked together by a communication network, and a special protocol guarantees safety of data exchange.

3.2 The Safety Nucleus

For the Safety Nucleus (SN) an architecture based on three parallel connected computer sections was adopted, to create a "2-out-of-3" majority logic (triple modular redundancy - TMR) with a fail-safe external hardware device detecting any disagreement of a section; when such a disagreement is detected, the section is cut off from the configuration (Exclusion Logic -EL).

Each TMR section is electrically insulated from the others, with independent power supply, and consists of two microprocessor units: the first one is dedicated to the safety logic functions and the operator interface management (Logic Unit), the second manages exchanges with the trackside peripheral units (Peripheral Unit). The Safety Nucleus consists therefore of 6 units, interconnected by means of dedicated serial links, with optically insulated transmitting and receiving drivers (Figure 2).

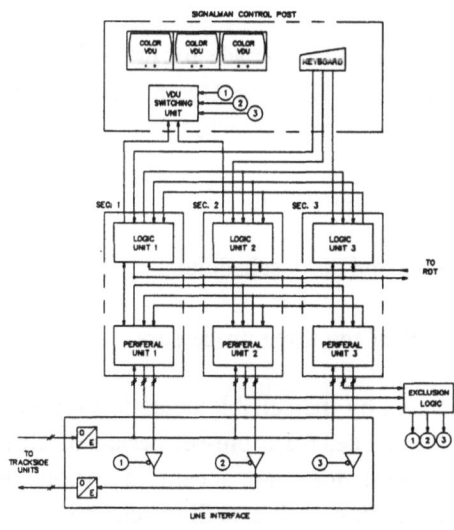

Figure 2: SN - Architecture diagram

The three Safety Nucleus sections independently perform the same functions: the software architecture and the operating environment are exactly the same, while the diversity has been adopted, however, for the development of application software modules. Following software diversity criteria were adopted:

- different programming teams;

- use of different programming language for the replicas (assembly and C language);

- different coding of data;

- data and code loading at different memory location.

The operating system, application programs and data are stored in the read-only memory of each unit, as a stand-alone program, in such a way that when the Safety Nucleus is activated the operating system starts controlling the operations.

This software structure was custom-made for the Safety Nucleus; its operation is based on the characteristics of the Operating System contained therein, which is custom-developed as well. It includes the following functions:

- the memory addressing management;

- the processing cycle and tasks scheduler;

- the failures' control of the three processors.

All the tasks are activated by the suitable operating system function according to a constant cyclic sequence, for a pre-set period of CPU time. The Safety Nucleus operating system does not accept interrupts, but controls tasks' activation sequence according with a polling scheme. Operations cycle has a constant length and on each of the three sections operating system checks the time still available; when the cycle time is over, operating system runs data and processing results exchange and majority voting software function; it detects disagreements, performs recovery actions and prepares a common internal status for the 3 software replicas, when the case occurs. Internal data exchange is used to synchronise the Safety Nucleus sections.

The TMR architecture with the safety Exclusion Logic combines the safety characteristics of "two-out-of-two" systems with the availability resulting from the redundancy already existing in the system. In the case of failures in a unit, the Exclusion Logic cuts off the unit in disagreement so that a further failure in one of the two units which are still operating cannot be summed up to the first and cause majority error; the unit cut off by the Exclusion Logic can be tested for the identification and removal of the faulty component and then be restored to normal operation.

The Exclusion Logic consists of three identical modules, that receive and process the analog signals generated by Safety Nucleus sections, after software voting operation; each of these signals is set to the active/passive, value, saying if each section agrees/disagrees from the other two (Figure 3).

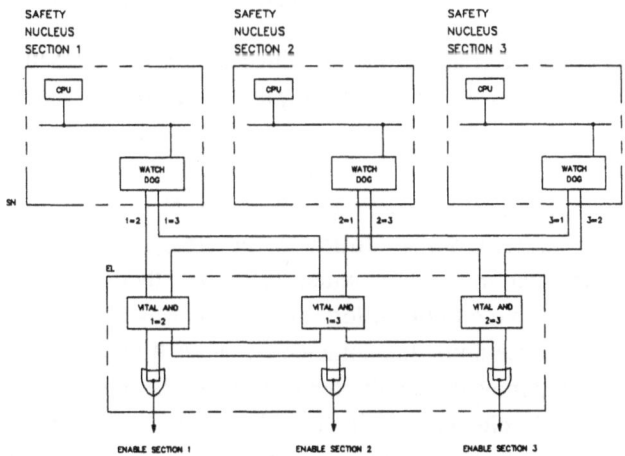

Figure 3: The Exclusion Logic

Each card verifies that no disagreement exists between the two units connected thereto and gives active value to its output signals, if the check is successful. By combining the signals generated by the three cards, enable/disable signals are obtained for the three Safety Nucleus sections. These signals are used as a direct power supply for the transmitter drivers off all serial links among TMR sections and between it and the Trackside Units. When a TMR unit disagrees, the Exclusion Logic sets the corresponding signal to disabling value; the serial drivers, no more powered, insulate that unit from both the other two in the Safety Nucleus and the Trackside Units. The Exclusion Logic is designed in such a way that any single internal fault therein cannot change the value of the enable signals, altering in such a way the Safety Nucleus configuration running at that cycle time.

Last, some hints about the operator interface safe management.

Vital section of CBI operator interface is made with:

- a functional key-board for the input of commands by the operator;

- colour VDUs, for the continuous display of the logic and yard devices, as well as for the display of the alarm signals and control sequence

All the listed devices are controlled by the three sections of the Safety Nucleus, according to the 2-out-of-3 majority logic, so as to guarantee fail-safe operation. The command acquisition is performed using three contacts for each push buttons or key on the key-board, independently connected to the hardware modules of the three sections. The fail-safe information display on colour VDU's is carried out as follows: the images for each VDU are independently generated in the three units; for each of them a CRC polynomial is calculated on the video-controller; through voting operations of these polynomials on the three units, the coincidence of such images is checked and any possible discrepancy detected; the results of the voting operation are transmitted to the Exclusion Logic, together with the results of further diagnostic operations; the Exclusion Logic enabling signals are used to control an external switching unit which selects the video signals from a correctly operating unit to each of the VDUs.

3.3 The Trackside Units

Also for the Trackside Units a vital architecture was adopted, equipped with fail-safe hardware circuits that force the Trackside Unit outputs into a safe status, in the case of discrepancy between the two units (Figure 4).

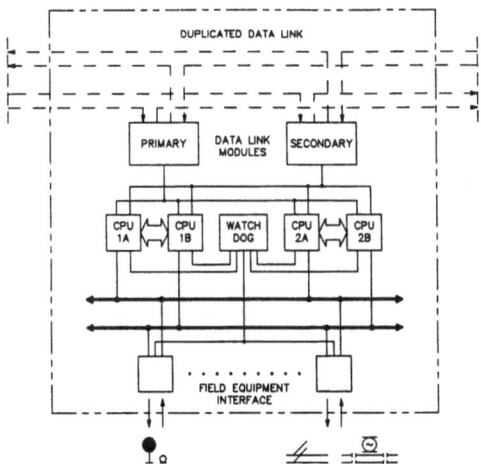

Figure 4: The Trackside Unit

Each Trackside Unit (TU) controls field devices of a yard's sector and consists of the following:

- the fail-safe processing section;

- the field device controllers;

- the interface equipment;

it is connected to the Safety Nucleus by means of the serial communication system described here below.

The TU processing section consists of two microprocessors in a fail-safe "2-out-of-2" ("duplex") configuration; a further pair of microprocessors can be added in accordance with availability requirements, as "stand-by". A system watch-dog module checks that the operation of the two microprocessors of each pair is correct, while selecting the accesses to the control modules. In the case of incorrect operation of one of the pairs, the watch-dog module enables module connection to the standby pair; should this operate incorrectly as well - or should no standby pair be provided - all field device controllers are no more addressed, so they force the output circuits of all modules to a safe status and stop their operation.

Field equipment and physical interfaces are controlled by means of interface controller modules, connected to the microprocessors by means of serial buses. Each controller consists of the following:

- field data collection circuits; these are redundant to ensure independent data acquisition of the two microprocessors in the 2-out-of-2 configuration;

- fail-safe comparison circuits, controlled by the two microprocessors, for command actuation;

- fail-safe module operation supervision circuit (controller watch-dog).

The watch-dog circuit is carried out in such a way as to disable controller operation, whenever a fault in its circuits is detected, while keeping the rest of the Trackide Unit operating.

Safety and availability requirements are thus fulfilled also with reference to yard equipment: in each Trackside Unit safety is ensured by the 2-out-of-2 configuration and by the fail-safe operation of the comparators, provided for each command; as far as service regularity is concerned, in the case of failure in any component of the "duplex" architecture, stand-by pair is activated; in the case of one module failure, the module is disabled while the Trackside Unit operation of the other field controllers is regularly continued; a device controller stand-by is provided, should a field device be critical.

In the Trackside Unit the functions relevant to the control of yard equipment are implemented, including the control of yard equipment status variation. In each unit of the "duplex" configuration, active programs are organised in such a way as to be run according to a strict schedule, based on a pre-set activation priority. The two units operate in loose synchronism, checking each other's correct operation via data exchange through a devoted optocoupled channel, at a fixed time rate. Cross-checking of status data collection and diagnostic test results on hardware modules is carried out. Status data collection is checked to ensure that processing is effected on correct data. Diagnostic test on I/O modules and microprocessor auto-test verify correct behaviour of all vital modules, before they are used in each operation cycle. According with tests' results, controller watch-dog and system watch-dog set value of their output signals, that are used to enable/disable controller circuits (the former) and to switch access of the I/O modules to the first or second pair of microprocessors (the latter).

The whole set of the Trackside Unit's programs has reduced size; then techniques to guarantee absence of faults were adopted (structured design and exhaustive testing), but no diversity criteria.

3.4 The Communication System

The Trackside Units are connected to the Safety Nucleus by means of a vital communication system consisting of high-speed serial links (optical fibers) which irradiate from the centre to the peripheral areas along the main direction lines where the Trackside Units are located; each "multidrop" link consists of a primary line (P) and a secondary line (S) (Figure 5).

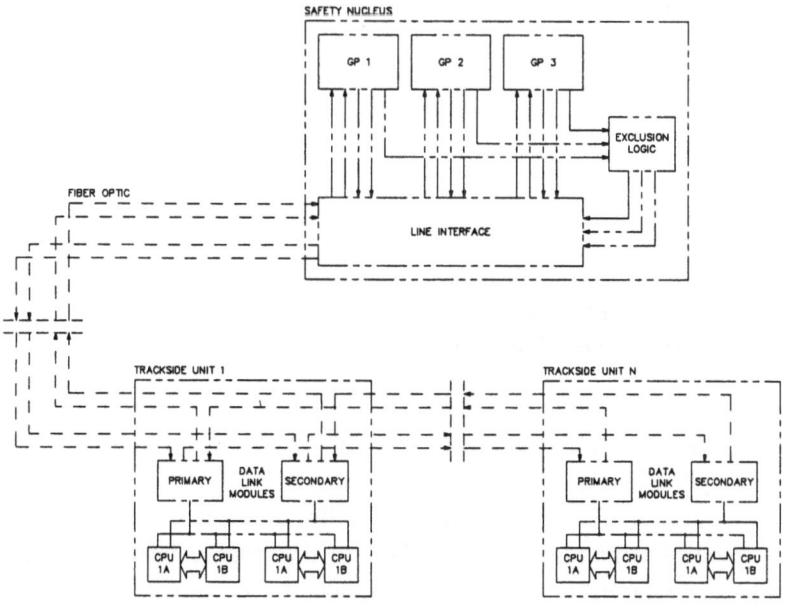

Figure 5: The Communication System

The three units are connected in such a way that, according with a regular rotation, each of them can send data on both P and S lines; in the opposite direction each line is connected to each of the three sections, that simultaneously can receive data on the different data links. Within each Trackside Unit, both P and S lines are connected to the microprocessors of the two pairs (normal and stand-by). At each operation cycle the Safety Nucleus section in charge of each multidrop link is selected by a software agreement among the three sections, to send data to all Trackside Units. Further, all transmitter drivers of the three units, corresponding to each multidrop line, are supplied by the outputs of the Exclusion Logic; so, in the case of failure and subsequent exclusion of a section, its access to the serial lines is disabled, thus avoiding any potentially wrong commands to be sent there to.

The communication protocol provides two steps, in both direction, during which Safety Nucleus and all Trackside Units exchange data and protection codes. The control of operations and the check of CRC codes added to each packet perform a complete diagnostic test of all working components, in each operation cycle. Usually both P and S lines are used; should a fault occur on one of the two lines, provided steps are executed on the other, following the same sequence. Such a way, the vital communication between the Safety Nucleus and the Trackside Units, as well as maximum availability, are ensured.

3.5   Recording and Diagnosis unit

According with general requirements, a processing unit dedicated to Event Recording, Diagnosis and connection to remote automation or Telecontrol systems (RDT) was provided (Figure 6).

The RDT Unit is linked to the Safety Nucleus via synchronous serial communication links; the communication is based on a dedicated protocol, where the Safety Nucleus always acts as a master. The RDT Unit can consist of an individual unit or of two units in a standby configuration, whenever availability requirements make it necessary. In that case connection to the Safety Nucleus and the RDT peripherals is provided by an automatic switching device, controlled by watch-dog circuits of RDT units.

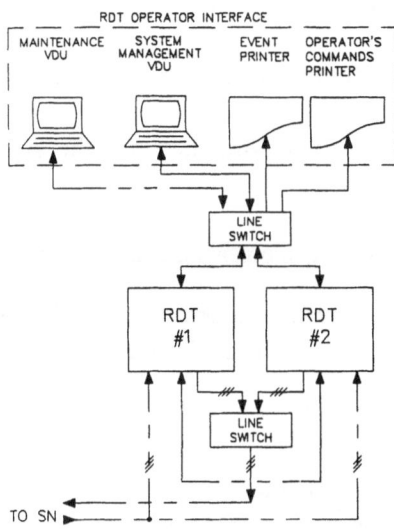

Figure 6: RDT - Architecture diagram

The RDT Unit directly controls the VDU in the operator desk and the following peripherals, as a Maintenance Post:

- the plant console;

- the event recording and operator commands printer

- the modem for the connection to the remote supervision system.

4 A traffic control application

The specific application of this fail-safe system architecture to the railway station interlocking required an in-depth analysis of the requirements and constraints to be complied with. Methods and time of the application design and of the functional testing, as well as the needs of modifications after commissioning were carefully considered; as a result, the following aspects were specially outlined:

- software characteristics of the interlocking logic;

- procedures and tools for the system application design;

- the methodology and the environment for the system verification and validation.

4.1 Application software: requirements and design procedures

The software elements required to guarantee fail-safe system behaviour were separated from those performing the application.

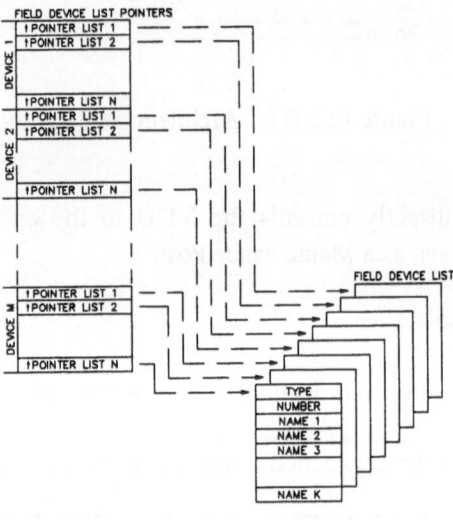

Figure 7: Structure of interlocking logic data base

The first group includes programs for communication management, diagnostic test and voting operation, system activation and disabling and recovery of individual units. These elements are unchanged regardless of the application and are verified and validated once. The other group includes interlocking logic.

A data driven structure was adopted in this case. The interlocking logic is made by a lot of logic operations, that include a list of verifications, on variables corresponding to physical or logical elements, and a list of value assignments to other variables (operation's outputs). The verification and assignment statements are coded as alphanumeric strings, following few simple rules; a set of strings, that realises an operation, is structured as a table. Other tables include pointers to memory locations where the value of variables referred in strings is written (Figure 7). So, the interlocking logic assumes the form of a data base, where tables are linked together according with operations meaning.

The executable code of this software item is a controller of data base accesses; again, it does not change regardless of the application and is validated once.

The compilation of the data base implementing logic rules and describing station layout is a repetitive and time consuming task, but nevertheless critical; then a custom software tool has been designed - the "configurator". Another inspecting tool is provided to check the compilation correctness.

4.2 System design

4.2.1 The design environment

The system design environment is composed of the following:

- the configurator

- the verificator

- an EPROM programmer

- a target processing unit.

The configurator and the verificator run on a workstation equipped with a colour graphic VDU terminal for the layout input and with a printer or plotter as output, in addition to the mass memory units required for data base and file storage.

The target unit is used during two stages:

- for the generation of the software system, following successful verification, during which the data bases and all the software elements are linked together;

- for the final software testing.

4.2.2 The design procedure (configuration)

The CBI application design includes the following steps:

- the station layout is entered via a graphic VDU, together with functional information characterising the system;

- the configurator obtains the system physical configuration, including number and location of variable parts, and providing reference data for their subsequent CAD customisation (Trackside Units, operator interface devices);

- the configurator subsequently obtains the data base for the simulator (see below);

- it then processes and draws up the interlocking data bases.

The configurator is now ready to produce an "intermediate design document" for the first inspections to be carried out by the Railway Signal man; then, connected to the simulator, it can also simulate logic operation so as to allow the designer to check whether the requirements have been properly met.

Once these inspections have been completed:

- the configurator generates data files to be used by the video controller modules for operator interface management;

- all the generated files are transferred to the target unit where the link is performed, in the final operational environment, with all the software elements and the relevant unchanging data structures.

The complete software is then transferred to the PROM programmer, which loads it onto the EPROM chips, to be subsequently put in the memory boards.

A last checking of the data is carried out at this stage by the verificator, which reads EPROM contents, decompiling it and generating a description of the system layout and of the functional requirements so that they can be compared - both automatically by the tool itself and visually by the designer - with the "intermediate document" generated during the configuration stage.

Finally, a complete functional test of the interlocking logic is carried out by operating the generated and verified software system in the target unit, serially connected to the environment simulator (see below).

4.3 Verification and validation

The software correctness and reliability are obtained through the application of structured and rigorous methods, both for development and verification and validation. The Verification and Validation (V&V) methodology adopted is a "step-by-step" one [3]; this allows for the timely identification of any errors introduced during development phases and results in a higher level of confidence in the system accuracy and in a higher overall quality and reliability of the final software product, together with a reduction in the time and cost of development.

The development cycle adopted is described in Figure 8: the activities of design and V&V are shown for each phase; the adopted V&V methods and techniques differ according to each phase.

After the choice of methods to be adopted [4], a dedicated V&V environment was defined, populated with software tools designed and implemented to support or perform the required activities:

- the Decision Table Tool, for the verification and validation of functional specifications;

- the Test/Case Specification Tool, for the test case generation;

- the Test Driver, for the test execution;

- the Flagger, for the program instrumentation;

- the Environment Simulator.

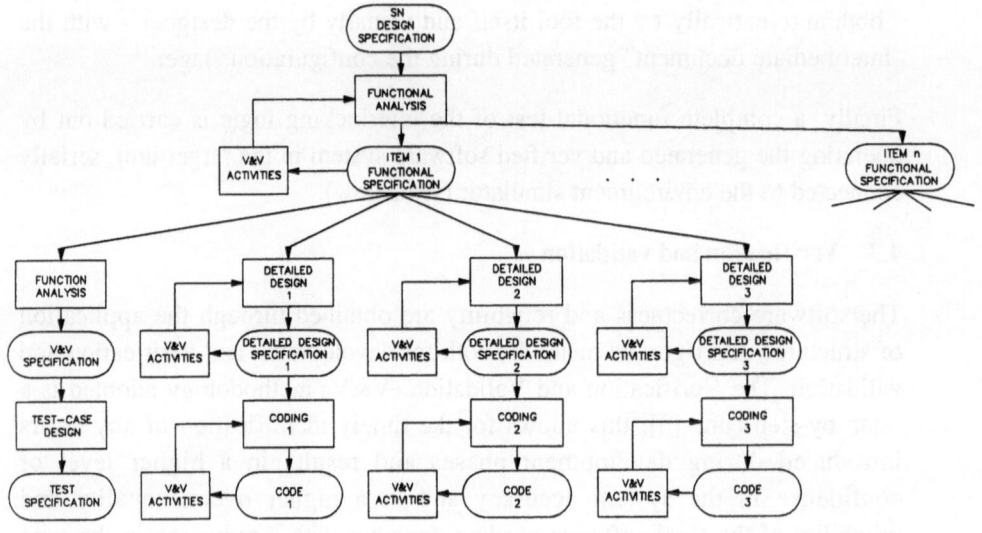

Figure 8: Safety Nucleus software development cycle

The Decision Table Tool is based on the Complete Decision Table formalism [12] in order to:

- formalise the software functional specifications, so as to make verification and approval easier;

- identify and select the functional test cases;

- guarantee that a given function is fully covered by the Test-Cases and prevent Test-Case redundancy.

The Test Case Specification Tool derives the individual Test-Cases from the

Decision Tables, generating the relevant test specifications.

The Test Driver allows a guided execution of tests, including the automatic recording of all the operations carried out by the operator, as well as of test results, so as to allow any test campaign to be subsequently automatically repeated.

The Flagger is used to insert assertion statements in specific points of the programming code. Once a test campaign has been carried out, the execution of these statements makes it possible to evaluate the obtained coverage, in terms of a percentage of code branches.

The Decision Table and the Test-case Specification Tools were developed and run in a DEC-VMS environment. The Test Driver and the Flagger run on a CBI target unit.

The real-time environment simulator is a software tool capable of real time simulating the controlled process. It can be customised to the features of a given station or of a section of the railway line including more than one station; it simulates the behaviour of field devices, reacting to commands originating from the interlocking or from its own operator interface. Moreover, by varying the free/occupied status of the track circuits, it can move trains onto the simulated yard. Lastly, it can also simulate all the typical field devices abnormal operations.

5 Field experience

The first CBI prototypes have been implemented within the framework of the system validation process, carried out in co-operation with the Italian National Railways, in two stations belonging to the railway network in the Genoa area.

At the former, on the main railway line connecting Genoa to La Spezia, the CBI Central Post was installed, connected to the existing relay interlocking by means of a comparison and recording system. All the operations carried out by the operator from his actual Control Desk are automatically transferred to the CBI, so as to allow it to process commands in a parallel mode.

CBI became operational in June 1988 and underwent a number of tests until

April 1990. Afterwards it was supervised along a year by Railway engineers in order to identify any discrepancy between its operation and the relay interlocking. All recorded data have been stored. At the end of the experimental stage, the system was kept operational in order to collect statistical data on reliability.

No wrong side failure was ever recorded; failures were observed in a few hardware modules and in all cases the exclusion logic took over, reconfiguring the Safety Nucleus to a 2-out-of-2 scheme, but without any interruptions.

At the latter station, on the side line connecting Genoa to Turin and Milan, a complete system was installed - a Central Post with four Trackside Units - as well as a comparison and recording system. In co-operation with the Railway engineers a special connection between the interlocking relay outputs and the CBI outputs was also established, so as to allow the latter to control the field devices directly, even though fail-safe commands are carried out by the former.

The system was commissioned in July 1991. After a test campaign carried out by Railway engineers, in October it was formally handed over to the Italian Railways local office.

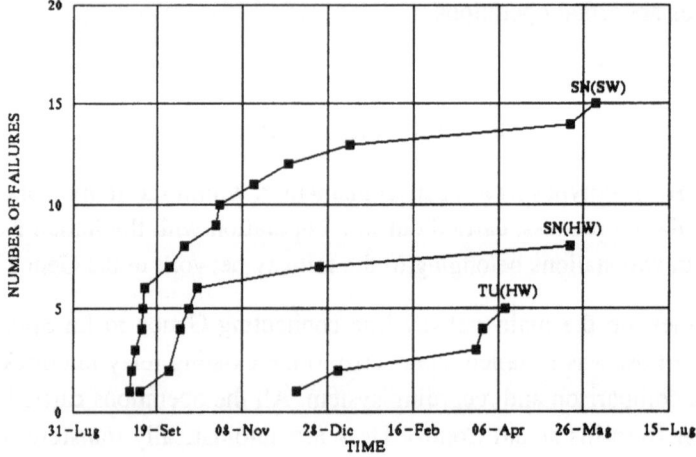

Figure 9: Failures trend

The trial period is now closed. A diagram shows failures trend between October 21, 1991 and June 30, 1992, classified under type and system section or component (Figure 9).

Some failures occurred in the Safety Nucleus, at the high speed serial communication controller; this component had already been detected as critical and has been substituted on the new release of the printed circuit board.

Few failures occurred in the Trackside Units, at the opto-coupler used for the inter processor link; the Customer declared it defeating and redesigned it.

As software faults, none has been revealed in the Trackside Units software, simple and well tested. As far as Safety Nucleus is concerned, three categories of faults occurred, because the field trial ran contemporaneously with factory testing:

- functional diversity between the CBI and the relay interlocking, mostly due to the latter;

- coding errors;

- errors in the application data base.

Now those faults have been removed, new software releases have been installed and the railway station daily operations have not revealed anyone more.

All examined faults did not cause any wrong side failure, being masked by the redundancy and fault-tolerance techniques; sometimes, yet, they caused interrupts to the normal station service, due to the existing connection between the CBI outputs and the relay interlocking ones (i.e. a signal lamp kept red aspect, when the green was right). Then, the CBI operational availability trend has been considered (Figure 10), evaluating the ratio between the effective service time and the total time, the latter being equivalent to the total calendar time minus periods spent for the proofs on new devices and the installation of new software releases.

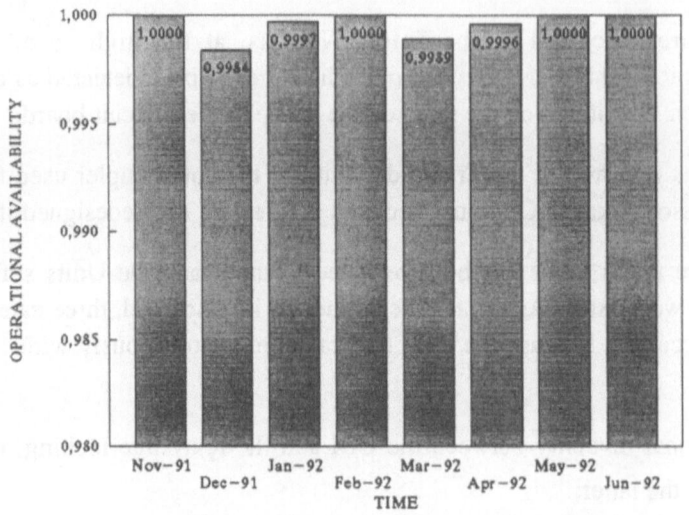

Figure 10: Operational availability

6 Conclusions

The fault-avoidance and fault-tolerance techniques have been described, used to obtain dependability in the Computer Based Interlocking; their effectiveness is demonstrated by the results of the first campaign of factory testing and of the field testing, performed through a comparison with a relay interlocking.

The Italian Railways expressed good evaluations about the CBI testing, so application studies started for renewing of actual station's signalling systems with the CBI technology: Rome main station is the most important one, owing to its size and critical position in the Italian railway network. Moreover, the same technology will be applied to the recently designed high speed lines and to the signalling system of the first underground line in Naples as well.

References

[1] J. B. Balliet, J. R. Hoelscher. Microprocessor based interlocking control - concept to application. *Proc. APTA Rail Transit Conf.*, Miami, Fl., 1986, pp. 13.

[2] A. H. Cribbens, M. J. Furniss, H. A. Ryland. The solid state interlocking project. *Proc. IRSE Symposium "Railways in the Electronic Age"*, London, UK, 1981, pp. 1-5.

[3] M. S. Deutsch. *Software verification and validation*. Prentice-Hall of Software Engineering, 1982.

[4] EWICS. Guideline for verification and validation of safety related software. *EWICS TC7 333*, 1983.

[5] E. K. Holt. The application of microprocessors to interlocking logic. *Proc. APTA Rail Transit Conf.*, Miami, Fl., 1986, pp. 13.

[6] IEC 65A. Software for computers in the application of industrial safety-related systems. IEC 65A (Secretariat) 122 Std.

[7] IRSE 92. Safety system validation. *IRSE - International Technical Committee*, Rep.n. 1, 1992.

[8] D. Nordenfors, A. Sjoeberg. Computer controlled electronic interlocking system, ERILOCK 850. *ERICSSON Review*, n. 1, 1986, pp. 1-12.

[9] I. Okumura. Electronic interlocking to be tried in Japan. *Railway Gazette International*, n. 12, 1980, pp. 1043-1046.

[10] ORE A155, Office de Recherches et d'Essais de l'Union Internationale de Chemins de Fer, Question A155.2. Use of electronics in railway signalling - Software for safety systems. Utrecht, NL, 1985-87.

[11] H. Strelow, H. Uebel. Das sichere Mikrocomputersystem SIMIS. *Signal und Draht*, n.4, 1978, pp. 82-86.

[12] A. M. Traverso. A tool for specification analysis: complete decision tables. *Proc. SAFECOMP 85*, Como, Italy, 1985, pp. 53-56.

[13] G. Wirthumer. VOTRICS - Fault tolerance realised in software. *Proc. SAFECOMP 1989*, Vienna, Austria, 1989, pp. 135-140.

EXPERIMENTAL EVALUATION

A HYBRID MONITOR ASSISTED

FAULT INJECTION ENVIRONMENT

Luke T. YOUNG, Carlos ALONSO
Integrity Systems Division, Tandem Computers, Inc.
Austin, Texas, USA

Ravi K. IYER, Kumar K. GOSWAMI
Center for Reliable and High Performance Computing
Coordinated Science Laboratory, University of Illinois at Urbana-Champaign,
Urbana, Illinois, USA

Abstract

This paper describes a hybrid (hardware/software monitor) fault injection environment and its application to a commercial fault tolerant system. The hybrid environment is useful for obtaining dependability statistics and failure characteristics for a range of system components. The Software instrumentation keeps the introduced overhead small so that error propagation and control flow are not significantly affected by its presence. The Hybrid environment can be used to obtain precise measurements of instruction-level activity that would otherwise be impossible to perform with a hardware monitor alone. It is also well suited for measuring extremely short error latencies. Its utility is demonstrated by applying it to the study of a Tandem *Integrity* S2 system. Faults are injected into CPU registers, cache, and local memory. The effects of faults on individual user applications are studied by obtaining subsystem dependability measurements such as detection and latency statistics for cache and local memory. Instruction-level error propagation effects are also measured.

1 Introduction

Fault Injection is well known for its successful use in system validation and in the extraction of dependability statistics such as latency and fault coverage. However, conventional fault injection environments depend on software monitors alone to measure the effects of faults. The use of software instrumentation to record as well as observe a system may cause the control flow and timing to be substantially different from those of the uninstrumented system. On the other hand, passive hardware monitors lack the ability to discriminate between software-defined, system components (e.g. sections of a given application).

This paper describes a hybrid fault injection environment, wherein faults are injected via software and the impact is measured by both software and hardware. The environment is useful for evaluating system dependability, and it has the advantage in that it introduces minimal perturbation, and provides a high degree of control over the location of faults to be injected. It also measures latencies with high precision. We illustrate the environment by applying it to the study of the Unix-based, *Tandem Integrity S2* computing system. Faults can be injected into cache and any location that has a physical address, e.g., CPU control registers, local memory, mass storage and network controllers. Faults can also be injected into locations allocated to individual user programs or even into the kernel, and propagation can be characterized down to the instruction level.

The term *error detection latency* is defined as the time that elapses between the activation of an error and its discovery. Similarly, fault latency[1] is the time delay between when a fault comes into existence and when it becomes active by producing an error. In computer systems, failure rates can be elevated during a burst of system activity because errors may remain undiscovered until then. For this reason, it is generally believed that long fault and error latencies are undesirable and can have a significant impact on a computing system's reliability.

Most early fault latency experiments were conducted by emulating the systems under study. Studies of CPU fault latency are described in [9], [12] and [15]. A methodology for studying error latency characteristics of medium to large computer systems in a full production environment was developed in [1]. The technique is applied to the memory subsystem and employs periodic sampling by

---

[1] also referred to as *fault dormancy* [10]

a hardware monitor. In [13], this technique was extended to a shared-memory multiprocessing system and used to calculate the risk of encountering multiple latent errors. A failure acceleration method for determining fault detection characteristics is discussed in [2]. Because this study used periodic sampling, the discovery times of only **permanent faults** could be measured. In [17], a hybrid monitor approach to measuring error latency was applied to TI Explorer II Lisp machine. The method is based on simulation of the error discovery process taken from a continuous trace of software-selected locations.

Software fault injection studies have been used to study reliability as well. FIAT is an automated environment for injecting faults in a distributed system [14], [5], and utilizes *software implemented fault injection* to emulate various hardware faults. The emulation of hardware failure manifestations by automatic instruction substitutions is described in [18]. Another automated fault injection environment, FERRARI, is able to emulate faults in hardware components such as opcode decoding circuitry, program control units, data registers, ALU, and address and data buses [8]. In [11], a simulation environment was used to study error propagation from the gate to chip level. FOCUS, a simulation environment to conduct fault sensitivity analysis of chip-level designs, is described in [3] and [4]. Another simulation environment, DEPEND, studies the effects of faults at the system level [6]. Instruction-level simulations are used to supplement SWIFI in [5]. Such methods are useful at the design stage, but they fail to provide a complete environment for fault propagation. They cannot, for example, include the effects of paging, various interrupts, scheduling, and I/O.

This paper describes a hybrid fault injection environment that can be used to evaluate the dependability of computing systems. The environment combines the versatility of software injection and monitoring with the accuracy of hardware monitoring. The environment consists of a fault injection system, an external assistant system, a software monitor, a hardware monitor, and a supervisory system to coordinate the other systems and automate the measurements. Details of the environment follow.

2 Experimental environment

2.1 The hybrid fault injection environment

Figure 1 illustrates how the subsystems of the environment are physically configured. The *supervisor, fault injector* and *software monitor* execute on the *test system,* while the *assistant program* executes on the *control host.* The assistant interprets high-level commands from the supervisor and performs low-level communication with the hardware monitor. Probes attach the hardware monitor to the address/data backplane of the test system so that the monitor can analyze and record the signals generated. Communication between the control host and the hardware monitor takes place over an RS-232 or GPIB connection.

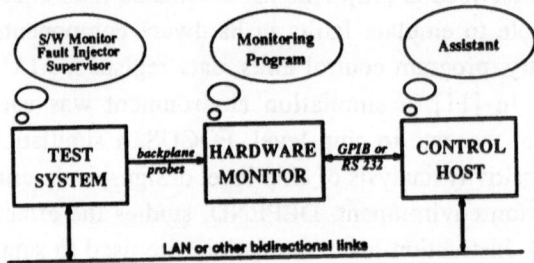

Figure 1: Physical layout of hybrid fault injection system.

The function of the environment is to perform experiments that repeatedly inject faults and record observations. It introduces faults into the test system during the execution of a *target program,* measures the effects of that fault, and returns the test system to preset conditions. These operations form a single *observation loop.* Multiple observation loops form one *experiment* (the control flow of an experiment is illustrated in Figure 2).

To use the environment, one must specify which *target program* to run (with data), the number of times to repeat an observation loop, the number of faults to generate per observation, and termination conditions (typically a time limit or CPU fail). The *target program* can be any user program desired. After the supervisor starts the target program, it randomly generates physical addresses and commands the assistant to reconfigure the hardware monitor with them.

While the hardware monitor reconfigures (this takes about 15 seconds), the software monitor determines which virtual addresses will both map to a physical address and be allocated to the target program. If no match is found at the time when the assistant reports reconfiguration completion, the supervisor must generate another random set of physical addresses and restart the hardware monitor and target program. Thus, the *assistant* and the *software monitor* work in parallel over the network and under the *supervisor's* control to generate locations in the test system to fault.

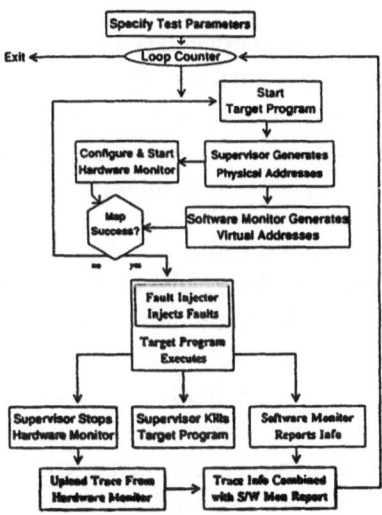

Figure 2: Fault injection control flow.

Next, the supervisor invokes the *fault injector* . Although typical experiments constrain fault injections to portions of memory allocated to the target program, the fault injector can generate faults with **any** location that has a physical address. When termination conditions such as CPU crash, application completion, or timeout occur, the supervisor stops the hardware monitor and kills the target program. The supervisor then commands the assistant to collect measurement reports from both the software and hardware monitors and to reconcile the two reports before appending a summary to the *tracefile* . The observation loop is complete at this point. The remainder of this section examines the component systems of the environment in greater detail.

2.2 Hybrid environment subsystems

The supervisor runs on the test system and serves to synchronize experiments and communicate with the control host via NFS sockets. At the beginning of an observation loop, it passes physical addresses to both the software and hardware monitors. At the end of an observation loop, it receives virtual addresses back from the software monitor and acquisition data back from the hardware monitor. For synchronization purposes, the supervisor controls when to start and stop both the target program and the hardware monitor.

The **assistant** serves as both an intermediary between the supervisor and the hardware monitor and as a data analyzer. The analysis role of the assistant involves taking hardware monitor measurements in the form of a timestamped trace and parsing it according to measurements taken by the software monitor. Further details of the analysis are given in subsection 2.3.

The hardware monitor can record and timestamp any activity, addresses, or data present on the address/data backplane of the system under test. In our hybrid environment, the hardware monitor is a Tektronix DAS 9200 - a programmable, digital analysis tool. Transfers of acquisition data and instrument setup data between the hardware monitor and supervisor and supported through a Programmatic Command Language (PCL) [16]. Transfers are physically supported over an RS-232 connection that uses a software-reconfigured TTY port. The PCL commands allow the assistant to reprogram the DAS, start and stop acquisitions, and upload acquisition files. The 92A90 data acquisition module of the DAS 9200 permits general-purpose state analysis for up to 90 channels. The 20 MHz buffer probe accompanying the 92090 can be retargeted for a wide variety of test systems and can store up to 32,768 samples, where each sample is timestamped with a resolution of 20 ns.

The **software monitor** is designed to function within a Unix operating system. It assists the supervisor by determining which virtual addresses can be used for fault injection during an experiment. Virtual-to-physical address translations are performed by accessing the system page table. If no match can be found, the software monitor notifies the supervisor. Otherwise, it provides the fault injector with the matching physical addresses and a randomly generated bit vector that specifies where, within the word, to inject a fault. After a fault has been injected, the software monitor performs periodic, relatively unobtrusive checks of the system status to determine whether termination conditions have been reached. At the end of an observation cycle, the software monitor reports information such as virtual page frame numbers and cause of termination to the supervisor.

The **fault injector** is a device driver and was partially implemented through the addition of a small, special-purpose kernel routine. During a fault injection, the content of a physical address is read and then written back to a dummy register with a fixed, uncached physical address. The original value is the XORed with the bit vector provided by the software monitor, and the new value is written back to the original address. By this scheme, every fault injection is immediately preceded by a write to a fixed, physical address. Thus, by programming the hardware monitor to detect and record all write activity to this address, we can obtain a record of the precise moment of fault injection.

2.3 Analysis

One of the roles of the assistant is to analyze and merge the data it obtains from the hardware and software monitors. Figure 3 shows the information reported by both the software and hardware monitors.

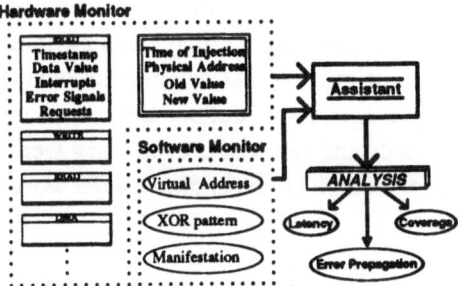

Figure 3: Analysis of hardware and software monitor reports.

For each injected fault, the software monitor reports a virtual address, an XOR bit vector, and the effect of the fault at termination time (e.g., latencies, interrupts, and cpu shutdowns). For the same fault, the hardware monitor reports the time of fault injection, a physical address, and the contents of that address before and after that fault. This information is followed by a timestamped list of all reads, writes, and DMA accesses to the physical address (complete with data values and signals such as interrupts, bus errors, and interrupt requests). The assistant must translate and merge the information it receives so that hardware-level activity and timing can be connected to software specific information. The result of the merge is a trace file that can be analyzed at many levels.

Analysis of a trace file can yield a number of dependability statistics. Latency measurements can be derived from the difference between two timestamps, where the first timestamp corresponds to the moment of fault injection and the second timestamp corresponds to the moment of fault detection (as indicated by the appropriate read or interrupt signal). By examining the application task image, it is possible to determine which instruction is being faulted, and what immediate error propagation effect that fault will have. An example of a typical trace file and its interpretation is provided in subsection 3.2.

2.4 Target programs tested

In all the experiments described in this paper, two C applications were executed under this fault injection environment. They are PRIME and ANAGRAM, and are 3,926 and 7,302 instructions long, respectively. PRIME is a CPU and memory intensive program that generates the first half million prime integers. ANAGRAM is a program that finds all three-word anagrams of a string of letters. Since it must access a local dictionary, it must perform extensive I/O in the early stages of execution. Both were tested under heavy and light multiuser workloads and selected because they were each able to take greater than half an hour to complete, depending on the workload.

3 The Tandem Integrity S2 system

3.1 S2 architecture

The Tandem *Integrity S2* is a fault-tolerant computing platform for Unix-based applications. Figure 4 depicts the architecture of the S2 system used in these experiments. The principal feature of the S2 is its RISC-based, TMR processing core, making it a highly-available, fault-tolerant system. Each of the three CPU boards contains a traditional CPU with cache and 8 MB of local memory. The tree CPUs act as one CPU, performing identical operations and accessing the duplexed Tripple Modular Redundant Controllers (TMRCs) through the *Reliable System Bus* (RSB). Each TMRC provides a 32 MB global memory and dual-rail voters. The Reliable I/O Bus (RIOB) interconnects the TMRCs with the I/O Packetizers (IOPs), which handle all system I/O (including mirrored disks, tape drives, ethernet controllers, terminals, etc...). Further information on the S2 architecture may be found in [7].

The voters are the key error detection mechanisms present on the S2. Whenever a CPU attempts to access the duplexed global memory and I/O systems, it must issue requests through the voters and, if necessary, wait for voting to complete and a result to be returned. An error is detected if two CPUs do not issue the same request within a fixed time period. Such requests are issued in the form of address, data, control, and interrupt values. For instance, if only two CPUs generate an interrupt, the voters will time out and signal that an error has occurred.

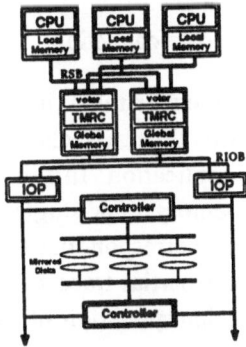

Figure 4: Architecture of the TANDEM *Integrity S2*.

The nature of the local memories provides incentive for studying detection and error propagation. In the local memories, parity checking is not performed, and errors are free to propagate to the CPU and cache without detection. Once inside the CPU, errors can propagate to registers and other local memory locations until the CPU is forced to access the TMRCs. These system features also point out a need to examine **error latency** on the S2. Such examinations can reveal the degree to which the amount of time spent by a voter in detecting an error affects overall error latency. Measuring error latency within local memory can tell us how long errors are free to propagate within a single CPU board before being detected.

Knowing the time associated with isolating a fault is also important to the study of the S2. An intricate chain of interrupts and tests are performed on the S2 in order to isolate faults. If a fault stems from a voter, the effect is the same as if a fault had stemmed from a CPU. To isolate the source of an error, an exception handler running on all the CPUs uses a collection of registers on the CPU and TMRC to determine whether the voter or one of the CPUs contains the fault. It should be useful to know how long fault isolation takes, because additional errors could appear during this chain of tests and prevent successful isolation.

The issues of detection ratio, latency, error propagation, and fault isolation times are addressed by this study. To better illustrate fault injection and data collection, the following example demonstrates an application of the hybrid fault injection environment to the S2 system.

3.2 Example: fault injection and data collection

This example is take from an experiment in which we measured error detection latency and fault isolation times associated with single bit faults in the instruction-stream of a given application. In this example, single-bit faults were injected into the local memory of a single CPU. The locations were randomly selected from memory corresponding to the code section of a test application. Figure 5 illustrates a partial trace file from a single observation generated by the assistant. The interpretation of this example is as follows: Information recorded includes a timestamp, a memory address, direction of data transfer, data, interrupt flags, and state analysis values. Reads are denoted by "=>", and writes are denoted by "<=".

```
Phys Addr = 00xxx994

        time       location      data     flags   comments (added)
    ------------------------------------------------
1    189.84 us  |  (00567994)  => 0043C821  F 3  ← addu t9,v0,v1 (before)
2    196.40 us  |  (00567994)  <= 0043C801  F 4  ← reserved instruction
                |
3    56.06224 ms |  (00567994)  => 0043C801  F 3  ← error activated
4    56.38838 ms |  (007E07C0)     8017B3D0  B 1  ← error detected
5    57.99304 ms |  (1FD10030)     B844FF0E  8 1
6   103.31650 ms |  (1FC00000)  => 0BF00082  F 3  ← fault isolated
```

Figure 5: Example of hardware monitor trace.

The fault corresponds to a flip of bit 5 in the instruction addu *t9, v0, v1*, which turned it into a reserved instruction. An attempt to execute the instruction 56.06 msecs later forced a cache miss and caused the error to migrate to the cache. Upon an attempt to execute this illegal instruction, the CPU generated a reserved instruction exception. This is a los-level exception and does not trigger voting. Whenever the kernel discovers that no instruction interpretation has been implemented, it generates an illegal instruction fault signal which presents a high-level exception to the voters. The generation of this last exception indicates the moment of error detection at the system level. The error detection latency (line 4 minus line 3) in this example was 326.14 usecs. Because the other two CPUs were not faulted, they did not report similar activity. After a brief signal timeout, the TMRC performed testes and determined that DPU B (the one that was faulted) needed to be taken offline. The fault isolation time (line 6 minus line 4) in this example was 46.92812 msecs (+/- 20 ns).

4 Experimental results

4.1 Fault chronology

Figure 6 depicts the life of a fault in the local memory subsystem of the S2. In this example, an error occurs when the CPU obtains incorrect data from the memory. The error is detected at the system level when the voters detect a request mismatch. The *fault latency* is the time that elapses from the moment a fault occurs to the moment it is accessed by CPU as data. The *error detection latency* is the time between CPU access of this data and system-wide acknowledgment of a problem (via the voters). Within a fault tolerant system such as the S2, there is a delay associated with locating the source of the error. This delay is the *fault isolation time*. If the fault is transient, it will be corrected by performing a "power-on self test" (POST) and reintegrating the subsystem. The time required to perform POST and reintegration is the *fault correction time*.

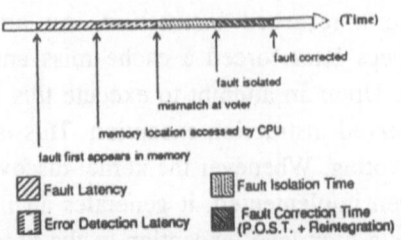

Figure 6: Fault Timeline.

In the following subsections, we present the results of performing several different forms of fault injection. The S2 is used as a case study and the results here may not be representative of tandem systems as a whole. We begin in subsection 4.2 by characterizing **fault** latency in the local memory subsystem of the S2. In subsection 4.3, we examine issues of fault detection percentage and the error propagation effects due to single bit faults on instructions. Then we move on to detection at the system level by studying error detection latency by error propagation type in subsection 4.4. Finally, in subsection 4.5, we characterize the time taken by the system to perform **fault isolation.**

4.2 Fault latency

In this experiment, fault locations were selected from all sections of memory allocated to the ANAGRAM test program (data, stack, and text), for the purpose of measuring fault latency. Since the virtual-physical address mappings are constantly changing in an executing application, the hardware monitor usually had to be reconfigured several times before a virtual address would map into the configured physical address. As a result, it was difficult to observe many fault injections. The ANAGRAM program was executed under a moderately heavy workload, and a total of 750 fault injections were observed. Trial runs indicated that the general distribution forms did not change by extending observation times, so we restricted observations to ten minutes each. Table 1 lists the results obtained:

(fate of fault)		count	(%)
Fault never accessed:		687	91.6%
Activated	CPU Failure:	27	3.6%
(4.3%)	no effect:	5	.07%
Undetected	Overwritten:	16	2.1%
(4.1%)	Deallocated:	15	2.0%

Table 1: Fault injection results.

These results tell us what can happen to a fault, though may (91.6 percent) were never accessed in the first ten minutes of observation. Of those faults that were accessed (8.4 percent), about one half were read-accessed first (51 percent), thus becoming active errors. Most of the active errors cause CPU failure (84 percent) and the rest had no effect (16 percent). The remainder of the accessed faults were undetected, either because the application overwrote them with new values or because the pages containing them were deallocated. Fifty-two percent of the accessed, undetected faults were overwritten and 48 percent were within pages that were deallocate.

Figures 7a and 7b show the latency associated with the removal of undetected faults, i.e., those faults that were removed before they could become errors. Figure 7a shows the latency associated with the deallocation of the clean page containing a fault, and Figure 7b shows the latency associated with overwriting a fault. We observe that faults within clean pages were deallocated after an average of about 210 seconds, which is more than the average time spent before a fault is overwritten by the application (about 164 seconds). In Figure 7b, we observe that many of the fault overwrites occurred within the first minute. These observations tell us that early fault removals are due mostly to overwrites by the application and that later removals are due mostly to page deallocation. This information characterizes a natural fault removal process and could assist in the design of efficient scrubbing techniques.

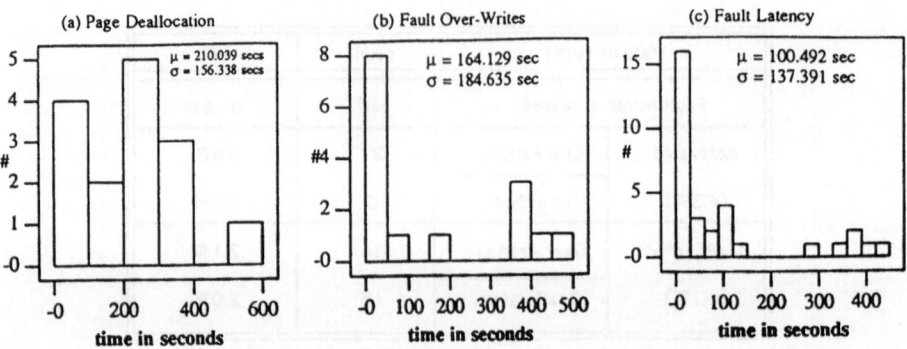

Figures 7a-7c: Histograms of overwrite, remap, and fault latencies.

In Figure 7c, we again see a sharp decay and long-tailed distribution for detected faults. In this example, faults were detected after an average of about 100 seconds, but half were detected in the first 35 seconds. This distribution is very similar to the distribution of latency associated with fault overwrites (Figure 7b). The similarity in distributions implies that the probability that a local memory location will be read before it will be written does not significantly change with increasing dormancy. This finding can be used to support memory management decisions.

4.3 Instruction level fault detection

In examining the fixed-width, Mips RISC instruction set, we found that a single bit fault will immediately impact a given instruction in one of several fashion - we also found that the first-order (immediate) impact could be determined automatically by a disassembler-like program. Table 2 lists the eleven manifestations into which all instructions and faulted bits were mapped for this study. The multiplier column in this table indicates the potential error propagation fanout. The NIL case is that in which a flipped bit is incapable of altering normal operation. Cases USV and ILI trigger error detection mechanisms on a CPU, which in turn alert the system that an unacceptable instruction has been encountered. In both cases, no further corruption occurs. Execution of the wrong instruction (WIE) can be caused by an improperly specified opcode. A branch to a wrong address (BWA) can be caused either by an improperly specified address or base register, or by misdirected calculations. Both WIE and BWA cases are capable of causing multiple corruptions. Writes to wrong registers (WWR) and addresses (WWA) can cause double corruptions: the intended location fails to be updated with new data, and a misintended location is corrupted. The process of reading a wrong register or address does not, in itself, cause error propagation. But when such actions cause wrong values to be passed on to operations that update registers (RWR, RWA, and BID) or memory (BVM), error propagation results. The case of RWA is a potential exception - if the read of a wrong address causes a page fault at the wrong time, then flow of control is altered, causing more than just data corruption to occur.

Id	Description	Multiplier
NIL	Fault has no impact whatsoever on execution	0
USV	(Unassigned Space Violation) Address Exception	0
ILI	ILlegal Instruction	0
WIE	Wrong Instruction Executed	many
BWA	Branch to Wrong Address	many
WWR	Write to Wrong Register	2
WWA	Write to Wrong Address	2
RWR	Read from Wrong Register (reg. corrupted)	1
RWA	Read from Wrong Address (reg. corrupted)	1+
BID	Bad Immediate Data (reg. corrupted)	1
BVM	Wrong register (Bad Value) transferred to Memory	1

Table 2: Error propagation types.

To study fault detection at the instruction level, faults were injected randomly within the code sections (local memory and cache) of executing programs. Measured values were obtained from 2,400 fault injections, 95 percent of which caused CPU failure. Using the tradefile data obtained, we computed first-order error propagation effects from the instruction type and XOR bit vector associated with each injected fault. The fault detections were then tabulated by effect and analyzed. Table 3 shows the results of this analysis. Roughly five percent of the instruction single-bit errors had no observable impact on program execution. Two of the forms trigger error detection without error propagation, but they are caused by only 9.13% of the instruction single-bit errors. This means that over 90 percent of all detectable single-bit errors in instructions cause propagating errors. Comparing ILI with WIE, we observe that a single alteration in an instruction's opcode bit is 3.44 times as likely to produce another legal (but wrong) instruction than it is to produce an illegal instruction. We also observe that the two most common error forms (*wrong instruction execution* and *branch to wrong address*) occur

36.2 percent of the time and are also the two with the highest propagation fan-out. This finding means that the expected propagation fan-out can be quite high.

Effect of error on instruction execution:	Portion
wrong instruction executed (multiple effects)	21.18%
branch to wrong address (multiple effects)	14.98%
read from wrong address (register corrupted)	13.10%
bad immediate data (register corrupted)	9.83%
write to wrong register (double fault)	9.83%
write to wrong address (double fault)	8.12%
read from wrong register (register corrupted)	6.90%
illegal instruction (triggers detection)	6.16%
no observable impact on program execution	4.97%
(USV) address exception (triggers detection)	2.97%
value of wrong register transferred to memory	1.97%

Table 3: Effects of instruction single bit errors.

A comparison of fault detection breakdowns by propagation effect between two applications (Table 4) shows that several propagation effects are sensitive to application. An instruction single bit error was more likely to produce wrong instruction execution (WIE) and bad immediate data (BID) propagations within the PRIME application than within ANAGRAM. Also, WWR, RWR, and RWA errors were more likely to result within the ANAGRAM application than in PRIME. For the rest of the error types, the odds were about even.

Description	ANAGRAM	PRIME
wrong instruction executed	18.56%	24.36%
write to wrong register	11.13%	8.31%
read from wrong register	8.23%	5.35%
read from wrong address	14.52%	11.46%
bad immediate data	8.47%	11.46%

Table 4: Selected error manifestations probabilities.

4.4 Error detection latency

This experiment reuses the tracefile obtained in the previous subsection, where faults were injected into the code (text) sections of memory allocated to the target programs. Each program was tested to determine how long it would take the system to detect an error through its voters. The cache was selectively flushed after a fault injection to ensure that faults would propagate to the CPU and cache upon a cache miss. The error detection latency was measured for each fault, and the identity of each faulted instruction was obtained from the disassembler by comparing listed virtual addresses with those obtained from the software monitor. Corresponding error propagation type was then computed by the method of subsection 4.3, and latency distributions were tabulated according to error propagation type.

Figure 8a illustrates the overall error detection latency distribution. Latencies range from under 50us to over 10 seconds. The mean error detection latency was 40.61 msecs, but several distinct peaks can be seen in the distribution and may be due to distinct error detection mechanisms.

Type	Median	Mean
USV	252.2 μs	464.0 μs
ILI	280.1 μs	1.425 ms
WIE	381.0 μs	30.08 ms
BWA	543.5 μs	21.35 ms
WWR	654.9 μs	85.82 ms
WWA	609.0 μs	40.67 ms
RWR	481.8 μs	47.55 ms
RWA	596.8 μs	84.37 ms
BID	690.6 μs	20.43 ms
BVM	1.179 ms	43.06 ms
all	492.9 μs	40.61 ms

Table 5: Error detection latency by type.

Table 5 lists error detection latency means and medians for each of the error types. The first two, USV and ILI cause immediate detection at the CPU level and short error detection latencies at the **system** level. These system-level latencies are considerably longer than the CPU-level error detection latencies found in past simulation-based studies [5], but part of the system-level latency comes from the time taken during voting. As will be seen later, these latencies are usually negligible when compared to fault isolation times. In examining each of the error types, we observe that while the means are on the order of tens of milliseconds, most errors are detected in under 493 us. This corresponds to about 5,900 instruction cycles and includes at least one round of exception handling.

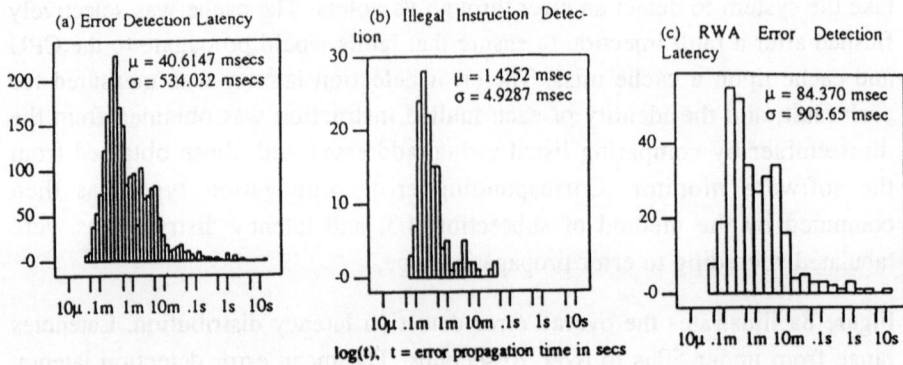

Figures 8a-8c: Histograms for error detection latency and ILI & RWA error detection.

Figures 8b and 8c illustrate distributions of error detection latency for error types ILI and RWA, respectively. For illegal instructions, the number of cycles required to cause the CPU to issue an interrupt is small and fairly constant. Therefore, what we see in Figure 8b is primarily the latency imposed by the voters in detecting the error of a lone CPU-generated interrupt. As expected, the main peak seen in an Illegal Instruction error is narrower than what is seen in the distributions for errors that propagate.

By contrast, the error detection latency distribution for errors that corrupt registers by reading from a wrong memory location (RWA) is quite wide. In figure 8c, we observe that the error detection latency introduced by the voter plus the latency due to error propagation can be as large as 10 seconds. This finding shows that propagating errors can remain undetected for millions of instructions and highlights the need to better characterize error propagation.

4.5 Fault Isolation

As defined in subsection 4.1, fault isolation time is the delay associated with locating the source of an error. Such a source may be faulty CPU board, memory board, voter, bus, etc... In this experiment, faults were injected only into the local memory of CPU B. Consequently, fault isolation identified CPU B as the component to be shut down every time.

As can be seen in Figure 9, the amount of time required to achieve fault isolation was roughly constant (47 msecs), with a standard deviation of only 3.88%. Times were fairly independent of workload level and the application tested. The fault isolation time corresponds to approximately 560,000 instructions.

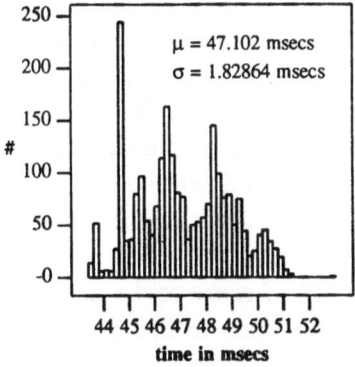

Figure 9: Histogram of fault isolation times.

From the previous subsection, we know that error propagation times can well exceed 47 mscec (3.45 percent of the error detection latencies exceeded 47 msecs). The potential for propagating errors to interfere with the fault isolation process is of much concern, since the propagation of multiple errors can continue undetected throughout fault isolation. This finding points to the need to investigate the effects of multiple errors.

5 Conclusion

In this paper, a hybrid fault injection environment was described, wherein faults were injected via software and the impact was measured by both software and hardware. The environment is useful for evaluating system dependability, and it has the advantage in that it introduces minimal perturbation, and provides a high degree of control over the location of faults to be injected. The injection system is not limited to just user application space. It can be used to inject faults in the Kernel, in CPU registers, cache, local memory, mass storage, network controllers, and any other subsystem that is mapped into physical address space. The fault injection environment was applied to the fault tolerant, Unix-based Tandem Integrity S2.

Using our hybrid monitor and fault injection environment, we obtained several key results: Our findings support the design decision to preserve error correction coding in cache but not in local memory. A fault injected in the local memory subsystem of the S2 caused a CPU divergence/shutdown only 3.6% of the time. But in the cache, CPU divergence/shutdown occurred 95.0% of the time (in only the first minute). Early fault removals were due mostly to overwrites by the application; later removals were due mostly to page deallocation. This information characterizes a natural fault removal process and could assist in the design of efficient scrubbing techniques. In cache, most detected faults were found in less than 493 microseconds. By comparison, it took, 51.2 seconds to scrub all of cache at the rate local memory uses. This finding supports the design decision to not perform memory scrubbing in cache. We characterized immediate error propagation effects caused by injecting faults into the instruction stream of the S2's Mips RI"SC processor. Single-bit errors in instruction code had a significant impact. During runtime, we found they propagated additional errors 85.9 percent of the time. Empirical measurements of all the dependability statistics discussed can further serve as input data for any simulation-based study of the long-term effects of faults on this system.

Acknowledgments

The authors wish to thank Howard Alt (Sun Microsystems) for his help in the virtual to physical address conversion routines, Rob Reinauer (Tandem) for supplying workloads, and especially Doug Jewett (Tandem) for supplying profile-parsing tools, S2 specifics, and many useful suggestions. This research was supported in part by Tandem Computers, Inc., in part by NASA Grant NAG-1-613, and in part by the Department of the Navy, Office of the Chief of Naval Research under Grand N00014-91J-1116. The content of this paper does not necessarily reflect the position or policy of the government and no official endorsement should be inferred.

References

[1] R. Chillarege, R. K. Iyer. Measurement-Based Analysis of Error Latency. *IEEE Trans Computers*, vol. C-36, No.5., May 1987, pp. 529-537.

[2] R. Chillarege, N. S. Bowen. Understanding large system failures - A fault injection experiment. *Proc. 19th International Symposium on Fault-Tolerant Computing*, June 1989, pp. 355-363.

[3] G. Choi, R. K. Iyer, V. A. Carreno. Simulated Fault Injection: A Methodology to Evaluate Fault Tolerant Microprocessor Architectures. *IEEE Transactions on Reliability-Special Issue on Experimental Evaluation*, Vol. 39, No. 4, October 1990, pp. 486-491.

[4] G. S. Choi, R. K. Iyer, V. Carreno. FOCUS: An Experimental Environment for Fault Sensitivity Analysis. To appear in *IEEE Transactions on Computers*.

[5] E. Czeck. On the Prediction of Fault Behavior based on Workload. *Ph.D. dissertation*, Electrical and Computer Engineering Department, Carnegie Mellon University, Pittsburgh, PA, April 19, 1992.

[6] K. Goswami, R. Iyer. A Simulation-Based Study of a Triple Modular Redundant System using DEPEND. *Proc. 5th International FTRS Conference*, Nurnberg, Germany, Sept. 25-27, 1991.

[7] D. Jewett. Integrity S2: A Fault-Tolerant Unix Platform. *Proc. 21st International Symposium on Fault-Tolerant Computing*, Montreal, June 25-27, 1991, pp. 512-519.

[8] G. Kanawati, N. Kanawati, J. Abraham. FERRARI: A Fault and ERRor Automatic Real-time Injector. *Proc. 22nd International Symposium on Fault-Tolerant Computing*, Boston, 1992.

[9] J. H. Lala. Fault detection, isolation and reconfiguration in FTMP: Methods and experimental results. *Proc. 5th Avionics Systems Conference*, Seattle, WA, Nov. 1983, pp. 21.3.1-21.3.9.

[10] J. C. Laprie. Dependable Computing and Fault-Tolerance: Concepts and Terminology. *Proc. 15th International Symposium on Fault-Tolerant Computing*, Ann Arbor, MI, USA, June 1985, pp. 2-11.

[11] D. Lomelino, R. Iyer. Error propagation in a digital avionic processor: A simulation-based study. *NASA CR-176501*, University of Illinois, 1986.

[12] J. G. McGough, F. L. Swern, S. Bavuso. New results in fault latency modeling. *Proc. IEEE EASCON Conf.*, Washington, D.C., Aug. 1983, pp. 882-889.

[13] S. G. Mitra, R. K. Iyer. Measurement-based Analysis of Multiple Latent Errors and Near-coincident Fault Discovery in a Shared Memory Multiprocessor. *Proc. 1988 International Conference on Parallel Processing*, St. Charles, IL, August 15-19, 1988, pp. 404-409.

[14] Z. Segall, D. Vrsalovic, et al.. FIAT - Fault Injection Based Automated Testing Environment. *Proc. 18th International Symposium on Fault-Tolerant Computing*, 1988, pp. 102-107.

[15] K. G. Shin, Y. H. Lee. Measurement and Application of Fault Latency. *IEEE Trans. Computers*, Vol. C-35, No. 4., April 1986, pp. 307-375.

[16] *DAS 9200 92A60/90 User's Manual* (8-/16-/32-Bit Microprocessor Support Modules). Tektronix, Inc., Beaverton, OR, May 1988.

[17] L. Young, R. Iyer. Error Latency Measurements in Symbolic Architectures. *Proc. AIAA Computing in Aerospace 8*, Baltimore, Maryland, October 22-24, 1991, pp. 786-794.

[18] C. Yount, D. Siewiorek. Automatic Generation of Instruction-Level Error Manifestations of Hardware Failures. *(pending technical report)*, Center for Dependable Systems, Dept. of Electrical and Computer Engineering, Carnegie Mellon University, Pittsburgh, PA, 1992.

SPACE/TIME OVERHEAD

ANALYSIS AND EXPERIMENTS

WITH TECHNIQUES FOR

FAULT TOLERANCE

Luiz A. LARANJEIRA, Miroslaw MALEK, Roy JENEVEIN
Department of Electrical and Computer Engineering
The University of Texas at Austin
Austin, Texas 78712, USA

Abstract

This paper presents the results of practical experiments, implemented on a multiprocessor platform, with several techniques for fault tolerance. We analyze and contrast the time and space overheads incurred by each technique. Three iterative algorithms were used in the experiments: solution of Laplace equations, the calculation of the invariant distribution of Markov chains, and the solution of systems of linear equations. Fault-tolerant versions of these algorithms were implemented with two general techniques for fault tolerance (triplication with voting, and checkpointing and rollback) and three application-specific techniques for fault tolerance (self stabilization, algorithm-based fault tolerance, and natural redundancy). The experimental results show that the approach based on natural redundancy, for applications possessing that property, presents the most attractive cost/benefit ratio when only single faults

This research was supported in part by CAPES (Coordenacao de Aperfeicoamento de Pessoal de Ensino Superior - Brazil) under fellowship 7099/86-2, ONR under Grant N00014-88-K-0543 and NASA under Grant NAG9-426.

are likely to occur. The execution of the three algorithms implemented with this technique required less than a 15% time redundancy for significantly sized problems in fault-free situations, only one extra iteration to recover from a fault, and no extra processors. These capabilities are ideal for critical applications that must deliver reliable and timely service. Surprisingly, time overheads for some other methods were much higher than expected.

1 Introduction

A system may incur space overhead (extra hardware and/or software) and/or a time overhead in order to be fault tolerant (see Figure 1a).

In the design of critical applications that must deliver continuous service under real-time constraints, it is essential to achieve fault tolerance with as little time redundancy as possible. Furthermore, in embedded systems, space may also be a limited resource. Therefore, for critical applications, we must provide reliable computing with low space *and* time redundancy. This situation can be represented by a point, such as P_o, in Figure 1b. Consequently, the quantification of time and space redundancy (overhead) in the implementation of fault tolerance is a necessity.

The literature proposes several techniques for fault tolerance, some general and others application-specific. However, little experimental data reflects the cost (in terms of space and time overheads) of actually implementing these techniques.

This paper intends to partially fill this gap by presenting the experimental results of implementing three algorithms on a multiprocessor platform using different techniques for fault tolerance. We present measurements of the space and time overheads required for each technique. Concerning time redundancy (overhead), we examine redundancy in fault-free situations and redundancy necessary to execute recovery when a fault occurs. We also present the fault coverage of the fault detection procedures.

We use the following algorithms in our experiments: the solution of Laplace equations by Jacobi's method; the solution of systems of linear equations, also by Jacobi's method; and the computation of the invariant distribution of Markov chains.

We analyze general techniques for fault tolerance, such as triplication with voting, and checkpointing and rollback; and application-specific techniques, such as algorithm-based fault tolerance, self stabilization, and natural redundancy.

The experimental results show that application-specific techniques are more likely to provide the low space and time overheads required by critical applications. More specifically, for situations in which only single faults are likely to occur, the approach based on natural redundancy, provided that this property is present in the application, offers the most attractive cost/benefit ratio. This technique requires no extra processor and very low time overhead in fault-free situations, as well as when a fault occurs. The implementation of the Laplace algorithm with this technique caused a time overhead of less than 15% in fault-free situations for significantly sized problems. The Markov algorithm implementation incurred less than an 8% time overhead in fault-free situations for all data sets utilized, while the linear equations algorithm implementation incurred less than an 11% time overhead under the same conditions for all data sets utilized. In all cases one extra iteration was necessary to execute fault recovery. One disadvantage of this approach is that it applies only to algorithms possessing natural redundancy.

This paper is divided as follows. Section 2 describes the computation model we utilized. Section 3 presents the target architecture and the assumed fault model. Section 4 lists the techniques for fault tolerance and algorithms used in the experiments. Section 5 presents a discussion of the testbed and experimental results quantifying the space/time overhead. Finally, Section 6 contains concluding remarks.

2   The parallel model of computation

This work adopts a computation model, based on the bulk-synchronous model of parallel computation proposed by Valiant [1] with some convenient modifications. In the adopted model the execution of a parallel algorithm proceeds in *supersteps* and only one process runs on each processor. So, we use the words process and processor interchangeably. The processes participating in a superstep are initially given a *step* of L time units to execute a given amount of processing. After each period of L time units, a global check is made to determine if the superstep has been completed by all participating processes. If it is completed, the computation advances to the next superstep. Otherwise, the next period of L units is allocated to the unfinished superstep. The model assumes the existence of facilities for a barrier synchronization of processes at regular intervals of L time units where L

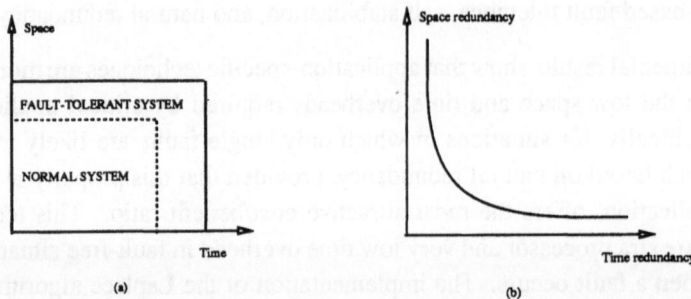

Figure 1: (a) Time and space overhead needed for fault-tolerant system implementation. (b) An example of a graph illustrating a tradeoff between space and time overhead required for fault tolerance.

is the *periodicity parameter*. The value of L may be controlled by the program, even at runtime. This synchronization mechanism captures quite simply the idea of global synchronization at a controllable level of coarseness.

3 Target architecture and fault model

We consider as our target architecture a time-bounded, asynchronous, shared memory MIMD machine such as the Sequent Symmetry, where a common bus links a number of processors. The parallelism in such an architecture is considered to be coarse-grained.

We adopt a fault classification scheme, similar to the one in [2], that proposes several nested fault classes (see Figure 2). We assume that processors operate by responding to triggering events, such as acquiring a lock or reaching a synchronization point. A crash fault occurs when a processor systematically stops responding to its triggering events. An omission fault occurs when a processor either systematically or occasionally does not respond to a triggering event. A timing fault occurs when, in response to a triggering event, a processor gives the right output too early, too late, or never. An incorrect computation fault may be a timing fault or a computation delivered on time containing incorrect or corrupted output values. Since ours is a nested fault model, incorrect computation faults include timing faults; timing faults include omission faults; and omission faults include crash faults. Another aspect of our fault model is that faults may

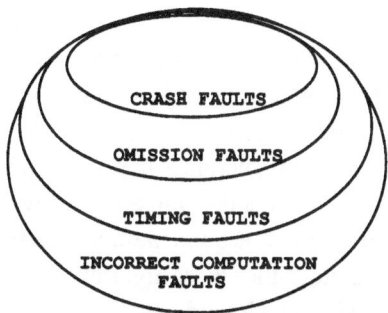

Figure 2: A nested fault classification scheme.

be permanent or temporary (transient or intermittent). Crash faults are always permanent, whereas omission, timing, and incorrect computation faults can be either permanent or temporary.

Our experiments addressed single hardware faults (either temporary or permanent) in the computational paths. The software was assumed to be fault-free. It was also considered that bus and memory faults, as well as faults in the address generation logic of processors and in the address decoding circuits of the memory system, are tolerated by error correcting hardware schemes.

4   Techniques for fault tolerance and algorithms used in the experiments

The techniques for fault tolerance used in the experiments are: (i) general techniques such as triplication with voting [3] and checkpointing and rollback [4]; and (ii) application-specific techniques such as algorithm-based fault tolerance [5], self stabilization [6], and natural redundancy [7]. Triplication with voting and checkpointing are well known techniques. Self stabilizing algorithms do not need explicit procedures to implement fault tolerance. Once a fault occurs, a valid state is eventually reached as the execution of the algorithm rolls forward. Naturally redundant algorithms possess inherent redundancy which must still be exploited by the addition of explicit procedures for fault detection, location and recovery. Algorithm-based fault tolerance (ABFT) explicitly inserts "artificial" redundancy (such as checksums) and the procedures to exploit it to the algorithm.

We experimented with three iterative naturally redundant algorithms: (i) the solution of Laplace equations by Jacobi's method; (ii) the computation of the invariant distribution of Markov chains; and (iii) the solution of systems of linear equations by Jacobi's method. Each of these algorithms was implemented with each of the techniques for fault tolerance mentioned above. The exceptions were: (i) the Laplace algorithm was not implemented with the ABFT technique because the implementation would be quite inefficient; and (ii) the Markov chains algorithm is not self stabilizing. Since these algorithms are iterative, each iteration corresponds to a superstep in the adopted computation model. Details about the basic algorithms and their fault-tolerant implementations are found in [7] and [8].

A general strategy for designing fault-tolerant algorithms is to look first for natural redundancy or self stabilization properties. If these properties are not present, or if the fault tolerance they provide does not meet system specifications (tolerable time and/or space overhead, expected fault coverage), other techniques such as replication, checkpointing, or algorithm-based fault tolerance must be used.

5   Experimental results

5.1   Testbed description and discussion of main issues

We ran our experiments on a Sequent Symmetry bus-based MIMD architecture with a 12-processor configuration. We implemented normal and fault-tolerant versions of the three algorithms with the various techniques for fault tolerance using the programming language "C" with a library of parallel extensions called PPL (Parallel Programming Language). For the sake of simplicity, we call the algorithm for solving Laplace equations by Jacobi's method *the Laplace algorithm*; the algorithm for computing the invariant distribution of a Markov chain *the Markov chains algorithm*; and the algorithm for solving systems of linear equations *the linear equations algorithm*. The main issues involved in our experiments are fault insertion, the effects of finite precision arithmetic, error coverage, and the space/time overheads introduced by the schemes for fault tolerance (relative to the non fault-tolerant version of the algorithms).

A scheme similar to the one in [9] was used to test the fault-tolerant implementations, with the difference that we included fault detection, fault location, and fault recovery, while only fault detection was addressed in that work. Hardware faults were simulated using software techniques. We used C statements to manipulate bits and words, turning correct data into erroneous data at specific points in the computation paths. We used normal floating point precision. Bit and word errors, both transient and permanent, were inserted in the calculation of multiplications and additions during the algorithms' execution. These inserted errors simulated faults in the adder and multiplier of the processors' ALUs (Arithmetic Logic Unit). In the case of the Laplace algorithm, we assumed that a transient (or intermittent) fault lasts for a period of time equal to or greater than the computation time needed to calculate one row or column of points (whichever is larger) in a grid's partition. This assumption is not necessary for the Markov chains and linear equations algorithm. We considered a fault permanent if it lasts more than three iterations. Since we simulated faults by introducing errors in the computational paths, we use the terms error checking and fault detection interchangeably.

The effects of finite precision arithmetic can influence the error detection procedures. The error detection procedures verify whether the computation's output meets a certain relation or is equal to the output of a replicated computation. When replicated computations are compared, finite precision arithmetic causes no problem. Roundoff errors due to finite precision arithmetic are the same in the executions of equivalent computations (homogeneous processors are assumed). So, if an error is introduced and affects one of the computations, the comparison detects the error.

If error checking is done by verifying whether the computation results meet a certain relation or property, then the effects of finite arithmetic are important and need to be considered.

In addition to fault coverage, the quantities measured in the experiments with the fault-tolerant algorithms were: the extra computation time per superstep (iteration) in fault-free situations; the extra number of supersteps necessary to execute fault recovery when a fault occurs; and the space overhead incurred by each technique.

5.2   Discussion of the experiments and comparative analysis

First we discuss the fault coverage of the fault detection schemes. For detection schemes based on the comparison of quantities obtained by equal computation paths (like the ones used with the Laplace algorithm and in the triplication with voting versions of the Markov chains and linear equations algorithms), we observed 100% coverage for the simulated faults.

FAULT DETECTION COVERAGE - MARKOV ALGORITHM					
Percentage of detectable and recoverable faults					$\epsilon = 1.0 * 10^{-7}$
		Data set 1 $\delta = 5.74 * 10^{-7}$ 6 processors	Data set 2 $\delta = 4.04 * 10^{-7}$ 8 processors	Data set 3 $\delta = 1.12 * 10^{-7}$ 10 processors	Data set 4 $\delta = 0.73 * 10^{-7}$ 12 processors
Error type	Unit				
Transient bit error	adder multiplier	78.13 78.13	81.25 81.25	81.25 84.38	78.13 71.86
Transient word error	adder multiplier	100.00 100.00	100.00 100.00	100.00 100.00	100.00 100.00
Permanent bit error	adder multiplier	93.75 87.50	93.75 87.50	87.50 90.63	84.38 84.38
Permanent word error	adder multiplier	100.00 100.00	100.00 100.00	100.00 100.00	100.00 100.00

Table 1: Fault coverage of the fault detection scheme used with the Markov algorithm (except the triplication with voting version).

FAULT DETECTION COVERAGE - LINEAR EQUATIONS ALGORITHM					
Percentage of detectable and recoverable faults					$\epsilon = 1.0 * 10^{-5}$
		Data set 1 $\delta = 3.82 * 10^{-6}$ 6 processors	Data set 2 $\delta = 7.64 * 10^{-6}$ 8 processors	Data set 3 $\delta = 1.15 * 10^{-5}$ 10 processors	Data set 4 $\delta = 1.54 * 10^{-5}$ 12 processors
Error type	Unit				
Transient bit error	adder multiplier	84.38 75.00	84.38 75.00	87.50 68.75	81.25 75.00
Transient word error	adder multiplier	100.00 100.00	100.00 100.00	100.00 100.00	100.00 100.00
Permanent bit error	adder multiplier	100.00 100.00	100.00 100.00	100.00 100.00	100.00 100.00
Permanent word error	adder multiplier	100.00 100.00	100.00 100.00	100.00 100.00	100.00 100.00

Table 2: Fault coverage of the fault detection scheme used with the linear equations algorithm (except the triplication with voting version).

To measure the fault coverage of the fault detection procedures used with the Markov chains and linear equations algorithms (except for the triplication with voting versions), we used four different data sets and four different sets of processors. Since the error checking procedure compares quantities calculated through different data paths, roundoff errors are important. The error checking is done by subtracting the two quantities and comparing the absolute value of the result

TYPE OF TECHNIQUE	REDUNDANCY				FAULTS TOLERATED
	SPACE		TIME		
	# of processors		# of supersteps		
	needed	extra	needed	extra	
Triplication with Voting	$3N$	$2N$	$T+1$	1	multiple temporary and permanent
Checkpointing and Rollback	N	—	$T+I$	I	multiple temporary
Algorithm-based Fault Tolerance	$N+1$	1	$T+1$	1	single temporary
Self Stabilization	N	—	?	?	multiple temporary
Approach Based on Natural Redundancy	N	—	$T+1$	1	single temporary

Table 3: Necessary space overhead, time overhead needed for recovery, and types of faults tolerated by the techniques for fault tolerance used in the experiments.

of this subtraction to a certain δ that accounts for roundoff errors. If δ is too small, *false alarms* will occur; that is, correct computations will be interpreted as erroneous. If this delta is too large, many errors will remain undetected. The employed solution was to determine experimentally, for each data set, the minimum value of δ that causes no false alarms in a fault-free execution of the algorithm. This δ represents the maximum value of the finite arithmetic error for that fixed data set and problem size. Using this value for δ, we measured the fault coverage of the schemes for fault tolerance.

Tables 1 and 2 give the percentage of detected faults. Notice that all faults causing word errors are covered. In particular, those causing floating exceptions are detectable through watchdog timers. In the floating point representation, bits 0 to 23 correspond to the mantissa, bits 24 to 30 define the exponent, and bit 31 is the number's sign bit. Most of the undetected faults were simulated by errors inserted in the lower order bits of the floating point representation, and caused no error in the algorithm's final outcome. This fact and the data in the tables confirm that, for practical considerations, the fault coverage of these schemes for fault tolerance is quite acceptable with exception of transient bit faults which had no impact on the correctness of the results. In these tables ϵ represents the convergence factor used to determine when convergence is achieved.

Now we analyze the space and time overheads observed in our experiments, as well as the types of faults tolerated by each technique. Table 3 shows the space

SELF-STABILIZING LAPLACE ALGORITHM							
$NIC_{ff} = 6325$				$\epsilon = 0.1$			
bit_{er}	NIC_{fc}	bit_{er}	NIC_{fc}	bit_{er}	NIC_{fc}	bit_{er}	NIC_{fc}
0	6325	8	6325	16	6325	24	6399
1	6325	9	6325	17	6325	25	6370
2	6325	10	6325	18	6325	26	6977
3	6325	11	6325	19	6325	27	11202
4	6325	12	6325	20	6325	28	20691
5	6325	13	6325	21	6327	29	39674
6	6325	14	6325	22	6327	30	FE
7	6325	15	6325	23	6326	31	6289

Table 4: Number of supersteps necessary for the self-stabilizing (normal) Laplace algorithm to converge when different transient faults occur.

overhead, in terms of the number of extra processors, and gives the time overhead, in terms of the number of extra supersteps, necessary for fault recovery when a fault occurs.

Replication with voting requires the largest amount of space overhead. Variables and processes are at least triplicated. Alternately, the time overhead for recovery is minimal in terms of the number of supersteps. If a fault occurs in one superstep, recovery is executed in the next superstep. As we discuss later, however, the time overhead incurred by this technique in fault-free situations is considerable for bus-based multiprocessor architectures, such as the Sequent. This technique covers a large set of faults, both temporary (transient and intermittent) and permanent.

The checkpointing and rollback technique requires no extra processors. This technique is usually used to tolerate temporary faults. The time overhead necessary for recovery may vary depending on the distance, in terms of the number of supersteps, of the superstep in which the fault occurred to the one in which the latest correct state was saved. An upper bound for this distance is I, which is the interval, in terms of the number of supersteps, between two checkpoints. Previous work in this area [10] reports that the optimal checkpointing interval for a given system is a function of the time required to perform a checkpoint and the system's failure rate. Since failure rate information was unavailable for our computer system, we chose values of I for our experiments that were reasonable with respect to the problems we worked with. Since the case studies we used to measure time overhead were run to completion with 500 to 1000 supersteps

SELF-STABILIZING LINEAR EQUATIONS ALGORITHM							
$NIC_{ff} = 51$				$\epsilon = 0.0001$			
bit_{er}	NIC_{fc}	bit_{er}	NIC_{fc}	bit_{er}	NIC_{fc}	bit_{er}	NIC_{fc}
0	51	8	51	16	49	24	65
1	51	9	51	17	51	25	81
2	51	10	51	18	56	26	93
3	51	11	51	19	62	27	117
4	51	12	51	20	50	28	165
5	51	13	51	21	55	29	261
6	51	14	51	22	61	30	438
7	51	15	55	23	61	31	72

Table 5: Number of supersteps necessary for the self-stabilizing (normal) linear equations algorithm to converge when different transient faults occur.

(iterations), we arbitrarily selected values of I equal to 50, 100, and 200.

Algorithm-based fault tolerance is accomplished with a small space overhead. The time overhead for recovery is one superstep. In our implementations, the computation of the fault detection procedure was spread over the original set of processors, but the recovery procedure was executed by an extra processor in order to minimize the time overhead in fault-free situations. Otherwise (that is, without the extra processor) this time overhead could become prohibitive.

Self stabilization requires no space overhead. Conversely, the time overhead necessary for the algorithm to converge after a fault's occurrence is unpredictable and may be quite large. Tables 4 and 5 present the results of an experiment in which transient bit errors were inserted during the execution of the Laplace and linear equations algorithms (which are self stabilizing) simulating faults. The number of supersteps necessary for convergence under a faulty condition, NIC_{fc}, is shown for different locations of bit errors, bit_{er}. The number of supersteps necessary for convergence in the fault-free execution of the algorithms is given by NIC_{ff}. Transient bit errors were inserted during iteration 3165 of the Laplace algorithm and iteration 25 of the linear equations algorithm, that is, when approximately half of the entire computation had been executed. In certain situations the inserted bit error caused a floating exception (see Table 4), terminating the corresponding process and causing, therefore, a crash fault. This permanent fault could not be tolerated because self stabilization can only tolerate temporary faults.

The large time overhead that may be necessary for recovery and the nonnegligible probability of permanent faults seem to point out that self stabilization may be insufficient to achieve the high levels of reliability and performance required by many critical applications.

The time overhead for recovery incurred by the approach based on natural redundancy was found to be only one superstep. This technique can be used to tolerate single temporary faults. It can also be easily extended to handle permanent single faults by the addition of a reconfiguration procedure and the availability of spare processors. In our experiments this extension was implemented and the algorithm versions coded with the extended technique were able to tolerate single permanent faults.

Figure 3 shows the time overhead measurements in fault-free situations for the three algorithms coded with the various techniques for fault tolerance. The experiments were carried out with three different sets of "basic" processors (four, six, and ten processors). By basic processors we mean those necessary for the execution of the algorithm's normal (non-fault-tolerant) version. The algorithm-based version of the Markov algorithm that runs with six "basic" processors needs, in fact, seven processors, because an additional processor is necessary to calculate the checksum. The triplication with voting version of the Laplace algorithm that runs with four "basic" processors needs, in fact, twelve processors. We also used different data sets in the experiments: various grid sizes with the Laplace algorithm; various numbers of states with the Markov algorithm; and various numbers of variables with the linear equations algorithm. Also, the checkpointing and rollback versions of the algorithms were executed with different checkpointing intervals ($I = 50$, $I = 100$, and $I = 200$) corresponding to the number of supersteps in between two subsequent checkpoints (see Figure 3).

In general, the time overhead in fault-free situations decreases as the problem size increases. This is because, for larger data sets, the amount of computation due to the schemes for fault tolerance represent a smaller portion of the total computation of a superstep. The triplication with voting scheme was the technique incurring the largest amount of time overhead in fault-free situations, followed by the checkpointing and rollback technique. The algorithm-based fault tolerance technique and the approach based on natural redundancy incurred very small time overheads in fault-free situations. The advantages of the natural redundancy scheme are that it does not require additional processors (unless permanent faults occur), and incurs smaller time overhead (in fault-free situations) than the algorithm-based approach.

The reasons why the replication with voting technique caused a higher time overhead in fault-free situations are related to the increased data traffic in the bus and the increased synchronization penalty implied by this technique. In a bus-based architecture machine, such as the Sequent Symmetry, the bus is a bottleneck. When one triplicates the number of processors used in a computation, the data traffic also triplicates, but only one bus handles the increased traffic. Furthermore, PPL synchronization primitives use a busy wait protocol. Processes needing to synchronize constantly check the value of a variable in shared memory. To access the value of this variable, each process must use the bus. Triplicating these accesses causes additional time overhead.

The time overhead in fault-free situations for the checkpointing and rollback technique depends on the size of the state to be stored in each checkpoint. In our experiments, this technique incurred larger time overheads with the implementation of the Laplace algorithm because the amount of data composing a checkpoint is larger for that algorithm than for the others. Another important parameter is the checkpointing interval given by I, the number of supersteps between checkpoints. Larger checkpointing intervals incur smaller time overhead values in fault-free situations, but require longer recovery times when faults do occur.

The natural redundancy version of the Laplace algorithm required a time overhead of less than 15% in fault-free situations for significantly large problems. The Markov chains algorithm implementation incurred a time overhead of less than 8% in fault-free situations for all data sets, while the linear equations algorithm implementation incurred a time overhead of less than 11% under the same conditions for all data sets.

The self-stabilizing versions of the algorithms incurred no performance overhead in the absence of faults. However, as discussed before, this technique may require a very high time overhead to recover from faults when they do occur.

Given the tradeoffs between the various techniques for fault tolerance, the approach based on natural overhead, when this property is present in the application, results in the most attractive cost/benefit ratio if only single faults are likely to occur (which is true in most situations). This approach requires no extra processors (unless it must handle permanent faults), one superstep of time overhead

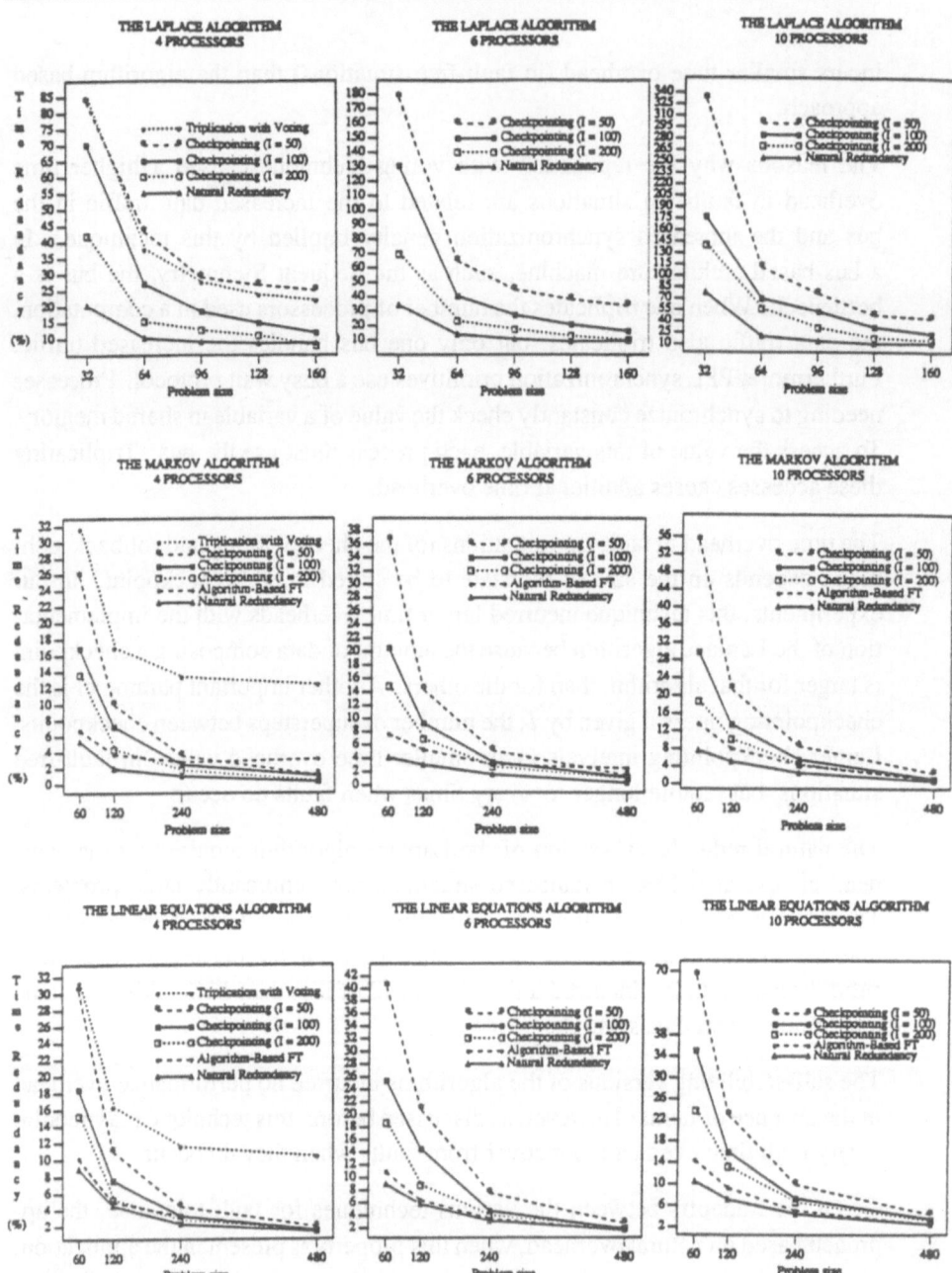

Figure 3: Comparison of time redundancy (overhead) in fault-free situations for the three algorithms implemented with several techniques for fault tolerance executing on four, six and ten "basic" processors.

for recovery, low time overhead in fault-free situations, and provides high fault coverage.

6 Conclusions

We presented and discussed the results of practical experiments with three different algorithms implemented with several techniques for fault tolerance. The algorithms were: the solution of Laplace equations, the computation of the invariant distribution of Markov chains, and the solution of systems of linear equations. The techniques used were: triplication with voting, checkpointing and rollback, self stabilization, algorithm-based fault tolerance, and natural redundancy. The implementations were realized on a bus-based shared memory architecture (Sequent Symmetry) and executed with several data sets and three different sets of processors.

The experimental results demonstrate that application-specific techniques are more likely to ensure fault-tolerant execution with the lower amounts of space and time overhead required by critical applications.

The fault detection schemes provided a fault coverage close to 100% for the simulated faults producing measurable errors in the final output.

More specifically, the approach based on natural redundancy proved to be especially attractive for single fault situations. The implementation of the Laplace algorithm with this technique incurred a run-time overhead in fault-free situations of less than 15% for significantly sized problems. The Markov chains algorithm implementation incurred a time overhead of less than 8% in fault-free situations for all data sets, while the linear equations algorithm implementation incurred a time overhead of less than 11% under the same conditions for all used data sets.

The experimental results confirm that fault-tolerant parallel/distributed systems can be successfully built by the development of an ultrareliable, formally-proved correct kernel and application-specific techniques for fault tolerance. This is especially valid in the case of critical applications that must deliver continuous service in a timely manner.

References

[1] L. G. Valiant. A bridging model for parallel computation. *Communications of the ACM,*

Vol. 33, No. 8, August 1990, pp. 103-111.

[2] F. Cristian, H. Aghili, R. Strong, D. Dolev. Atomic Broadcast: from simple diffusion to byzantine agreement. *15th Int. Conference on Fault-Tolerant Computing*, 1985, pp. 200-206.

[3] D. P Siewiorek, R. S. Swarz. *The theory and practice of reliable system design*. Digital Press, 1982.

[4] R. Koo, S. Toueg. Checkpointing and rollback-recovery for distributed systems. *IEEE Transactions on Software Engineering*, Vol. SE-13, No. 1, January 1987, pp. 23-31.

[5] K. H. Huang, J. A. Abraham. Algorithm-based fault tolerance for matrix operations. *IEEE Transactions on Software Engineering*, Vol. SE-33, No. 6, June 1984, pp. 518-528.

[6] E. W. Dijkstra. Self-stabilizing systems in spite of distributed control. *Communications of the ACM*, Vol. 17, No. 11, November 1974, pp. 643-644.

[7] L. Laranjeira, M. Malek, R. Jenevein. On tolerating faults in naturally redundant algorithms. *Proc. of 10th Symposium on Reliable Distributed Systems,* Pisa, Italy, September 1991, pp. 118-127.

[8] L. Laranjeira, M. Malek, R. Jenevein. Experimental evaluation of techniques for fault tolerance. *Technical Report TR-92-32*, Department of Computer Sciences, The University of Texas at Austin, July 1992.

[9] V. Balasubramanian, P. Banerjee. Tradeoffs in the design of efficient algorithm-based error detection schemes for hypercube multiprocessors. *IEEE Transactions on Software Engineering*, Vol. SE-39, No. 2, February 1990.

[10] K. M. Chandy. A survey of analytic models of rollback and recovery strategies. *IEEE Computer*, May 1975, pp. 40-47.

PROTOCOLS FOR DEPENDABILITY

PRIMARY–BACKUP PROTOCOLS:

LOWER BOUNDS AND OPTIMAL

IMPLEMENTATIONS

Navin BUDHIRAJA[1], *Keith MARZULLO*[1]
Fred B. SCHNEIDER[2], *Sam TOUEG*[3]
Department of Computer Science, Cornell University
Ithaca, New York 14853, USA

Abstract

We present a precise specification of the primary–backup approach. Then, for a variety of different failure models we prove lower bounds on the degree of replication, failover time, and worst-case blocking time for client requests. Finally, we outline primary–backup protocols and indicate which of our lower bounds are tight.

[1]Supported by Defense Advanced Research Projects Agency (DoD) under NASA Ames grant number NAG 2–593 and by grants from IBM, Siemens, and Xerox. Budhiraja is also supported by an IBM Graduate Fellowship. The views, opinions, and findings contained in this report are those of the authors and should not be construed as an official Department of Defense position, policy, or decision.

[2]Supported in part by the Office of Naval Research under contract N00014-91-J-1219, the National Science Foundation under Grant No. CCR-8701103, DARPA/NSF Grant No. CCR-9014363, and by a grant from IBM Endicott Programming Laboratory.

[3]Supported in part by NSF grants CCR-8901780 and CCR-9102231 and by a grant from IBM Endicott Programming Laboratory.

1   Introduction

One way to implement a fault-tolerant service is by using multiple servers that fail independently. The state of the service is replicated and distributed among these servers, and updates are coordinated so that even when a subset of servers fail, the service remains available.

Such fault-tolerant services are generally structured in one of two ways. One approach is to replicate the service state at all servers and to present all client requests, in the same order, to all non-faulty servers. This service architecture is commonly called *active replication* or *the state machine approach* [22] and has been widely studied from both theoretical and practical viewpoints (*e.g.*, [9], [11], [19]).

The other approach to building replicated services is to designate one server as the *primary* and all the others as *backups*. Clients make requests by sending messages only to the primary. If the primary fails, then a *failover* occurs and one of the backups takes over. This service architecture is commonly called the *primary-backup* or the *primary-copy* approach [1] and has been widely used in commercial fault-tolerant systems. However, the approach has not been analyzed nearly as extensively as the state machine approach. Little is known of its costs and tradeoffs, the degree of replication required, or the worst-case response time for various failure models. In this paper, we derive some of these tradeoffs. For example, in some primary–backup protocols [15] the number of servers used is more than twice the number of failures to be tolerated. We are now able to explain this phenomenon by showing that the number of servers needed depends on the failure model.

With both active replication and the primary-backup approach, the goal is to provide a client with the illusion of a service that is implemented by a single server, despite failures. The key difference between active replication and the primary-backup is how each handles failures. With active replication, the effects of failures are completely masked by voting, and the service implemented is indistinguishable from a single non-faulty server. With the primary-backup approach, a request to the service can be lost if it is sent to a faulty primary.[1] Thus, clients can observe the effects of failures. However, the periods during which requests can be lost are bounded by the length of time that can elapse between failure of the primary and

---

[1] Of course, the client can subsequently resend a copy of that request to the new primary.

takeover by a backup. Such behavior is an instance of what we call *bofo* (*bounded outage finitely often*).

To formulate the notion of a bofo server, define a *server outage* to occur at time t if some client makes a server request at that time but never receives a response to that request.[2] In a (k, Δ)–*bofo server*, all server outages can be grouped into at most k intervals of time, with each interval having length at most Δ. Accordingly, even though some requests made to a bofo service (that is, a service that implements the abstraction of a bofo server) will be lost, this number is limited. Note that if clients of a service are restricted to send requests only to one server, then it is not possible to implement a specification that is stronger than bofo. This is because if the client sends a request to a (single) server and that server subsequently crashes, then the request can be lost and will not be processed.

This paper gives lower bounds for various costs associated with implementing a bofo service by using the primary-backup approach. These lower bounds depend on message delivery delay and the class of failures to be tolerated. These bounds characterize the degree of replication, the time during which the service can be without a primary, and the amount of time it can take to respond to client requests (blocking time). In some cases, our results are surprising. For example, more than $f + 1$ servers are necessary to tolerate f failures of certain types (crash and link failures, receive-omission failures, or general-omission failures). Also, we have proved that if a majority of the servers can be faulty, then any primary–backup protocol for receive–omission failures will have a run in which a non-faulty primary is forced to let a faulty server become the primary in its place. Finally, we outline some primary–backup protocols. This allows us to determine which of our lower bounds are tight.

The paper is organized as follows. Section 2 gives a precise specification of the primary-backup approach. Section 3 describes the system model we consider. Section 4 discusses lower bounds, and in Section 5 we outline our protocols and state which of our bounds are tight. We conclude in Section 6.

---

[2]For simplicity, we assume in this paper that every request elicits a response.

2 Specification of the primary–backup approach

Since we wish to derive lower bounds, we must first give a precise specification of primary-backup that is general enough to satisfy any protocol one would characterize as being primary-backup. The following four properties do this.

The first property states that no more than one server can be the primary at any time.

Pb1: There exists a local predicate $Prmy_s$ on the state of each server s. At any time, there is at most one server s whose state satisfies $Prmy_s$.[3]

For brevity, whenever we say that "s is the primary (at time t)" we mean that the state of s satisfies $Prmy_s$ (at time t). We define the *failover time* of a service to be the longest period of time during which $Prmy_s$ is not true for any s.

Property Pb2 distinguishes the primary-backup approach from active replication, where each client broadcasts its request to all the servers.

Pb2: Each client i maintains a server identity $Dest_i$ such that to make a request, client i sends a message (only) to $Dest_i$.

We assume that requests sent to a server s are enqueued in a *message queue* at s.

Pb3: If a client request arrives at a server that is not the current primary, then that request is not enqueued (and therefore is not processed).

Properties Pb1–Pb3 specify a protocol for client interactions with a service, but not the obligations of the service. For example, the properties do not rule out a primary that ignores all requests. A fourth property eliminates such trivial implementations by stipulating that the service implements a single bofo server for some values of k and Δ:

Pb4: There exist fixed values k and Δ such that the service behaves like a single (k, Δ)–bofo server.

---

[3]The protocol of [15] allows concurrent primaries, but only for bounded periods. If one replaces Pb1 by this weaker property, then except for the bounds on failover times, the bounds shown in Section 4 continue to hold.

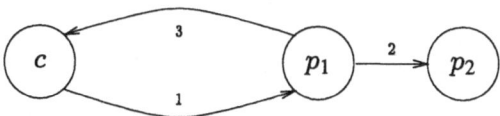

Figure 1: A simple primary–backup protocol.

We believe that the above four properties characterize a primary-backup approach and have checked that many primary-backup protocols in the literature (*e.g.* [1], [3], [4], [7]) do satisfy this characterization.

Note that Pb4 is not implementable if the number of failures (that is, the number of servers and communication components that fail) cannot be bounded *a priori*. This is because an unbounded number of servers would be required to implement the service. In a practical system, one can implement service outages of bounded lengths by bounding the rate of failures and allowing reintegration of recovered servers and communication links. We do not address failure rates or reintegration in this paper.

2.1 A simple primary–backup protocol

As an example of a service based on the primary–backup approach, consider the following protocol, which tolerates crash failures of a single server. Assume that all communication is over point-to-point non-faulty links and that each link has an upper bound δ on message delivery time.[4] Refer to Figure 1. There is a primary server p_1 and a backup server p_2 connected by a communications link. A client C initially sends all requests to p_1 (indicated by the arrow labeled 1 in the figure). Whenever p_1 receives a request, it

- processes the request and updates its state accordingly,

- sends information about the update to p_2 (message 2 in the figure),

- without waiting for an acknowledgement from p_2, sends a response to the client (message 3 in the figure).

---

[4]To simplify exposition, we assume that the maximum message delay between the clients and the servers is the same as the delay between the servers. However, our results can be easily extended to the case when the delays are different.

The order in which these messages are sent is important because it guarantees that, given our assumption about failures, if the client receives a response, then either p_2 will eventually receive message 2 or p_2 will crash.

Server p_2 updates its state upon receiving messages from p_1. In addition, p_1 sends dummy messages to p_2 every τ seconds. If p_2 does not receive such a message for $\tau + \delta$ seconds, then p_2 becomes the primary. Once p_2 has become the primary, it informs the clients (who update their copies of *Dest*) and begins processing subsequent requests from the clients.

We now show how this protocol satisfies our characterization of a primary–backup protocol. Property Pb1 requires that there never be two primaries. This is satisfied by the following definitions of *Prmy*:

$$Prmy_{p_1} \overset{\text{def}}{=} (p_1 \text{ has not crashed})$$
$$Prmy_{p_2} \overset{\text{def}}{=} (p_2 \text{ has not received a message from } p_1 \text{ for } \tau + \delta)$$

Predicate $Prmy_{p_1} \wedge Prmy_{p_2}$ is always false in a system executing our protocol, and hence Pb1 is satisfied. The *failover time* for this protocol is the longest interval during which $\neg Prmy_{p_1} \wedge \neg Prmy_{p_2}$ can hold, and it is $\tau + 2\delta$ seconds. Property Pb2 follows trivially from the description of the protocol. Property Pb3 is true because requests are not sent to p_2 until after p_1 has failed. Finally, Pb4 requires that the protocol implements a single bofo server for some values of k and Δ. Since p_1 sends message 2 before message 3, it will never be the case that p_1 sends a response to the client and p_2 does not get information about that response from p_1. In this protocol, there is at most one switch of the primary. So there is at most one outage period *i.e.* $k = 1$. To compute Δ, it suffices to compute the longest interval during which a client request may not elicit a response. Assume that p_1 crashes at time t_c. Thus any client request sent to p_1 at $t_c - \delta$ or later may be lost since p_1 crashes at t_c. Furthermore, p_2 may not learn about p_1's crash until $t_c + \tau + 2\delta$, and clients may not learn that p_2 is the primary for another δ. So, the total period during which a request may not elicit a response is $t_c - \delta$ through $t_c + \tau + 3\delta$: the protocol implements a single $(1, \tau + 4\delta)$–bofo server.

3   The model

We consider a system consisting of n servers and a set of clients. We assume that server clocks are perfectly synchronized with real time.[5] Clients and servers communicate by exchanging messages through a completely connected point-to-point network.[6] Each message sent is enqueued in a queue maintained by the receiving process, and a process accesses its message queue by executing a **receive** statement. We assume that links between processes are FIFO (*i.e.* if p_i sends message m followed by m' to process p_j, then p_j will never receive m after m') and there is a known constant δ such that if processes p_i and p_j are connected by a (non-faulty) link, then a message sent from p_i to p_j at time t will be enqueued in p_j's queue at of before $t + \delta$.

We are interested in identifying the costs inherent in primary–backup protocols, and so we assume that it takes no time for a server to compute a response. Our theorems characterize lower bounds; they are not invalidated by servers that require a substantial amount of time to compute a response.

We model an execution of a system by a *run*, which is a sequence of timestamped events involving clients, servers, and message queues. These events include sending messages, enqueuing messages, receiving messages, and internal events that model computation at processes. Two runs σ_1 and σ_2 of the system are defined to be *indistinguishable* to a process p if the same sequence of events (with the same timestamps) occur at p in both σ_1 and σ_2. We assume that if two runs σ_1 and σ_2 are indistinguishable to p and p has the same initial state in both runs, then at any time t the state of p at t in σ_1 is the same as the state of p at t in σ_2. It is not hard to extend the definition of indistinguishability to handle nondeterministic servers.

We assume that the clients can send any request at any time. If we impose restrictions on the behavior of the clients, then we can derive protocols that violate the lower bounds in this paper.

Define $\prec$ to be the *potential causality* relation [12] on server events e_1 and e_2. Thus $\prec$ is the transitive closure of the following relation $\rightsquigarrow$: $e_1 \rightsquigarrow e_2$ iff both e_1 and e_2 occur at the same server s and e_1 occurs before e_2, or e_1 is a send

---

[5]Our protocols can be extended to the case where clocks are only approximately synchronized [14].

[6]Another approach would assume that servers are interconnected with redundant broadcast busses [2], [8]. We have not pursued this approach.

event and e_2 is the corresponding receive event. Informally, we say a request m is an *update request* if it changes the state of the service in such a way that the responses to subsequent requests depend on m. More formally, let m be a request with associated response r (and $e(m)$ and $e(r)$ be the events in the run associated with the receipt of m and the sending of r respectively). Then m is an update request if all request/response pairs m'/r', where m' was sent after r, have $e(m) \prec e(r')$. We assume that update requests exist since otherwise the actions performed by the primary do not have to be communicated to the backups.

We assume that failures occur independently from each other. We consider the following hierarchy of failure models:

Crash failures: A server may fail by halting prematurely. Until it halts, it behaves correctly [13]. [7]

Crash+Link failures: A server may crash or a link may lose messages (but links do not delay, duplicate or corrupt messages).

Receive-Omission failures: A server may fail not only by crashing, but also by omitting to receive some of the messages directed to it over a non-faulty link.

Send-Omission failures: A server may fail not only by crashing, but also by omitting to send some of the messages over a non-faulty link [10].

General-Omission failures: A server may exhibit send-omission and receive-omission failures [20]).

Note that crash+link failures and the various types of omission failures are quite different. Although all of these failure models concern loss of messages, each class of failures is dealt with by a different masking technique. In particular, crash+link failures can be masked by adding redundant communication paths, while omission failures can only be masked by adding redundant servers so that faulty processes can detect their own failure and halt. We return to these masking techniques in Section 5.

Failures are counted by the number of failing components (either servers or links). We say that a protocol tolerates f failures if it works correctly despite the failure

---

[7]The lower bounds we derive for crash failures also hold for fail-stop failures [21] except for the bound on failover time. The lower bound on failover time depends on the maximum duration between when a server p_i fails and when *failed_i* becomes true.

of up to f faulty components (note that each faulty component may fail many times during an execution).

4 Lower bounds

For each failure model, we now give lower bounds for implementing a single (k, Δ)–bofo server using the primary–backup approach.

4.1 Bounds on replication

The first bound is obvious. However, to introduce our notation and the proof technique that will be used later in the section, we give a formal proof of the theorem.

Theorem 1 *Any primary–backup protocol tolerating f crash failures requires $n \geq f + 1$.*

Proof: We prove the result by contradiction. Suppose there is a protocol P for $n < f + 1$. Thus, P satisfies Pb4. Consider a run in which all n servers are crashed initially and clients submit $R > k\lceil \Delta/d \rceil$ requests, where d is the minimum time between the sending of any two requests $(d > 0)$. By Pb4, at least one of these requests must elicit a response. This is because the number of requests that cannot have responses must fall into at most k intervals of length at most Δ, and each interval of Δ can contain at most $\lceil \Delta/d \rceil$ requests. However, such a response is impossible since, by assumption, all servers have crashed. □

The following lemma is used in the rest of the theorems in this section.

Lemma 4.1 *Consider any protocol that satisfies Pb4. Suppose two disjoint and nonempty sets of servers A and B can be found that meet the following three properties:*

1. There exists a run σ_a containing $R > 2k\lceil \Delta/d \rceil$ requests where d is the minimum time between the sending of any two client requests $(d > 0)$. Furthermore, in this run the servers in A do not crash and all other servers crash at time 0.

2. *There exists a run σ_b containing R requests. Furthermore, in this run the servers in B do not crash and all other servers crash at time 0.*

3. *There exists a run σ_{ab} containing R requests. Furthermore, the servers in A and B do not crash, σ_{ab} is indistinguishable from σ_a to all servers in A, and σ_{ab} is indistinguishable from σ_b to all servers in B.*

At least one of the above runs violates Pb2.

Proof: Suppose for contradiction that the lemma is false and runs σ_a, σ_b and σ_{ab} all satisfy Pb2.

For σ_a, by Pb4 at least $R - k\lceil\Delta/d\rceil$ of the requests must have been received by servers in A. Similarly, for σ_b, at least $R - k\lceil\Delta/d\rceil$ of the requests must have been received by servers in B. Finally, since σ_{ab} is indistinguishable from σ_a to servers in A, they must execute the same number of receive events in both runs. The same holds for the servers in B. By Pb2, each request is sent to at most one server and so at least $2(R - k\lceil\Delta/d\rceil)$ requests must have been sent in σ_{ab}. Since only R requests were sent, we must have $R \geq 2(R - k\lceil\Delta/d\rceil)$, or $R \leq 2k\lceil\Delta/d\rceil$, which contradicts the assumption that $R > 2k\lceil\Delta/d\rceil$.

$\square$

Theorems 2 and 3 depend on two parameters of primary–backup protocols. Let Γ be the maximum time that can elapse between any two successive client requests (possibly from different clients), and let D be a duration such that if some server s becomes the primary at time t_0 and remains the primary through time $t \geq t_0 + D$ when a client c_i sends a request, then $Dest_i = s$ at time t. Hence, D is the minimum delay until all clients know the identity of a new primary. For simplicity of notation, we write $D < \Gamma$ to mean that D is bounded and Γ is either unbounded or bounded and greater than D. Note that when $D < \Gamma$ the service must be able to detect the failure of a primary and disseminate the new primary's identity to the clients without using any messages from clients.

With both send-omission failures and crash+link failures, messages may fail to reach their destinations. The following theorem shows that crash+link failures are more expensive to tolerate, as they require more replication.

Theorem 2 *Suppose there is at most one link between any two servers. Then any primary–backup protocol tolerating f crash+link failures and having $D < \Gamma$ requires $n \geq f + 2$.*

Proof: For contradiction, assume the existence of a protocol P with $n < f + 2$. We will show that P has three runs σ_a, σ_b and σ_{ab} that satisfy the conditions of Lemma 4.1. From the lemma, at least one of these runs violates Pb2, which implies that P cannot be a primary–backup protocol.

Let A be a set containing the one server s_a and let B be the set of remaining servers. Since $|A| = 1$ and $|B| = n - 1 \leq f$, A and B can become disconnected by link failures.

We first construct the run σ_{ab} in which no server crashes, postulating that the links between the servers in A and B are faulty and do not deliver any messages. As required by Lemma 4.1, clients send a total of $R > 2k\lceil \Delta/d \rceil$ requests. Let $0 < d \leq \Gamma - D$ be the minimum interval between any two such requests. We postulate that a request will be sent at time t iff no request has been sent during the interval $[t - d..t)$ and one of the following rules hold.

1. A server s is the primary during the interval $[t - D..t]$. This request arrives immediately and is enqueued (at s, by Pb3 and the definition of D).

2. There is no primary at time t. This request arrives immediately and by Pb3 will never be enqueued at any server.

3. A server s is the primary at time t but another server s' is the primary immediately after time t. If this request is sent to s, then it arrives after t, and if it is sent to any other server, then it arrives immediately. In both cases, it arrives at a server that is not the primary, and so will not be enqueued (again by Pb3).

Note that, by construction, the maximum interval between any two client requests is $D + d$. This interval occurs when a server s becomes the primary just before d after a client message is sent, and s remains the primary for at least D. Hence, the client will be able to send R requests within time $R(D + d)$. This completes the construction of σ_{ab}.

We now construct σ_a and σ_b, recalling that in σ_a all of the servers except s_a crash at time 0, and in σ_b server s_a crashes at time 0. The clients send the same requests and at the same times in σ_a and in σ_b as in σ_{ab}. Furthermore, by construction these requests will arrive at the servers according to the same rules used in constructing σ_{ab}. Of course, a client request may not be delivered to the same servers in σ_a or σ_b as in σ_{ab}, since different servers are operational in these runs.

Since s_a does not receive any messages from servers in B in either σ_{ab} or σ_a, these two runs are indistinguishable to s_a as long as it receives the same client requests at the same times in both runs. We show that this is the case by contradiction: let t be the earliest time that s_a can distinguish between these two runs.

Thus, at time t either s_a received a request m in σ_{ab} but not in σ_a or it received a request m in σ_a but not in σ_{ab}. We will assume the former; the proof for the latter is similar. The request m must have been enqueued at some time $t' \leq t$ at s_a in σ_{ab}. Since m was received by s_a, m must have been sent by rule 1. By rule 1, s_a must have been the primary through $[t' - D..t']$ in σ_{ab} and therefore, by indistinguishability, in σ_a as well. By the definition of D, m would have been enqueued at s_a at time t' in σ_a as well.

Since s_a cannot distinguish between the runs before t, s_a cannot receive m before t in σ_a, and s_a must execute a receive in both σ_a and σ_{ab} at time t. So, it must be the case that s_a receives another request $m' \neq m$ at time t in σ_a. Assume that m' was enqueued at time t''. By an indistinguishability argument similar to above, m' must be enqueued at time t'' at s_a in σ_{ab} as well. Therefore, if s received m' in σ_a at time t, it must receive m' in σ_{ab} as well, a contradiction.

A similar argument can be used to show that the servers in B receive the same requests in σ_b and σ_{ab}, and so these two runs are indistinguishable to the servers in B. Thus, by Lemma 4.1 P cannot be a primary–backup protocol.   □

The assumption in this theorem that $D < \Gamma$ is significant. As we discuss in Section 5, when $D \geq \Gamma$ protocols that tolerate f crash+link failures can be constructed that use only $f + 1$ servers.

The next theorem states that additional replication is required in order to tolerate receive-omission failures. The proof is similar to that of Theorem 2, and so it is omitted.

Theorem 3 *Any primary–backup protocol tolerating f receive-omission failures and having $D < \Gamma$ requires $n > \lfloor \frac{3f}{2} \rfloor$.*

The next lower bound holds independent of the relation between D and Γ.

Theorem 4 *Any primary–backup protocol tolerating f general-omission failures requires $n > 2f$.*

Proof: Assume for contradiction that there is a protocol for $n \leq 2f$. Partition the servers into two disjoint sets A and B of size at most f each. We will construct two runs σ_1 and σ_2. In each run, one set of servers will be faulty and the other set will be non-faulty.

σ_1: The servers in A are faulty and fail to communicate with all servers in B, but behave correctly otherwise. Clients send update requests until the first response is sent (this must happen, by Pb4). Assume that the first response r to an update request m is sent at time t. Say that this response is sent by server s.

σ_2: The same as σ_1 up to time t, but if s is in B, then in σ_2 it is the servers in B that are faulty and fail to communicate with all servers in A rather than the servers in A that are faulty. In either case, r is sent by a faulty server. Furthermore, no server can distinguish σ_1 from σ_2 through time t and therefore, the first response r is sent at time t in σ_2 as well.

Let all of the faulty servers in σ_2 crash immediately after r is sent and have clients continue to send requests until another response r' is sent. This response must have been sent by a non-faulty server which implies that $\neg(e(m) \prec e(r'))$. However this violates the fact that m is an update request.   □

4.2   Bounds on blocking

Informally, a *blocking* primary-backup protocol is one in which the primary must, after receiving a request m, either receive a message from another server or simply wait an interval before it can respond to m. Consider a failure-free run of a primary-backup protocol that is handling a request. Let the time that the request is received be t_m and the time that the response is sent be t_r. We say that this protocol is C–*blocking* if it is guaranteed that $t_r - t_m \leq C$ holds. For example, any primary-backup protocol in which the primary sends information about a request to the backups and waits for acknowledgement before sending the response to the client will be at least 2δ–blocking.

As shown in Section 5, 0–blocking primary-backup protocols can be built for crash and crash+link failure models. For servers that take no time to compute the response to a request, the simple protocol tolerating crash failures presented in Section 2 is 0–blocking. We call such protocols *nonblocking* because the primary can send a reply to the client as soon as the reply has been computed. Nonblocking protocols tolerating receive-omission failures also exist as long as $n > 2f$, but there is can be no nonblocking primary–backup protocol tolerating send-omission or general–omission failures.

Theorem 5 *Any primary–backup protocol tolerating f receive-omission failures with $f > 1$, $n \le 2f$ and $D < \Gamma$ is C–blocking for some $C \ge 2\delta$.*

Proof: For contradiction, suppose there is a primary–backup protocol for $n \le 2f$ and $f > 1$ that is C–blocking where $C < 2\delta$. Partition the servers into two sets A and B where $|A| = f$ and $|B| = n - f \le f$. We construct three runs. In all three runs, assume that all server messages take δ to arrive.

σ_1: There are no failures and all client requests take δ to arrive. Moreover, clients send update requests until some update request m evokes a response r. Let m be received at time t_m by server $p \in A$ and r be sent at time t_r by a server $q \in A$ (q could be the same as p). Notice that since the protocol is C–blocking where $C < 2\delta$, $t_r - t_m < 2\delta$. Also since, by construction, all requests take δ to arrive, all client requests sent after time $t_m + \delta$ will be received after time t_r.

σ_2: Identical to σ_1 until p receives m at time t_m. At this point in σ_2, all servers in A are assumed to crash and clients are assumed to send no request during the interval $[t_m + \delta..t_r]$. Finally, after time t_r clients are assumed to repeatedly send requests at intervals of at least d where $0 < d \le \Gamma - D$ as follows. A request is sent at time t if no request has been sent in $[t - d..t)$ and one of the following rules hold.

1. A server $s \in B$ is the primary during the interval $[t - D..t]$. This request arrives immediately and is enqueued (at s, by Pb3 and the definition of D).

2. There is no primary in B at time t. This request arrives immediately by Pb3 will never be enqueued at any server.

3. A server $s \in B$ is the primary at time t but another server $s' \in B$ is the primary immediately after time t. If the request is sent to s, then it arrives after t, and if it is sent to any other server it arrives immediately. In both

cases, it arrives at a server that is not the primary, and so will not be enqueued (again, by Pb3).

Notice that eventually, there will be a response (say r') in σ_2 because the protocol satisfies Pb4, and by construction it must be from a request sent by rule 1.

σ_3: The same as σ_2, except that the servers in A *do not* crash at time t_m. Instead, the servers in B commit receive failures on all messages sent after t_m by servers in A. Clients send requests at the same times as in σ_2 which arrive using the same rules as σ_2.

Now, consider these three runs. By construction, the runs are identical up to time t_m. Since all server messages take δ to arrive, clients cannot distinguish σ_1 and σ_3 through $t_m + \delta$, and so clients send the same requests to the same servers in both σ_1 and σ_3. Similarly, since all server messages take δ to arrive, the servers in B cannot distinguish between σ_1 and σ_3 through $t_m + \delta$. Therefore, since $t_r - t_m < 2\delta$, p (the server that received request m at time t_m in σ_1) and q (the server that sent response r at time t_r in σ_1) cannot distinguish between σ_1 and σ_3 through time t_r, and so q sends response r in σ_3 as well. Then, using an argument similar to the one in Theorem 2, servers in B cannot distinguish σ_2 and σ_3, and so response r' also occurs in σ_3. However, $\neg(e(m) \prec e(r'))$ which violates the assumption that m is an update request. $\qquad\square$

In run σ_3 of the above proof, a correct primary (p in set A) becomes the backup, while a faulty server from set B becomes the primary in p's place. It is always possible to construct such a run. This is a disconcerting property: there does not exist a primary–backup protocol that tolerates receive-omission failures with $n \leq 2f$ in which a primary cedes only when it fails. Moreover, this lower bound is tight—in [6], we give a receive-omission primary–backup protocol with $n = 2f + 1$ in which a primary cedes only when it fails.

And, if $f = 1$, then the following theorem holds: its proof is similar to the proof of Theorem 5 (and is therefore omitted), except that $p = q$.

Theorem 6 *Any primary–backup protocol tolerating receive-omission failures with $f = 1$ and $n \leq 2f$ and having $D < \Gamma$ is C–blocking for some $C \geq \delta$.*

Primary–backup protocols tolerating send-omission or general–omission failures exhibit the same blocking properties as those tolerating receive-omission failures, except that the restriction $D < \Gamma$ is no longer necessary. Here we prove just the results for send–omission failures. The results for general–omission failures then follow.

Theorem 7 *Any primary–backup protocol tolerating f send-omission failures with $f > 1$ is C–blocking for some $C \geq 2\delta$.*

Proof: For contradiction, suppose there is a primary–backup protocol that is C–blocking where $C < 2\delta$. We consider the following two runs in which all server messages take δ to arrive.

σ_1: There are no failures and all client requests take δ to arrive. Moreover, clients send update requests until some update request m evokes a response r. Let m be received at time t_m by server p and r be sent at time t_r by a server q (again q could be p). Notice that since the protocol is C–blocking where $C < 2\delta$, $t_r - t_m < 2\delta$. Also, since by construction all requests take δ to arrive, all client requests sent after time $t_m + \delta$ will be received after time t_r.

σ_2: Identical to σ_1 through t_m. After t_m, p and q fail and omit to send all messages to all servers except each other. Since, by construction, all messages take δ to arrive, servers and clients cannot distinguish between σ_1 and σ_2 through $t_m + \delta$ and, as a result, p and q cannot distinguish the two runs through $t_m + 2\delta$. Therefore, since $t_r - t_m < 2\delta$, q sends the response r at time t_r in σ_2 as well. Now let p and q crash at time t_r and the clients send requests after time t_r. By Pb4, there eventually must be some request m' that results in a response r'. However, $\neg(e(m) \prec e(r'))$, which violates the assumption that m us an update request. □

Again, if $f = 1$, then the following theorem can be proved using a proof similar to Theorem 7, except that $p = q$.

Theorem 8 *Any primary–backup protocol tolerating send-omission failures with $f = 1$ is C–blocking for some $C \geq \delta$.*

4.3 Bounds on failover times

Recall that the failover time is the longest interval during which *Prmy_s* is not true for any server s. In this section, we give lower bounds for failover times.

In order to discuss these bounds, we postulate a fifth property of primary–backup protocols.

Pb5: A correct server that is the primary remains so until there is a failure of *some* server or link.

This is a reasonable expectation and it is valid for all protocols that we have found in the literature.

Theorem 9 *Any primary–backup protocol tolerating f crash failures must have a failover time of at least $f\delta$.*

Proof: The proof is by induction on f.

Base case $f = 0$: trivially true since a failover time cannot be smaller than zero.

Induction case $f > 0$: Suppose the theorem holds for at most $f - 1$ failures, but (for a proof by contradiction) there is a protocol P for which the theorem is false when there are f failures. From the induction hypothesis, there is a run σ with at most $f - 1$ failures and an interval $[t_0..t_1]$ at least $(f - 1)\delta$ during which there is no primary. Let p_1 be the server that becomes the primary at t_1. Consider the two runs σ_1 and σ_2 that extend σ as follows:

σ_1: Assume p_1 crashes at time t_1. By assumption, there exists a new primary (say p_2) at time $t_2 < t_1 + \delta$. Since p_1 crashes at time t_1, p_2 does not receive any messages from p_1 that were sent after time t_1.

σ_2: Assume that p_1 is correct, there are no other crashes at or after t_1 and all messages sent by p_1 after time t_1 take δ to arrive.

Since p_2 cannot distinguish σ_1 from σ_2 through time t_2, p_2 becomes the primary at time t_2 in σ_2. By Pb5, however, p_1 remains the primary at time t_2 in σ_2. This violates Pb1, and so P is not a primary–backup protocol.   □

Failover times for all other failure models have a larger lower bound:

Theorem 10 *Any primary–backup protocol tolerating f crash+link failures has a failover time of at least $2f\delta$.*

Proof: The proof is by induction on f.

Base case $f = 0$: trivially true.

Induction case $f > 0$: Suppose the theorem holds for at most $f - 1$ failures, but (for a proof by contradiction) there is a protocol P for which the theorem is false when there are f failures.

From the induction hypothesis, there is a run σ with at most $f - 1$ failures and an interval $[t_0..t_1]$ at least $2(f - 1)\delta$ during which there is no primary. Let p_1 be the server that becomes the primary at t_1. Consider the three runs σ_1, σ_2 and, σ_3 that extend σ as follows:

σ_1: Assume that p_1 crashes at time t_1 and all messages sent after t_1 take δ to arrive. Furthermore, the crash of p_1 is the only failure at or after t_1. By assumption, there exists a new primary (say p_2) at time $t_2 < t_1 + 2\delta$. Since p_1 crashes at time t_1, p_2 does not receive any messages from p_1 that were sent after time t_1. Furthermore, since all messages take δ to arrive, any message that was sent after $t_1 + \delta$ can be received by p_2 only after time t_2.

σ_2: Assume that p_1 is correct, there are no other failures at or after t_1, and all messages sent after time t_1 take δ to arrive. Since there are no failures at or after time t_1, by Pb5 p_1 continues to be the primary through time t_2.

σ_3: The same as σ_2 except that the link between p_1 and p_2 is faulty and does not deliver any message sent by p_1 to p_2 after time t_1.

By construction, p_2 cannot distinguish σ_1 from σ_3 through time t_2, and so p_2 becomes the primary at time t_2 in σ_3. Similarly, p_1 cannot distinguish σ_2 from σ_3 through time t_2 and so p_1 remains the primary until time t_2 in σ_3. This violates Pb1, and so P is not a primary–backup protocol.   □

We omit the proofs of the following two theorems because they are similar to Theorem 9.

Theorem 11 *Any primary–backup protocol tolerating f receive-omission failures has a failover time of at least $2f\delta$.*

Theorem 12 *Any primary–backup protocol tolerating f send-omission failures has a failover time of at least $2f\delta$.*

5 Outline of the protocols

In order to establish that the bounds given above are tight, we have developed primary–backup protocols for the different failure models [6]. In this section, we outline these protocols and discuss which of our lower bounds are tight.

Our protocol for crash failures is similar to the protocol given in Section 2. Whenever the primary receives a request from the client, it processes that request and sends information about state updates to the backups before sending a response to the client. All servers periodically send messages to each other in order to detect server failures. This protocol uses $(f + 1)$ servers and so the lower bound in Theorem 1 is tight. Furthermore, it is nonblocking and so incurs no additional delay. It has the failover time $f\delta + \tau$ for arbitrarily small and positive τ, and so the lower bound in Theorem 9 is tight.

In order for the protocol to tolerate crash+link failures, we add an additional server. By Theorem 2, this server is necessary. The additional server ensures that there is always at least one non-faulty path between any two correct servers, where a path contains zero or more intermediate servers. The protocol for crash failures outlined above is now modified so that a primary ensures any message sent to a backup is sent across at least one non-faulty path. This protocol uses $(f + 2)$ servers and so Theorem 2 is tight. Furthermore, it is nonblocking and so incurs no additional delay. It has the failover time $2f\delta + \tau$ for arbitrarily small and positive τ, and so Theorem 10 is tight.

Most of our protocols for the various kinds of omission failures apply translation techniques [17] to the protocol for crash failures outlined above. These techniques ensure that a faulty server detects its own failure and halts, thereby translating a more severe failure to a crash failure. The translations of [17] assume a round-based protocol. Since our crash failure protocol is not round-based, we must modify the translations so that a server can send and receive messages at any time

failure model	degree of replication	amount of blocking	failover time
crash	$n > f$	0	$f\delta$
crash+link	$n > f + 1$ [†]	0	$2f\delta$
receive omission	$n > \lfloor \frac{3f}{2} \rfloor$ [* †]	δ if $n \leq 2f$ and $f = 1$ [†] 2δ if $n \leq 2f$ and $f > 1$ [* †] 0 if $n > 2f$	$2f\delta$
send omission	$n > f$	δ if $f = 1$ 2δ if $f > 1$	$2f\delta$
general omission	$n > 2f$	δ if $f = 1$ 2δ if $f > 1$	$2f\delta$

[*] Bound not known to be tight.

[†] $D < \Gamma$.

Table 1: Lower bounds.

rather than just at the beginning or the end of a round. All of these resulting omission–failure protocols have failover time $2f\delta + \tau$, and thus Theorems 11 and 12 are tight. The protocol for send-omission failures uses $f + 1$ servers and is $2\delta + \tau$–blocking. Furthermore, we also have a send-omission protocol for $f = 1$ that is δ–blocking. Thus, Theorems 7, 8 and 12 are tight. The protocol for general-omission failures also uses $2f + 1$ servers and is $2\delta + \tau$–blocking, and so Theorem 4 is tight, and Theorems 7 and 12 are tight for general-omission failures as well.

We have not been able to determine whether Theorems 3 and 5 are tight. Our protocol for receive-omission failures uses $2f + 1$ servers whereas the lower bound in Theorem 3 only requires $n > \lfloor \frac{3f}{2} \rfloor$. We have constructed protocols for $n = 2$, $f = 1$ and $n = 4$, $f = 2$ but are unable to generalize these protocols. We can also show that any protocol for $n \leq 2f$ has the following odd property: there is at least one run of the protocol in which a non-faulty primary is forced to relinquish control to a backup that is faulty. However, the protocol for $n = 2$, $f = 1$ is δ–blocking and so Theorem 6 is tight.

Table 1 summarizes all of our results.

6   Discussion

We give a precise characterization for primary–backup protocols in a system with synchronized clocks and bounded message delays. We then present lower bounds on the degree of replication, the blocking time, and the failover time under various kinds of server and link failures. We finally outline a set of primary–backup protocols that show which of our lower bounds are tight.

We now briefly compare our results to existing primary–backup protocols. The protocol presented in [3] tolerates one crash+link failure by using only two servers. This appears to contradict Theorem 2 which states that at least three servers are required to tolerate one failure. However, the protocol in [3] assumes that there are two links between the two servers, effectively masking a single link failure. Hence, only crash failures need to be tolerated, and this can be accomplished using only two servers (Theorem 1).

A more ambitious primary–backup protocol is presented in [15]. This protocol works for the following failure model (quoted from [15]):

> The network may lose or duplicate messages, or deliver them late or out of order; in addition it may partition so that some nodes are temporarily unable to send messages to some other nodes. As is usual in distributed systems, we assume the nodes are fail-stop processors and the network delivers only uncorrupted messages.

This failure model is incomparable with those in the hierarchy we presented. However, the protocol does tolerate general-omission failures and has optimal degree of replication for general-omission failures as it uses $2f + 1$ servers.

In Theorem 2, we assumed that $D < \Gamma$. This assumption is crucial: we are able to construct a two-server primary–backup protocol tolerating one crash+link failure for which $D \geq \Gamma$. Recall that link failures are masked by adding redundant paths between the servers. Our two-server crash+link protocol essentially uses the path from the primary to the backup through the client as the redundant path. Thus, there appears to be a tradeoff between the degree of replication and the time it takes for a client to learn that there is a new primary.

The lower bounds on failover times given in Section 4.3 assume Pb5. This is necessary as we have constructed protocols that have failover times smaller than the lower bounds given in Section 4.3 and these protocols do not satisfy Pb5. This smaller failover time is achieved at a cost of an increased variance in service response time.

Finally, in this paper we have attempted to give a characterization of primary–backup that is broad enough to include most synchronous protocols that are considered to be instances of the approach. There are protocols, however, that are incomparable to the class of protocols we analyze [5], [16], [18] since they were developed for an asynchronous setting. Such protocols cannot be cast in terms of implementing a (k, Δ)–bofo server for finite values of k and Δ. We are currently studying possible characterizations for a primary–backup protocol in an asynchronous system and hope to extend our results to this setting.

Acknowledgements

We would like to thank Lorenzo Alvisi, Mike Reiter and the anonymous conference referees for their helpful comments on earlier drafts of this paper.

References

[1] P. A. Alsberg, J. D. Day. A principle for resilient sharing of distributed resources. *Proc. Second International Conference on Software Engineering*, October 1976, pp. 627–644.

[2] Ö. Babaoğlu, R. Drummond. Streets of Byzantium: network architectures for fast reliable broadcasts. *IEEE Transactions on Software Engineering*, 11(6), June 1985, pp. 546–554.

[3] J. F. Barlett. A nonstop kernel. *Proc. Eighth ACM Symposium on Operating System Principles, SIGOPS Operating System Review*, vol. 15, December 1981, pp. 22–29.

[4] A. Bhide, E. N. Elnozahy, S. P. Morgan. A highly available network file server. *USENIX*, 1991, pp. 199–205.

[5] K. P. Birman, T. A. Joseph. Exploiting virtual synchrony in distributed systems. *Eleventh ACM Symposium on Operating System Principles*, November 1987, pp. 123–138.

[6] N. Budhiraja, K. Marzullo, F. Schneider, S. Toueg. Optimal primary–backup protocols. *Proc. Sixth International Workshop on Distributed Algorithms*, Haifa, Israel, November 1992. To appear.

[7] IBM International Technical Support Centers. IBM/VS extended recovery facility (XRF) technical reference. *Technical Report GG24-3153-0*, IBM, 1987.

[8] F. Cristian. Synchronous atomic broadcast for redundant broadcast channels. *Journal of Real-Time Systems*, 2, September 1990, pp. 195–212.

[9] F. Cristian, H. Aghili, H. R. Strong, D. Dolev. Atomic broadcast: from simple message diffusion to Byzantine agreement. *Proc. Fifteenth International Symposium on Fault-Tolerant Computing*, Ann Arbor, Michigan, June 1985. A revised version appears as *IBM Technical Report RJ5244*, pp. 200–206.

[10] V. Hadzilacos. Issues of fault tolerance in concurrent computations. *PhD thesis*, Harvard University, June 1984. *Technical Report 11-84*, Department of Computer Science.

[11] T. Joseph, K. Birman. *Reliable broadcast protocols*. ACM Press, New York, 1989, pp. 294–318.

[12] L. Lamport. Time, clocks, and the ordering of events in a distributed system. *Communications of the ACM*, 21(7), July 1978, pp. 558–565.

[13] L. Lamport, M. Fischer. Byzantine generals and transaction commit protocols. *Op. 62, SRI International*, April 1982.

[14] L. Lamport, P. M. Melliar-Smith. Synchronizing clocks in the presence of faults. *Journal of the ACM*, 32(1), January 1985, 52–78.

[15] B. Liskov, S. Ghemawat, R. Gruber, P. Johnson, M. Williams. Replication in the Harp file system. *Proc. 13th Symposium on Operating System Principles*, 1991, pp. 226–238.

[16] T. Mann, A. Hisgen, G. Swart. An algorithm for data replication. *Technical Report 46*, Digital Systems Research Center, 1989.

[17] G. Neiger, S. Toueg. Automatically increasing the fault-tolerance of distributed systems. *Proc. Seventh ACM Symposium on Principles of Distributed Computing*, ACM SIGOPS-SIGACT, Toronto, Ontario, August 1988, pp. 248–262.

[18] B. Oki, B. Liskov. Viewstamped replication: a new primary copy method to support highly available distributed systems. *Seventh ACM Symposium on Principles of Distributed Computing*, ACM SIGOPS-SIGACT, Toronto, Ontario, August 1988, pp. 8–17.

[19] M. Pease, R. Shostak, L. Lamport. Reaching agreement in the presence of faults. *Journal of the ACM*, 27(2), April 1980, pp. 228–234.

[20] K. J. Perry, S. Toueg. Distributed agreement in the presence of processor and communication faults. *IEEE Transactions on Software Engineering*, 12(3), March 1986, pp. 477–482.

[21] R. D. Schlichting, F. B. Schneider. Fail-stop processors: an approach to designing fault-tolerant computing systems. *ACM Transactions on Computer Systems*, 1(3), August 1983, pp. 222–238.

[22] F. B. Schneider. Implementing fault tolerant services using the state machine approach: a tutorial. *Computing Surveys*, 22(4), December 1990, pp. 299–319.

A LINGUISTIC FRAMEWORK

FOR DYNAMIC COMPOSITION

OF DEPENDABILITY PROTOCOLS

Gul AGHA, Svend FRØLUND, Rajendra PANWAR, Daniel STURMAN
Department of Computer Science, University of Illinois at Urbana-Champaign
1304 W. Springfield Avenue, Urbana, Illinois 61801, USA

Abstract

We present a language framework for describing dependable systems which emphasizes *modularity* and *composition*. Dependability and functionality aspects of an application may be described separately providing a separation of design concerns. Futhermore, the dependability protocols of an application may be constructed bottom-up as simple protocols that are composed into more complex protocols. Composition makes it easier to reason about dependability and supports the construction of general reusable dependability schemes. A significant aspect of our language framework is that dependability protocols may be loaded into a running application and installed dynamically. Dynamic installation makes it possible to impose additional dependability protocols on a server as clients with new dependability demands are integrated into a system. Similarly, if a given dependability protocol is only necessary during some critical phase of execution, it may be installed during that period only.

The research described has been made possible by support provided by a Young Investigator Award from the Office of Naval Research (ONR contract number N00014-90-J-1899) and by an Incentives for Excellence Award from the Digital Equipment Corporation Faculty Program.

1 Introduction

This paper describes a software methodology for supporting dependable *services* in *open systems*. Provision of a dependable service involves the *servers* implementing the service as well as the communication channel to the servers. We make no assumptions about the behavior of the customers, called *clients*, using the service. For our purposes, the most significant characteristic of an open system is *extensibility*: new services and new clients may be integrated into an open system while it is functioning.

In many existing methodologies for programming dependable applications, the dependability characteristics of an application are fixed statically (i.e., at compile time). This is unsatisfactory in many computer systems, which are required to function for a long period of time, yet are fault-prone due to the uncertain environment in which they operate. An example of such a system is the control system embedded in an orbiting satellite. Furthermore, in open systems the addition of new services and clients may impose new requirements for dependability of a service. For example, a file server may start only addressing safety by checkpointing the files to stable storage. In an open system, new clients added to the system may require the server to also provide security, e.g., by encrypting the files they transfer to the clients. Our method includes *dynamic installation* of dependability protocols which allows a system to start with a "minimal" set of dependability protocols and later be extended with more protocols where and when the need arises. As the file server example illustrates, not all dependability protocols that may be needed at runtime can necessarily be predicted at compile time.

Our methodology incorporates object-oriented programming methods and as a result offers the following advantages:

- *Separation of design concerns*: an application programmer need not be concerned with the particular dependability protocols to be used when developing an application.

- *Reusability*: code for implementing dependability protocols and application programs can be stored in separate libraries and reused.

We employ *reflection* as the enabling technology for dynamic installation of dependability protocols. Reflection means that a system can reason about, and manipulate, a representation of its own behavior. This representation is called the system's *meta-level*. In our case, the meta-level contains a description of the dependability characteristics of an executing application; reflection thus allows dynamic changes in the execution of an application with respect to dependability. Reflection in an object based system allows customization of the underlying system independently for every object as compared to customization in a micro kernel based system [1] where changes made to the micro kernel affect all the objects collectively. This flexibility is required for implementing dependability protocols since such protocols are mostly installed on very specific subsets of the objects in a system.

The code providing dependability is specified independently from the code which specifies the application specific functionality of a system. As we show later, our reflective model allows *compositionality* of dependability protocols. Compositionality means that we can specify and reason about a complex dependability scheme in terms of its constituents. Thus, logically distinct aspects of a dependability scheme may be described separately. Compositionality supports a methodology in which dependability protocols are constructed in terms of general, reusable, components. Dynamic composition is particularly useful; it allows software for additional dependability protocols to be constructed and installed without knowledge of previously installed protocols. It may not be possible to describe a protocol in general terms. In such cases, composition may not be possible. For example, the composition of the two-phase commit protocol with security mechanisms may not be done naively [11].

A number of languages and systems offer support for constructing fault tolerant systems. In Argus [12], Avalon [8] and Arjuna [16], the concept of nested transactions is used to structure distributed systems. Consistency and resilience is ensured by atomic actions whose effect are checkpointed at commit time. The focus in [14], [7] and [6] is to provide a set of protocols that represent common communication patterns found in fault tolerant systems. None of the above systems support the factorization of fault tolerance characteristics from the application specific code. In [21] and [15], replication can be described separate from the service being replicated. Our approach is more flexible since fault tolerance schemes are not only described separately but they can also be attached and detached dynamically. Another unique aspect of our approach is that different fault tolerance schemes may be composed in a modular fashion. For example, checkpointing may be composed with replication without having either method know about the other.

Reflection has been used to address a number of issues in concurrent systems. For example, the scheduling problem of the Time Warp algorithm for parallel discrete event simulation is modeled by means of reflection in [24]. A reflective implementation of object migration is reported in [22]. Reflection has been used in the Muse Operating System [23] for dynamically modifying the system behaviour. Reflective frameworks for the Actor languages MERING IV and Rosette have been proposed in [9] and [19], respectively. In MERING IV, programs may access *meta-instances* to modify an object or *meta-classes* to change a class definition. In Rosette, the meta-level is described in terms of three components: a *container*, which represents the acquaintances and script; a *processor*, which acts as the scheduler for the actor; and a *mailbox*, which handles message reception.

The paper is organized as follows. Section 2 gives a more detailed description of the concept of reflection. Section 3 introduces the Actor model which provides a linguistic framework for describing our methodology. Note that our methodology is not dependent on any specific language framework; we simply use an Actor language as a convenient tool to describe our examples. Section 4 describes our *meta-level architecture for ultradependability* (MAUD). Section 5 gives an example of a replicated service implemented using MAUD. Composition of methods for dependability is addressed in Section 6 and illustrated in Section 7 by means of an example composing replication with a two-phase commit protocol. The final section summarizes our conclusions and research directions.

2   Reflection

Reflection means that a system can manipulate a causally connected description of itself [18], [13]. Causal connection implies that changes to the description have an immediate effect on the described object. The causally connected description is called a *meta-level*. In a reflective system, implementation of objects may be customized within the programming language. The customization can take place on a per *object* basis in the form of *meta-objects*. In this paper, we use the term *object* as a generic term for clients and servers (when it is not necessary to distinguish between them). The object for which the meta-object represents certain aspects of the implementation is called the *base object*. A meta-object may be thought of as an object that logically belongs in the underlying runtime system.

A meta-object might control the message lookup scheme that would map incoming messages to operations in the base object. Using reflection, such implementation level objects can be accessed and examined, and user defined meta-objects may be installed, yielding a potentially customizable runtime system within a single language framework.

The reflective capabilities which are provided by a language are referred to as the *meta-level architecture* of the language. The meta-level architecture may provide variable levels of sophistication, depending on the desirable level of customization. The most general meta-level architecture is comprised of complete interpreters, thus allowing customization of all aspects of the implementation of objects. In practice, this generality is not always needed and, furthermore, defining a more restrictive meta-level architecture may allow reflection to be realized in a compiled language. The choice of a meta-level architecture is part of the language design. Customizability of a language implementation must be anticipated when designing the runtime structure. Although a restrictive meta-level architecture limits flexibility, it provides greater safety and structure. If all aspects of the implementation were mutable, an entirely new semantics for the language could be defined at runtime; in this case, reasoning about the behavior of a program would be impossible.

We limit our meta-level to only contain the aspects that are relevant to dependability. Application specific functionality is described in the form of base objects and dependability protocols are described in terms of meta-objects. Thus, dependability is modeled as a special way of implementing the application in question. Our methodology gives modularity since functionality and dependability are described in separate objects. Since meta-objects can be defined and installed dynamically, a system can dynamically switch between different dependability modes of execution. Dependability protocols could be installed and removed dynamically, depending on system need. Furthermore, new dependability protocols may be defined while a system is running and put into effect without stopping and recompiling the system. For example, if a communication line within a system shows potential for unacceptable error rates, more dependable communication protocols may be installed without stopping and recompiling the entire system.

Since meta-objects are themselves objects, they can also have meta-objects associated with them, giving customizable implementation of meta-objects. In this way, meta-objects realizing a given dependability protocol may again be subject to another dependability protocol. This scenario implies a hierarchy of meta-objects where each meta-object contributes a part of the dependability characteristics for the application in question. Each meta-object may be defined separately and composed with other meta-objects in a layered structure supporting reuse and incremental construction of dependability protocols.

Because installation of a malfunctioning meta-level may compromise the dependability of the a system, precautions must be taken to protect against erroneous or malicious meta-objects. To provide the needed protection of the meta-level, we introduce the concept of privileged objects called a *managers*. Only managers may install meta-objects. Using operating system terminology, a manager should be thought of as a privileged process that has *capabilities* to dynamically load new modules (meta-objects) into the kernel (meta-level). It should be observed that, because of the close resemblance to the operating system world, many of the operating system protection strategies can be reused in our design. We will not discuss particular mechanisms for enforcing the protection provided by the managers in further detail here. Because only managers may install meta-objects, special requirements can be enforced by the managers on the structure of objects which may be installed as meta-objects. For example, managers may only allow installation of meta-objects instantiated from special verified and trusted libraries. Greater or fewer restrictions may be imposed on the meta-level depending on the dependability and security requirements that a given system must meet.

3   The Actor model

We illustrate our approach using the *Actor model* [2], [3]. It is important to note that the idea of using reflection to describe dependability is not tied to any specific language framework. Our methodology assumes only that resources can be created dynamically, if needed to implement a particular protocol and that the communication topology of a system is reconfigurable. Our methodology does not depend on any specific communication model. In particular, it is immaterial to our approach whether synchronous or asynchronous communication is used.

Actors can be thought of as an abstract representation for multicomputer architectures. An actor is an encapsulated entity that has a local state. The state of an actor can only be manipulated through a set of *operations*. Actors communicate by asynchronous point to point message passing. A message is a request for invocation of an operation in the target actor. Messages sent to an actor are buffered in a *mail queue* until the actor is ready to process the message. Each actor has a system-wide unique identifier which is called a *mail-address*. This mail-address allows an actor to be referenced in a location transparent way. The *behavior* of an actor is the actions performed in response to a message. An actor's *acquaintances* are the mail addresses of known actors. An actor can only send messages to its acquaintances, which provides locality. In order to abstract over processor speeds and allow adaptive routing, preservation of message order is not guaranteed.

The language used for examples in this paper is HAL [10], an evolving high-level actor language which runs on a number of distributed execution platforms. HAL provides two other message passing constructs besides the asynchronous send. The first, SSEND, is a message order preserving send, or *sequenced send*. SSEND allows the sender to impose an arrival order on a series of messages sent to the same target. The second, BSEND, is a remote procedure call mechanism, or *blocking send*. The calling program implicitly blocks and waits for a value to be returned from the actor whose method was invoked.

4   A meta-level architecture

In this section we introduce MAUD (Meta-level Architecture for Ultra Dependability) . As previously mentioned, MAUD is designed to support the structures that are necessary to implement dependability. In MAUD, there are three meta-objects for each actor: *dispatcher, mail queue* and *acquaintances*. In the next three paragraphs we describe the structure of meta-objects in MAUD. Note that MAUD is a particular system developed for use with actors. It would be possible, however, to develop similar systems for most other models.

The dispatcher and mail queue meta-objects customize the communication primitives of the runtime system so that the interaction between objects can be adjusted for a variety of dependability characteristics. For instance, a dispatcher meta-object is a representation of the implementation of the (SEND...) action. Whenever the base object issues a message send, the runtime system calls the transmit method on the installed dispatcher. The dispatcher performs whatever actions are needed to send the given message. Installing new send behaviors makes it possible to implement customized message delivery patterns.

A mail queue meta-object represents the mail queue holding the incoming messages sent to an actor. A mail queue is an object with `get` and `put` operations. After installation of a mail queue meta-object, its `get` operation is called by the runtime system whenever base object is ready to process a message. The `put` operation on a mail queue is called by the runtime system whenever a message for the base object arrives. By installing a mail queue at the meta-level, it is possible to customize the way messages flow into the base object.

The acquaintances meta-object is a list representing the acquaintances of a base object. Information about an actor's acquaintances is necessary in order to check-point its state. For our current purposes, we assume that the acquaintances's meta-object is immutable. Otherwise, if customized acquaintance lists could be installed, static checking of legal names would be impossible.

Meta-objects are installed and examined by means of *meta-operations*. Meta-operations are defined in the class called `Object` which is the root of the inheritance hierarchy. All classes in the system inherits from `Object`, implying that meta-operations can be called on each actor in the system. The meta-operations `change-mailQueue` and `change-dispatcher` install mail queues and dispatchers for the object on which they are called. Similarly, the meta-operations `get-mailQueue`, `get-dispatcher` and `get-acquaintances` return the meta-objects of a given actor. If no meta-objects have been previously installed, an object representing the built-in, default, implementation is returned. Such default meta-objects are created in a lazy fashion when a meta-operation is actually called.

5   A replicated server

In a distributed system, an important service may be replicated to maintain availability despite processor faults. The type of faults that can be experienced are generally considered to be of two basic types: Byzantine failures or fail-stop failures [17]. In this section, we will give an example of how MAUD can be used in an actor domain to implement a modular and application-independent protocol which uses *replication* to protect against fail-stop failures.

Dependability of a server will be increased by creating several exact copies of the server on different nodes. We refer to the copies as *replicas*. The task of supporting the replicas is divided between two actors: a distributor, which broadcasts messages to the replicas, and a collector, which takes the replicas' responses and

```
(DEFINE-ACTOR serveMailq
  (SLOTS data members)
  (METHOD (get who) )
  (METHOD (put msg)
      ;; A bcaster actor broadcasts msg to members
      (BSEND bcast (NEW bcaster msg) members)))
(DEFINE-ACTOR bcaster
  (SLOTS msg)
  (METHOD (bcast l)
     (IF (not (null? l))
        ((SSEND (msg-type msg) (car l) (msg-data msg))
         (SEND bcast self (cdr l)))))))
```

Figure 1: Code for the mail queue which implements replication.

extracts one response to send to the client. If either the distributor or the collector crashes, the replicas will be unable to communicate with the rest of the world. Therefore, the applicability of the given solution is limited due to the need for more dependable processors for running the distributor and collector. However, because these two objects are computationally much less expensive than the application itself, the solution allows us to increase the dependability of a system given a few highly-dependable processors. It is possible to use MAUD for implementing replication schemes which do not require a centralized distributor/collector; however we restrict our discussion to the centralized distributor/collector scheme.

We assume that managers themselves install appropriate meta-objects realizing a given dependability protocol. Therefore, we specify the relevant dependability protocols by describing the behavior of the initiating manager as well as the installed mail queues and dispatchers. A manager in charge of replicating a service takes the following actions to achieve the state shown in Figure 2:

1. The specified server is replicated by a manager by creating actors with the same behavior and acquaintance list (using HAL's `clone` function).

2. A mail queue is installed for the original server to make it act as the *distributor* described above. Messages destined for the original server are broadcast to the replicas. A broadcast using SSENDs is done so that all replicas receive messages in the same order and thus solve the same task.

3. The dispatcher of the original server is modified to act as the *collector* described above. The first message out of each set of replica responses is

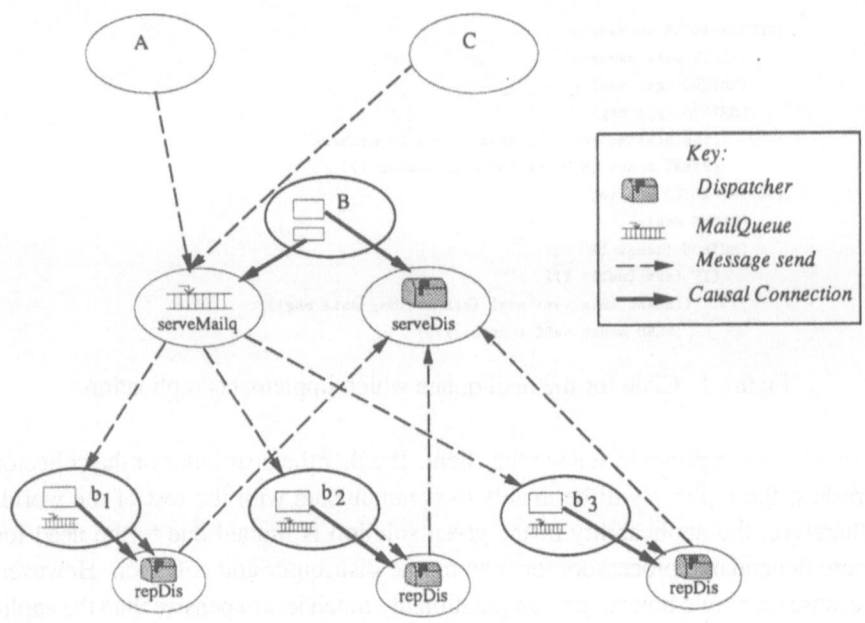

Figure 2: When a message is sent by the clients A or C to the replicated service B, the message is received by B's mail queue serveMailq (1). The message is then sent to each of the replicas (2).

selected to be passed to the destination. Since we assume only fail-stop failures, no complex voting scheme is necessary.

4. The dispatchers of the replicas are changed to forward all messages being sent to the original server's dispatcher. In addition, the messages are tagged so that the original server's dispatcher can eliminate multiple copies of the same message.

The dispatchers and mail queues are designed according to the specification in Section 4. The new mail queue for the original server is described in Figure 1. Note that message order is being preserved in the broadcast. We use HAL's SSEND function to guarantee consistent state at each replica. Figure 2 shows the resulting actions occurring when a message is sent to the replicated service. The original server is actor B. When a message is received by the distributor, $serveMailq$ (B's new mail queue), the message is broadcast to the replicas

```
(METHOD (add-mailq aMailq)
   (IF (null? mailq)
      (SEND change-mailq self aMailq)
      (SEND add-mailq mailq aMailQ)))
(METHOD (add-dispatcher aDispatcher)
   (IF (null? dispatcher)
      (SEND change-dispatcher self aDispatcher)
      (SEND add-dispatcher dispatcher aDispatcher)))
```

Figure 3: The additional methods which must be inherited to allow for protocol composition.

b_1, b_2, b_3. Each of the replicated actors has the same base-level behavior as B. Therefore, upon receipt of the message, each b_i responds in the same way B would have. However, if the replicas respond to the message, the message destinations would be rerouted by the dispatchers $repDis$ to the original server's dispatcher, $serveDis$ (serving as the collector). For each response, $serveDis$ gets three messages, one from each replica. It processes the three messages using some voting scheme and sends out a single response to the original destination. Note that the base-level actor B does not receive any messages now since all the incoming messages are redirected to the replicas by its mail queue $serveMailq$ and the outgoing messages are sent by the dispatchers of the replicas directly to its dispatcher $serveDis$.

Although this example is fairly simple, it does illustrate some of the benefits of our approach. The manager initiating the replication protocol needs no advance knowledge of the service to be replicated nor does the replicated service need to know that it is being replicated. Additionally, the clients using the replicated service are not modified in any way. These benefits give us the flexibility to dynamically replicate and unreplicate services while the system is running.

6   Composition of dependability characteristics

In some cases, dependability can only be guaranteed by using several different protocols. For example, a system employing replication to avoid possible processor faults may also need to guarantee consensus on multi-party transactions through the use of three-phase commit or some similar mechanism. Unfortunately, writing one protocol which has the functionality of multiple protocols can lead to very complex code. In addition, the number of possible permutations of protocols grows exponentially – making it necessary to predict all possibly needed combinations in a system. However, because the meta-components of an object are themselves objects in a reflective system, there is a general solution for composing two protocols using MAUD. A simple change to the meta-operations inherited from the `Object` class, along with a few restrictions on the construction of mail queues and dispatchers, allows us to layer protocols in a general way. Figure 3 shows how an *add-mailq* method could be expressed in terms of the other meta-operations to allow layering.

Because the current mail queue, `mailq`, and the current dispatcher, `dispatcher`, are objects, we can install meta-objects to customize their mail queue or dispatcher. By adding protocols in the above manner, the new mail queue functionality will be performed on incoming messages before they are passed on to the "old" mail queues. For the send behaviors, the process is reversed with the oldest send behavior being performed first and the newest behavior last, thereby creating an onion-like model with the newest layer closest to the outside world.

To preserve the model, however, several restrictions must be applied to the behavior of dispatchers and mail queues. We define the *partner* of a mail queue as being the dispatcher which handles the output of a protocol and the partner of a dispatcher as being the mail queue which receives input for the protocol. In Figure 4, B and C are partners as well as E and D. Each pair implements *one* protocol. It is possible for a meta-object to have a null partner.

The *owner application* of a meta-object is inductively defined as either its base object, if its base object is not a meta-object, or the owner application of its base object. For example, in figure 4, A is the owner application of meta-objects B, C, D, and E. With the above definition we can restrict the communication behavior of the actors so that:

- A mail queue or dispatcher may send or receive messages from its partner or an object created by itself or its partner.

- Dispatchers may send messages to the outside world, i.e. to an object which is not a mail queue or dispatcher of the owner application (although the message might be sent through the dispatcher's dispatcher). Dispatchers may receive `transmit` messages from their base object.

- Mail queues may receive messages from the outside world (through its own

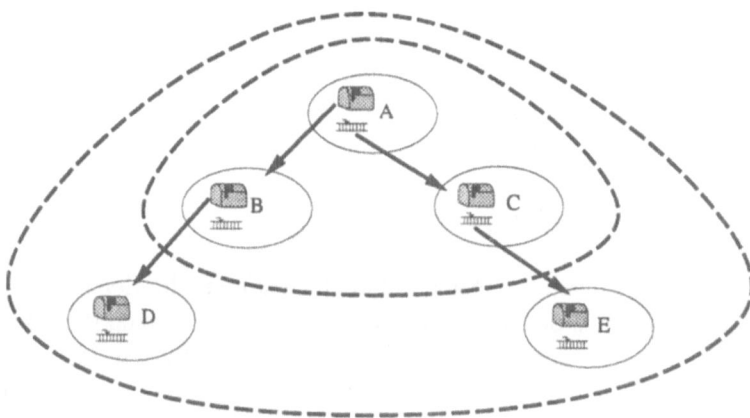

Figure 4: Partners and Owner relationships. A is the owner of all other actors in the figure. Dispatcher B and mail queue C are partners as well as dispatcher D and mail queue E.

mail queue) and send messages when responding to `get` messages from their base object.

- Objects created by a mail queue or dispatcher may communicate with each other, their creator, or their creator's partner.

Because of the above restrictions, regardless of the number of protocols added to an object there is exactly one path which incoming messages follow, starting with the newest mail queue, and exactly one path for outgoing messages in each object, ending with the newest dispatcher. Therefore, when a new dispatcher is added to an object, all outgoing messages from the object must pass through the new dispatcher. When a new mail queue is installed it will handle all incoming messages before passing them down to the next layer. Thus, a model of objects resembling the layers of an onion is created; each addition of a protocol adds a new layer in the same way regardless of how many layers currently exist. With the above rules, protocols can be composed without any previous knowledge that the composition was going to occur and protocols can now be added and removed as needed without regard not just to the actor itself, but also without regard to existing protocols. In figure 4, actors B and C are initially installed as one "layer". Messages come into the layer only through C and leave through B. Therefore,

D and E may be installed with the `add-mailq` and `add-dispatcher` messages as if they were being added to a single actor. Now messages coming into the composite object through E and are then received by C. Messages sent are first processed by B and then by D.

7   Composition example

In section 5, we demonstrated one method of implementing replication. We now build on that example to show how different protocols can be composed. Our system currently has three actors: A, B, and C, where B has been replicated and is currently represented by b_1, b_2, and b_3. The initial system configuration is shown in Figure 2.

Assume A initiates a transaction with databases B, and C. These transactions are implemented using a specific commit protocol. The commit protocol is chosen dynamically depending on the kinds of failures (site failures or communication failures [5]) that need to be tolerated by the system. Assume that a two-phase commit protocol is implemented by a mail queue called *tpcMailq* and dispatcher called *tpcDis* installed at A, B and C. Also assume that A acts as the coordinator [5] of the commit protocol. This scenario is depicted in Figure 5.

When A initiates a commit protocol, its *tpcDis* broadcasts a `vote-req` message to B and C, and waits for `vote` messages from them. The *tpcDis* of a participant sends a message to its *tpcMailq* informing it of the voted value. If the participant voted `no`, its *tpcMailq* assumes that the transaction is aborted and allows subsequent messages to proceed. Otherwise, it waits for a `commit` or `abort` message from the coordinator. The votes of B and C are received by the *tpcMailq* of A and forwarded to its *tpcDis*. Based on the `vote` values, the *tpcDis* of A decides to commit or abort the transaction and broadcasts its decision. If the *tpcMailq* of B or C receives a message to commit the transaction, it sends a message to its base actor to install the transaction.

The applied meta-objects are designed to follow the restrictions given in Section 6 to preserve the onion-layer model. Figure 5 shows the resulting system after all three actors have the two-phase commit protocol installed. Because *tpcMailq* does not enqueue a message on data until after two-phase commit is resolved, a message sent to B after the commit protocol starts, is not copied and sent to the replicates until the protocol is finished. There are no differences between the way

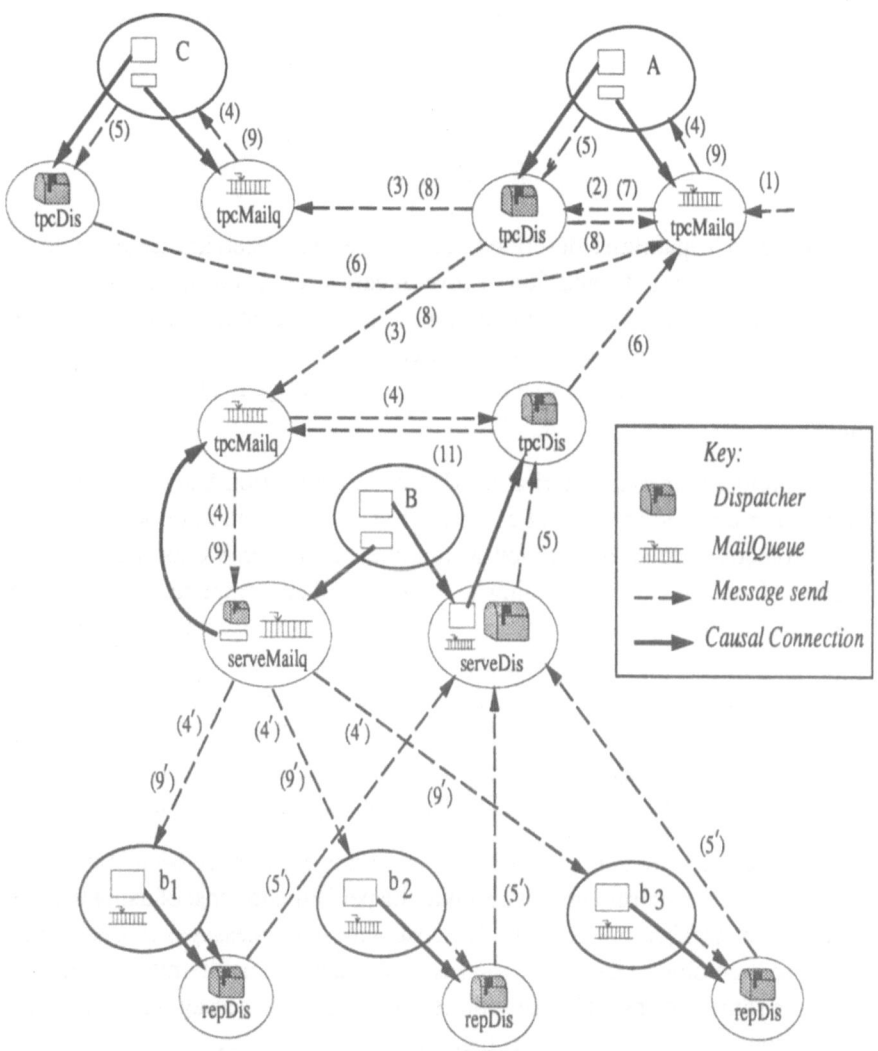

Figure 5: System resulting from the composition of two-phase commit with replication. When a transaction is received (1), *tpcMailq* notifies *tpcDis* (2). The *tpcDis* broadcasts a `vote-req` message (3). The vote is decided (4, 5) by the participants. The voting by the replicated actor *B* involves the messages (4', 5'). The decision of *B* and *C* is sent to the coordinator (6). The coordinator decides whether the transaction should be committed or aborted and sends its decision to the participants (7,8). If the final decision is to commit, the transaction is installed (9).

in which the two-phase commit protocol is installed on C and on the replicated B.

8   Discussion

The reflective capabilities described in this paper have been implemented in HAL [4]. Our work led to the addition of several categories of constructs to HAL. The structure for MAUD was built directly into HAL and the underlying system directs messages to dispatcher meta-objects, when sent, and to mail queue meta-objects when received. Additional functions to allow message manipulation were also added.

In our prototype implementation of MAUD, an application and its dependability protocols are linked together at compile time. It is currently not possible to load new protocols into an application at runtime. However, we are developing an execution environment containing a dynamic linker. Applications can invoke the dynamic linker and thereby be extended with new executable code: thus it is possible for an application to dynamically extend its repertoire of dependability protocols. Using dynamic linking instead of interpretation, dynamically constructed protocols have the same format and exhibit the same level of trustworthiness as statically linked protocols. In particular, it is possible to verify dynamically constructed protocols in the same way as statically linked protocols.

Several examples, including the ones given in this paper, have been implemented using HAL. Our experiments suggest that the performance cost of using the meta-objects is, in itself, not significant. The cost of the extra messages caused by meta-object to base-object communication is a constant factor; the meta-objects may be on the same node as their base-object, allowing these messages to be converted to function calls. The primary source of cost is that generalized protocols may be used instead of customized protocols which can exploit knowledge of a given application. An example of this cost is the inability of a general replication scheme to take advantage of commutable operations [14]. Unfortunately, the cost due to generalizing protocols is difficult to express quantitatively since it depends on the application and the protocol: in some cases, the cost is quite high and, in others, insignificant. However, even if a programmer wants to exploit knowledge of an application, the customized protocols can be handled as any other protocol. Our methodology still preserves code modularity, dynamic protocol installation and removal, and composability with other protocols.

The dependability of applications developed using our methodology is highly dependent on the installation of "correct" meta-objects. As we have already explained, installation of erroneous meta-objects may have dramatic consequences for the behavior of an application. An important part of our future work is to come up with ways in which it is possible to reason about the behavior of meta-objects. A formal semantic framework is needed to verify the adherence of MAUD components to a given specification. Preliminary work in this area has been done in [20]. Safety would be addressed by having managers only install meta-objects that conform to some specification. The approach is also flexible and open-ended since many meta-objects may implement a given specification.

Besides protecting the meta-level, a manager may initiate the installation of meta-objects. Because we have the flexibility of dynamic protocol installation, self-monitoring could be constructed using daemons that, without human intervention, initiate protocols to prevent faulty behavior. The above capability will be especially useful in the future as systems grow and become proportionally harder for human managers to monitor. For example, daemons may use sensors to monitor the behavior of a system in order to predict potentially faulty components.

Acknowledgments

The authors would like to thank Chris Callsen, Wooyoung Kim, and Anna Patterson for helpful discussions concerning the manuscript. We would also like to thank Takuo Watanabe for his insights into the use of reflection and Chris Houck for the modifications to the HAL compiler required for our implementations.

References

[1] M. Acceta, R. Baron, W. Bolosky, D. Golub, R. Rashid, A. Tevanian, M. Young. Mach: a new kernel foundation for UNIX development. *Proc. USENIX 1986 Summer Conference*, June 1986.

[2] G. Agha. *Actors: a model of concurrent computation in distributed systems*. MIT Press, 1986.

[3] G. Agha. Concurrent object-oriented programming. *Communications of the ACM*, 33(9), September 1990, pp. 125-141.

[4] G. Agha, W. Kim. Compilation of a highly parallel Actor-based language. *Proc. Workshop on Languages and Compilers for Parallel Computing*, U. Banerjee, D. Gelernter, A. Nicolau, and D. Padua, editors, Yale University, Springer-Verlag, 1992. *Lecture Notes in Computer Science*, to be published.

[5] P. A. Bernstein, V. Hadzilacos, N. Goodman. *Concurrency control and recovery in database systems*. Addison-Wesley, 1987.

[6] K. P. Birman, T. A. Joseph. Communication support for reliable distributed computing. *Fault-tolerant Distributed Computing*, Springer-Verlag, 1987.

[7] E. Cooper. Programming language support for multicast communication in distributed systems. *Tenth International Conference on Distributed Computer Systems*, 1990.

[8] D. L. Detlefs, M. P. Herlihy, J. M. Wing. Inheritance of synchronization and recovery properties in Avalon/C++. *IEEE Computer*, 21(12), December 1988, pp. 57–69.

[9] J. Ferber, J-P Briot. Design of a concurrent language for distributed artificial intelligence. *Proc. International Conference on Fifth Generation Computer Systems*, vol. 2, Institute for New Generation Computer Technology, 1988, pp. 755–762.

[10] C. Houck, G. Agha. HAL: a high-level actor language and its distributed implementation. *Proc. 21st International Conference on Parallel Processing (ICPP '92)*, vol. 2, St. Charles, IL, August 1992, pp. 158–165.

[11] S. Jajodia, C. D. McCollum. Using two-phase commit for crash recovery in federated multilevel secure database management systems. *Proc. 3rd IFIP Working Conference on Dependable Computing for Critical Applications*, Mondello, Sicily, Italy, September 1992, pp. 209–218. Preprint.

[12] B. Liskov, R. Scheifler. Guardians and Actions: linguistic support for robust, distributed programs. *Conference Record of the Ninth Annual ACM Symposium on Principles of Programming Languages*, Albuquerque, New Mexico, January 1982, pp. 7–19.

[13] P. Maes. Computational reflection. *Technical Report 87-2*, Artificial Intelligence Laboratory, Vrije University, 1987.

[14] S. Mishra, L. L. Peterson, R. D. Schlichting. Consul: a communication substrate for fault-tolerant distributed programs. *Technical report*, University of Arizona, Tucson, 1991.

[15] M. H. Olsen, E. Oskiewicz, J. P. Warne. A model for interface groups. *Tenth Symposium on Reliable Distributed Systems*, Pisa, Italy, 1991.

[16] G. D. Parrington, S. K. Shrivastava. Implementing concurrency control in reliable distributed object-oriented systems. *Proc. Second European Conference on Object-Oriented Programming, ECOOP88*, Springer-Verlag, 1988.

[17] F. B. Schneider. The State Machine approach: a tutorial. *Lecture Notes in Computer Science*, 448, 1990, pp. 18–41.

[18] B. C. Smith. Reflection and semantics in a procedural language. *Technical Report 272*, Massachusetts Institute of Technology, Laboratory for Computer Science, 1982.

[19] C. Tomlinson, V. Singh. Inheritance and synchronization with enabled-sets. *Proc. OOP-SLA*, 1989.

[20] N. Venkatasubramanian, C. Talcott. A MetaArchitecture for distributed resource management. *Proc. Hawaii International Conference on System Sciences*, IEEE Computer Society Press, January 1993. To appear.

[21] C. T. Wilkes, R. J. LeBlanc. Distributed Locking: a mechanism for constructing highly available objects. *Seventh Symposium on Reliable Distributed Systems*, Ohio State University, Columbus, Ohio, 1988.

[22] T. Watanabe, A. Yonezawa. An Actor-Based metalevel architecture for group-wide reflection. *Foundations of Object-Oriented Languages*, J. W. deBakker, W. P. deRoever, and G. Rozenberg, editors, Springer-Verlag, pp. 405–425, 1990. LNCS 489.

[23] Y. Yokote, A. Mitsuzawa, N. Fujinami, M. Tokoro. The Muse object architecture: a new operating system structuring concept. *Technical Report SCSL-TR-91-002*, Sony Computer Science Laboratory Inc., February 1991.

[24] A. Yonezawa, editor. *ABCL An Object-Oriented Concurrent System*, chapter Reflection in an Object-Oriented Concurrent Language, MIT Press, Cambridge, Mass., 1990, pp. 45–70.

USING TWO-PHASE COMMIT

FOR CRASH RECOVERY

IN FEDERATED

MULTILEVEL SECURE DATABASE

MANAGEMENT SYSTEMS

Sushil JAJODIA, Catherine McCOLLUM
The MITRE Corporation
7525 Colshire Drive, McLean, Virginia 22102-34811, USA

Abstract

In a federated database management system, a collection of autonomous database management systems (DBMSs) agree to cooperate to make data available for sharing and to process distributed retrieval and update queries. Distributed transactions can access data across multiple DBMSs. Securing such an environment requires a method that coordinates processing of these distributed requests to provide distributed transaction atomicity without security compromise. An open question is how much of its scheduling process an individual DBMS must expose to the federation in order to allow sufficient coordination of distributed transactions. In this paper, we address the application of the two-phase commit protocol, which is emerging as the dominant method of providing transaction atomicity for crash recovery in the conventional (single-level) distributed DBMS area, to the federated multilevel secure (MLS) DBMS environment. We discuss the limits of its applicability and identify the conditions that must be satisfied by the individual DBMSs in order to participate in the federation.

1 Introduction

A federated database management system (DBMS) is one in which a number of
independent database systems cooperate to share data without relinquishing
their autonomy [10]. The set of DBMSs involved may be heterogeneous. In an
environment under multilevel security, where systems may have evolved for
years in isolation, the federated paradigm is a particularly good model for the
type of database integration needs that arise. Standalone multilevel secure
(MLS) DBMSs are now available, and multilevel networks are beginning to
appear. Given the availability of these building blocks, the question that arises
is what would be required to allow them to cooperate to form a federated MLS
DBMS. One critical area that must be addressed is that of transaction
management.

In order to be usable, a federated DBMS must provide the capability to execute
distributed transactions and guarantee their atomicity even in the presence of
various types of system failures. By atomicity, we mean that either all parts of
a distributed transaction complete successfully, or all of its effects are erased
on all the DBMSs involved. If a system failure occurs, there must be a means
for unambiguously determining the status of all parts of the distributed
transactions so that they can be uniformly completed or discarded during crash
recovery. The methods for providing distributed transaction atomicity are
called *commit protocols*. The one most commonly implemented in DBMS
products is known as the *two-phase commit* protocol [8], [12]. Although others
have been developed (such as three-phase commit [18]) to be resistant to a
wider class of system crashes, two-phase commit is popular because it makes a
reasonable tradeoff between robustness in various types of failures and the
amount of overhead imposed on the routine execution of distributed
transactions [16].

The conventional (single-level) DBMS community has concluded [9] that, even
though it entails some loss of autonomy, DBMSs must make two-phase commit
primitives visible at the DBMS interface if they are to cooperate. No such
consensus has been reached among trusted DBMS developers, since the
problem of distributed transaction management in the multilevel environment

has yet to be thoroughly studied. Although there has been a recent flurry of activity in multilevel transaction management (for example, see [11], [1], [5], [14]), these algorithms are for centralized DBMSs. The only work pertaining to distributed transaction management in multilevel distributed systems [13], [15] raises some related questions but does not offer solutions.

Popular wisdom has long held that conventional transaction management techniques such as the two-phase commit protocol cannot be used in a multilevel secure distributed DBMS setting. However, there has been no systematic analysis of the conditions under which two-phase commit might apply in the multilevel environment and its limitations and vulnerabilities. In this paper, we show that, under certain circumstances, two-phase commit can be used to support distributed transactions in a federated MLS DBMS. This paper describes the use of the two-phase commit protocol for distributed transaction recovery with partitioned data in the multilevel federated distributed DBMS environment and argues that, used under the conditions we have laid out, it is secure.

In the remainder of this paper, we present our security model, distributed transaction model, and multilevel federated architecture model, then describe the use of two-phase commit in this environment and argue its correctness.

2 Two-phase commit protocol

Before discussing the application of the two-phase commit (2PC) protocol to the multilevel environment, we briefly review the ordinary 2PC protocol. 2PC is a method for ensuring the atomicity of distributed transactions by ensuring that all sites participating in the transaction either commit or abort the results of a transaction together.

We assume a federated DBMS architecture operating across multiple sites. Each site has a full-fledged DBMS with its own local transaction manager and a federated data management layer that includes a distributed transaction manager. Figure 1 shows our abstraction of a site participating in the federated system.

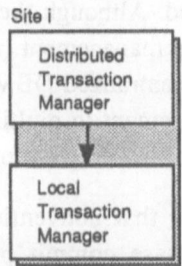

Figure 1: Federated DBMS site.

A distributed transaction may originate from any of the sites. In addition, each site may service local (single-site) transactions originating from local applications and users. As described in [3], 2PC basically works as follows. The site which first receives a distributed transaction request, the originating site, is responsible for breaking down the transaction into single-site subtransactions for distributed execution, as shown in Figure 2.

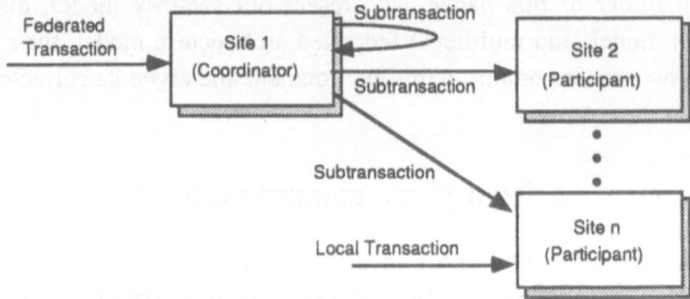

Figure 2: Distributed transaction execution.

2PC is used to ensure that these subtransactions either all commit or all abort together. That is, if a subtransaction fails for any reason, the 2PC protocol aborts all the other subtransactions and rolls back the overall federated transaction. The originating site, or *coordinator*, sends the subtransactions to the *participant* sites and requests their execution. When the user who initiated the transaction issues the commit command, the coordinator initiates the first phase of 2PC and asks each participate to send its vote. A participant responds with a Yes or No vote, depending on whether it was able to execute the subtransaction. In the second phase, based on the responses it has received, the coordinator tells the participants collectively whether to Commit or Abort their work.

The complete 2PC protocol also specifies when writes to nonvolatile storage must take place and provides for timeouts to ensure that transactions are recoverable in the event of site failures [16]. However, the preceding description will be adequate for our discussion.

3 Use of 2PC in a federated MLS DBMS

We now discuss how the 2PC protocol can be mapped to a federated MLS DBMS environment. We describe our security model, our model of the multilevel federated database system, and how 2PC is used within this environment.

3.1 Security model

In this paper, we address only mandatory access control. We assume a security model adapted from those of Bell and LaPadula [2] and Denning [6]. This security model characterizes the overall system formed by the union of the individual participating DBMSs and the federation layer.

The system is represented as a set D of *objects* (data elements), a set T of *subjects* (transactions), and a partially ordered set S of *security classes* with ordering relation $<$. A security class S_i is said to be *dominated* by another class S_j (denoted $S_i \leq S_j$) if and only if $i = j$ or $S_i < S_j$. There is a mapping L from $D \cup T$ to S; i.e., for every $x \in D$, $L(x) \in S$, and for every $T_i \in T$, $L(T_i) \in S$. In other words, every data element and every transaction has a security class associated with it.

The system is considered *secure* only if the following two conditions are satisfied at all times:

(1) Transaction T_i is not allowed to read data element x unless $L(T_i) \geq L(x)$.

(2) Transaction T_i is not allowed to write data element x unless $L(T_i) = L(x)$.

Intuitively, adhering to these conditions guarantees that no information is transferred directly from objects (i.e., data elements) with higher security classes to subjects (i.e., transactions) with lower classes. Property (2) is a restricted version of the *-property. The original *-property [2] allows T_i to write x when $L(T_i) \leq L(x)$. In the database context, however, it is undesirable to allow T_i to write x when $L(T_i) < L(x)$ (i.e., to write up) for integrity reasons. It can be argued that a subject at class $L(T_i)$ should not be trusted not to corrupt data at the higher class $L(x)$. Furthermore, aborting any T_i that violates an integrity constraint on x would open a potential covert channel. Thus, in our model of the federated system, we prohibit write-ups.

We must also be concerned with transferring information indirectly from objects with higher security classes to subjects with lower classes, through signalling channels. We distinguish signalling channels from covert channels [6]; we refer to channels that are inherent in the algorithm as signalling channels and those that may be artifacts of a particular implementation as covert channels. Signalling channels are those with which we are concerned here.

3.2 Transaction model

We define a database *transaction* as a sequence of atomic operations on data elements. An operation on a data element is either a Read (returns the value of the element) or a Write (updates the element with a specified new value). For simplicity, we assume that a transaction can read and write any data element at most once. (This is not a limiting assumption, because, as noted in [11], any transaction that reads or writes some data element more than once can be reduced to the necessary form without changing its semantic content.) Then, for any transaction T_i and data element x, $r_i[x]$ denotes a Read executed by T_i on x. Similarly, $w_i[x]$ denotes a Write executed by T_i on x. In general, a transaction need not be a totally ordered sequence. A Read and a Write on the same data element, however, must always be ordered.

Definition 1 A *transaction* T_i is a partially ordered set with ordering relation $<_i$ such that:

(1) T_i is a set of $\{r_i[x]: x \in D\} \cup \{w_i[x]: x \in D\}$;

(2)   For all $x \in D$, if $r_i[x] \in T_i$ and $w_i[x] \in T_i$, then either $r_i[x] <_i w_i[x]$

    or $w_i[x] <_i r_i[x]$.

                                         o

The above definition does not reflect any computations that a transaction may need to perform; it is concerned only with the transaction's requests for access to data. We cannot predict the exact effect of the transaction on the database, since we do not know what computed values may be used for updates.

Definition 2 A set $U_i = \{w_i[x]: w_i[x] \in T_i\}$ is called an *update projection* of transaction T_i. T_i is called a *parent transaction* of U_i.

                                         o

An *update* transaction can now be defined as a transaction with a non-empty update projection. Similarly, a *read-only* transaction is defined as a transaction with an empty update projection.

We now extend these definitions for distributed transactions and subtransactions. For now, we assume that there is no data replication, i.e., each x is stored at exactly one site. Definition 1, above, then suffices for a distributed transaction, requiring only the recognition that the data elements that the transaction reads and writes may be at different sites. A subtransaction is the projection of the transaction to contain only those reads and writes pertaining to data elements at one site while preserving the relative ordering of reads and writes to each individual data element.

Definition 3 Given a distributed transaction T_i with ordering relation $<_i$, a *subtransaction* $T_{i,j}$ for site N_j is a partially ordered set with ordering relation $<_{i,j}$ such that:

(1) $T_{i,j}$ is the set of all $\{r_i[x]: r_i[x] \in T_i \wedge x \in D_j\} \cup \{w_i[x]: w_i[x] \in T_i \wedge x \in D_j\}$, where D_j is the set of all $\{x: x \in D \wedge x$ exists at $N_j\}$;

(2)   For all $x \in D_j$, if $r_i[x] \in T_i$ and $w_i[x] \in T_i$, then either $r_i[x] <_i w_i[x]$ and $r_i[x] <_{i,j} w_i[x]$ or $w_i[x] <_i r_i[x]$ and $w_i[x] <_{i,j} r_i[x]$.

                                         o

We also refer to T_i as the parent transaction of $T_{i,j}$. An update projection of a subtransaction can be taken as for a parent transaction, by selecting only its write operations. Thus a subtransaction can also be characterized as being an update subtransaction or a read-only subtransaction. With respect to a particular transaction, T_i, a site N_j is referred to as a *read-only site* if the subtransaction for that site, $T_{i,j}$, is a read-only subtransaction.

3.3 Correctness criteria for distributed transactions

For the distributed execution of transactions to be correct, it is necessary to ensure consistency, isolation of transactions, and atomicity. Consistency means that two transactions cannot be allowed to interfere in such a way that they produce inconsistent effects in the database. Isolation means that one transaction cannot be allowed to observe the results of another transaction that has not completed and been committed, since these results may yet be rolled back if the partially complete transaction is later aborted. Atomicity is as we have discussed previously.

We can assume consistency, since each site maintains serializability for operations on its local resources, and there is no replication of data. The locally generated serializable schedules are for operations on disjoint domains, and thus any global interleaving of the local schedules is also automatically serializable [17].

Isolation is satisfied trivially. We assume that each local site provides isolation of uncommitted transactions. In 2PC, the sites are not told to commit their subtransactions until the overall distributed transaction is being committed. Hence, the distributed transactions' effects also remain isolated until they commit.

The correctness criterion for 2PC is that it maintain distributed transaction atomicity. In the absence of system failures, it is obvious that 2PC ensures that all subtransactions either commit or abort together. Taking failures into account, it has been shown [3] that 2PC is resilient to site and communication failures. That is, if sites are assumed to be fail-stop (i.e., cease processing when a failure occurs) rather than fail-insane (i.e., continue processing but behave incorrectly) [7], then 2PC with appropriate logging to stable storage will always eventually produce a uniform commit decision.

3.4 Multilevel federated architecture model

In our model of the multilevel federated data management system, we assume that each site hosts a local MLS DBMS. As shown in Figure 3, the multilevel site DBMS, supported by a Trusted Computing Base (TCB), has its own local transaction manager and a distributed transaction manager that is capable of originating or participating in distributed transactions and interfacing with the local transaction manager.

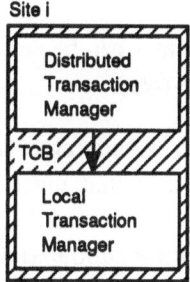

Figure 3: A site in the multilevel federated architecture.

The overall federated system architecture, shown in Figure 4, is like that of the federated data management system in the non-multilevel environment, with the addition of security classes associated with the sites, the transactions, and the data items. We represent the federated system as a set N of sites which contain data objects from the set D and where transactions from the set T can operate.

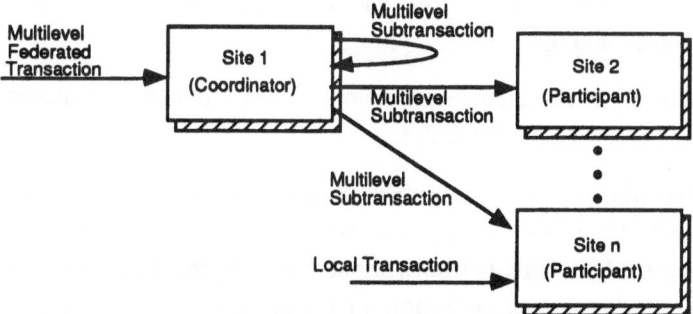

Figure 4: Multilevel federated system architecture.

Each site in the multilevel federated environment has an associated range of security classes defining the minimum and maximum levels at which subjects (transactions) and objects (data elements) can exist at that site. Thus, there is a mapping L' from N to S, in which each site is mapped to an ordered pair of security classes L'_{min} and L'_{max}. For every site $N_k \in N$, $L'_{min}(N_k) \in S$, $L'_{max}(N_k) \in S$, and $L'_{max}(N_k) \geq L'_{min}(N_k)$. This range constrains the data elements that N_k can contain as follows:

Constraint 1   A data element x can be stored in N_k only if $L'_{min}(N_k) \leq L(x) \leq L'_{max}(N_k)$.

In the multilevel environment, a federated transaction has an associated security class. The subtransactions that must be carried out as part of the federated transaction also have an associated security class. That is, both the federated transactions, T_i, and their subtransactions, $T_{i,j}$, belong to the set of transactions T which is associated with the set of security classes S through the mapping L. Transactions that can originate or execute at site N_k are constrained as follows:

Constraint 2a   A distributed transaction T_i can originate at N_k only if $L'_{min}(N_k) \leq L(T_i) \leq L'_{max}(N_k)$.

Constraint 2b   A subtransaction $T_{i,j}$ can execute at N_k only if $L'_{min}(N_k) \leq L(T_{i,j}) \leq L'_{max}(N_k)$.

Further, because the distributed transaction manager must be able to read the federated transaction and write the corresponding subtransactions, the following constraint applies:

Constraint 3   For all subtransactions $T_{i,j}$ of a federated transaction T_i, $L(T_{i,j}) = L(T_i)$

A federated transaction T_i is multilevel in the sense that it may read data at multiple security classes $\leq L(T_i)$; however, it can write data only at the single security class $L(T_i)$. This is true without regard to the sites in which data are found. Similarly, a subtransaction $T_{i,j}$ may read data at multiple security classes $\leq L(T_{i,j})$, but can write data only at its own security class $L(T_{i,j})$, within a single site.

3.5 Two-phase commit in multilevel architecture

We now describe how 2PC works in the multilevel environment and examine its operation with respect to security. Our goal is to identify the weakest set of constraints on transactions and on the operating ranges and capabilities of participating nodes under which 2PC can be used securely.

As in the single-level environment, the site which originates a distributed transaction T_i becomes the coordinator for 2PC. This site is responsible for resolving references in the submitted request to determine where the desired data are located. It uses this information to generate a set of subtransactions $T_{i,j}$, one for each site where data are to be accessed. Recall that the original request has a security class that is inherited by T_i and by all the $T_{i,j}$. We observe then that each transaction takes the form of a set of zero of more writes of data items at its own security class, i.e., $w_i[x]$ where $L(x) = L(T_i)$, and zero or more reads of data items at security classes dominated by its own, i.e., $r_i[x]$ where $L(x) \leq L(T_i)$. Each subtransaction has the same form, since it is created simply by removing any read or write operations not involving a data item at the site for which it is destined. The sites available to the coordinator for processing subtransactions are those N_j having $L(T_i)$ within their security class range: $L'_{min}(N_j) \leq L(T_{i,j}) \leq L'_{max}(N_j)$. Therefore, if any of the requested data are located at sites whose ranges do not contain the security class of the request, the transaction is rejected immediately.

If all the requested data are located at accessible sites, the coordinator dispatches each subtransaction $T_{i,j}$ to the appropriate site N_j for execution. These sites become participant sites in the 2PC protocol. The coordinator itself may also be one of the sites processing a subtransaction, even though it is not explicitly identified as a participant, so its own potential commit or failure status also affects the eventual decision to commit or abort the parent transaction. As we have noted, since there is no replication of data, there are no conflicts among operations in different subtransactions. Thus, the individual participants are able to schedule the operations of the subtransactions locally.

The coordinator and participants interact as shown in Figure 5 to process the distributed transaction. When the coordinator sends the subtransactions to the participants, each participant site is expected to perform the actions requested in the subtransaction but not commit it. Then, just as in the single-level environment, the two phases for which 2PC is named are carried out: first a voting phase, then a decision phase. In the voting phase, the coordinator requests votes from all the participants. A participant responds by voting Yes if it was able to carry out the subtransaction and is prepared to commit. If for any reason a participant is not able to complete the subtransaction, it votes No and then unilaterally aborts. If the coordinator is itself executing a subtransaction, it casts a vote of its own. (If any of the sites are read-only sites, however, they need not participate in the second phase of 2PC, since a read-only subtransaction will not have any effect on other operations at the site.) In the decision phase, each participant that voted Yes awaits instructions from the coordinator to commit or abort. If the coordinator has received Yes votes from all participants, it instructs all participants to commit their subtransactions. Otherwise, if the coordinator has received a No vote from any participant or has itself decided to vote No, it instructs all participants that voted Yes to abort.

Site 1 Coordinator	Site j Participant
Receives T_i, assumes coordinator role, decomposes T_i into $T_{i,j}$, $j = 1, ... ,n$, with $L(T_{i,j}) = L(T_i)$. Sends $T_{i,j}$ to site j.	
	Receives and executes $T_{i,j}$, and sends any output to the coordinator.
When user issues the commit command, the coordinator asks each participant to send its vote.	
	A participant votes "Yes" to commit $T_{i,j}$, or "No" to abort $T_{i,j}$. Sends vote to site 1
Receives votes from sites $j = 1, ..., n$. If all votes are "Yes," then sends "Commit," else sends "abort."	
	Receives message to "Commit" or "Abort" and executes.

Figure 5: Sequence of execution in two-phase commit.

3.6 Correctness of two-phase commit in the multilevel environment

The correctness of 2PC in single-level environments, in terms of consistency, isolation, and atomicity, is well established. Using 2PC as described here in the multilevel environment in no way violates these aspects of correctness, since it is merely being applied under more restrictive conditions. Thus, we need to show only that 2PC does not introduce security problems when used in the multilevel environment.

In this section, we sketch the proof that 2PC is free of channels when constrained as we have specified. Security concerns could arise in three areas of the multilevel 2PC scenario:

- the security classes of requests and responses sent back and forth between the coordinator and the participants

- interference among activities at different security classes within a site

- interactions among operations at different security classes across sites.

The first concern is easily addressed. All messages between the coordinator and the participants are assigned the same security class as that of the original transaction $L(T_i)$, including the messages from the coordinator to the sites requesting execution of the subtransactions, the messages back from the participants with their votes to commit or abort, the messages conveying the coordinator's commit decision to the participants, acknowledgments that are exchanged, and any data returned to the originating site as a result of read operations. Thus there is no security impediment to exchanging the necessary information as the distributed execution proceeds.

Each individual site is assumed to have a MLS DBMS that is able to manage read and write conflicts across security classes that may arise locally from interactions among different subtransactions and local transactions executing concurrently. As part of its own multilevel security responsibilities, the DBMS must provide the capability for reading down without signalling channels. At a given site, suppose that there are two transactions T_1 containing $w_1[x]$ and T_2 containing $r_1[x]$ such that $L(T_1) = L(x)$ and $L(T_2) \geq L(x)$. If $L(T_2) \neq L(T_1)$, then the site's multilevel DBMS must ensure that T_1 is never made to wait because of T_2. This stricture applies irrespective of whether the transactions involved are local transactions or are subtransactions spawned by distributed transactions controlled from elsewhere. (Only read-write conflicts are of concern, however, since two transactions can read x without delaying each other and any two transactions writing x must both have the same security class.) The multilevel DBMS is relied upon to ensure that no subtransaction writing a data item x is delayed because another transaction with a higher security class is reading x.

The remaining area for consideration is whether the delays across sites introduced by the 2PC protocol could be used as a signalling channel. In 2PC, the status of operations at one site can create delays that are propagated to other sites because the coordinator must await votes from all participants, and all participants must in turn await the coordinator's decision. However, because we can count on the MLS DBMS at each site to perform reads on behalf of higher level subtransactions without leaving "footprints" that are perceptible by write operations being executed for lower level transactions (or subtransactions), delays in one subtransaction are not detectable at a lower security class. This assertion means that there is no potential for using the propagated delays as a signalling channel across security classes.

3.7 Range restrictions

Clearly, 2PC can be used as we have described if we assume that all sites operate over the same security class range, i.e., for all $i, j, L'min(N_i) = L'min(N_j)$ and $L'max(N_i) = L'max(N_j)$. However, so strong a restriction would severely limit its usefulness, since we foresee a need for federated systems in which the security class range is heterogeneous. Observing that only the sites participating in the distributed transaction need have the same range allows only minor improvement. Though sites in the federated system could then operate over different ranges, with transactions at different security classes being processed by disjoint subsets of the sites, the system would in effect be partitioned by security class range, with no ability to coordinate distributed access across the partitions.

Fortunately, a weaker restriction is sufficient. We can relax the requirement as follows. For any distributed transaction T_i, the relative security class ranges of sites in the federated system must be such that it would be possible to read all x

for which $L(x) \leq L(T_i)$ and write all x for which $L(x) = L(T_i)$. To ensure that this will be the case for all possible transactions, we require that all sites have the same maximum security class, although the minimum security classes may vary; i.e., for all $i, j, L'_{max}(N_i) = L'_{max}(N_j)$.

To see why, consider a federated system composed of site N_1 operating over the range from Secret to Confidential and site N_2 operating from Confidential to Unclassified. If a Secret request were received by N_1, it would not be possible to read relevant Confidential and Unclassified data from N_2, because sending a Secret subtransaction to N_2 would require an illegal write-down. The only way to be sure of obtaining all data satisfying the request would be for the requester to issue the transaction once at the Secret level and a second time at the Confidential level (or more generally, once at maximum security class of every site in the system), without guaranteed atomicity of the entire operation.

In contrast, if the restriction given above is levied, there can be no security class at which a distributed transaction request is unable to access all data items for which it satisfies the basic security constraints.

4 Conclusion

It seems reasonable to require that DBMSs participating in a federated DBMS support the 2PC protocol. Although it sacrifices autonomy to a limited degree, 2PC is becoming the prevalent distributed transaction control method used in conventional DBMSs. Thus it offers a counterbalancing advantage for use in a federated DBMS in that it is not a vendor-specific approach. In this paper, we have shown that using 2PC in a multilevel environment within the limited circumstances we have identified does not introduce signalling channels. This result is not sensitive to the concurrency control methods of individual sites (e.g., locking, timestamping, or multiversioning) or to the architecture of the multilevel DBMS at the site (e.g., single-level untrusted schedulers or trusted schedulers), since we made no assumptions concerning these characteristics. It is necessary only to assume that the DBMS at each site meets the multilevel security requirement of reading without channels and that data are not replicated. We have also identified the minimal requirement on ranges of sites participating in a federation to ensure that data for which a transaction is authorized are accessible. It is worth noting, too, that the range restrictions could accommodate the use of an untrusted DBMS participating as a single-level site. The significance of our results is that, if trusted DBMS developers follow the lead of the single-level DBMS community in building externally accessible 2PC primitives into their systems, some degree of distributed transaction atomicity can be obtained virtually for free.

Acknowledgment

The authors would like to thank Joseph V. Giordano for his support of our work and LouAnna Notargiacomo for stimulating discussions on this topic.

References

[1] P. Ammann, S. Jajodia. A Timestamp Ordering Algorithm for Secure, Single-Version, Multi-Level Databases. In: *Database Security, V: Status and Prospects*, C. Landwehr, S. Jajodia eds., North-Holland, 1992, pp. 191-202.

[2] D. E. Bell, L. J. LaPadula. Secure Computer System: Unified Exposition and Multics Interpretation. *Technical Report MTR-2997*, MITRE Corp., Bedford, MA, July 1975.

[3] P. A. Bernstein, V. Hadzilacos, N. Goodman. *Concurrency Control and Recovery in Database Systems*. Addison-Wesley, 1987.

[4] S. Ceri, G. Pelagatti. *Distributed Databases; Principles and Systems*. McGraw-Hill, 1984.

[5] O. Costich. Transaction Processing Using an Untrusted Scheduler in a Multilevel Database with Replicated Architecture. In: *Database Security, V: Status and Prospects*, C. Landwehr, S. Jajodia eds., North-Holland, 1992, pp. 173-190.

[6] D. E. Denning. *Cryptography and Data Security*. Addison-Wesley, 1981.

[7] H. Garcia-Molina, R. K. Abbott. Reliable Distributed Database Management. *Proc. of the IEEE*, Vol. 75, No. 5, May 1987, pp. 601-620.

[8] J. N. Gray. Notes on Database Operating Systems. *Operating Systems: An Advanced Course*, Lecture Notes in Computer Science, Vol. 60, Springer-Verlag, 1978, pp. 394-481.

[9] Informal discussion with Paula Hawthorn. November 1991.

[10] D. Heimbigner, D. McLeod. A Federated Architecture for Information Management. *ACM Transactions on Office Information Systems*, Vol. 3, No. 3, July 1985, pp. 253-278.

[11] S. Jajodia, B. Kogan. Transaction Processing in Multilevel-Secure Databases Using Replicated Architecture. *Proc. IEEE Symposium on Research in Security and Privacy*, Oakland, CA, May 1990, pp. 360-368.

[12] B. Lampson, H. Sturgis. Crash Recovery in a Distributed Data Storage System. *Technical Report*, Computer Science Laboratory, Xerox, Palo Alto Research Center, Palo Alto, CA, 1976.

[13] G. H. MacEwen. Effects of Distributed System Technology on Database Security: A Survey. In: *Database Security: Status and Prospects*, C.E. Landwehr ed., North-Holland, 1988, pp. 253-261.

[14] C. D. McCollum, L. Notargiacomo. Distributed Concurrency Control with Optional Data Replication. In: *Database Security, V: Status and Prospects*, C. Landwehr, S. Jajodia eds., North-Holland, 1992, pp. 149-172.

[15] J. McHugh, B. M. Thuraisingham. Multilevel Security Issues in Distributed Database Management Systems. *Computers and Security*, Vol. 7, No. 4, August, 1988, pp. 387-396.

[16] C. Mohan, B. Lindsay, R. Obermarck. Transaction Management in the R* Distributed Database Management System. *ACM Transactions on Database Systems*, Vol. 11, No. 4, December 1986, pp. 378-396.

[17] M. T. Özsu, P. Valduriez. *Principles of Distributed Database Systems*. Prentice-Hall, 1991.

[18] D. Skeen. Nonblocking Commit Protocols. *Proc. ACM SIGMOD International Conference on Management of Data*, Ann Arbor, Michigan, 1981, pp. 133-147.

Author Index

Dependable Computing and Fault-Tolerant Systems

Edited by *A. Avizienis, H. Kopetz, J.-C. Laprie*

Volume 1

A. Avizienis, H. Kopetz, J.-C. Laprie (eds.)

The Evolution of Fault-Tolerant Computing

In the Honour of William C. Carter

1987. 52 figs., 35 portraits and 1 frontispiece.
X, 465 pages. Cloth DM 118,–, öS 830,–
ISBN 3-211-81941-X

Volume 2

U. Voges (ed.)

Software Diversity in Computerized Control Systems

1988. 41 figs. VII, 216 pages.
Cloth DM 75,–, öS 530,–
ISBN 3-211-82014-0

Volume 3

P.A. Lee, T. Anderson

Fault Tolerance

Principles and Practice

Third edition in preparation.

Volume 4

A. Avizienis, J.-C. Laprie (eds.)

Dependable Computing for Critical Applications

1991. 88 figs. XIII, 431 pages.
Cloth DM 162,–, öS 1134,–
ISBN 3-211-82249-6

Volume 5

J.-C. Laprie (ed.)

Dependability: Basic Concepts and Terminology

in English, French, German, Italian and Japanese
1992. 3 figs. XII, 265 pages.
Cloth DM 128,–, öS 896,–
ISBN 3-211-82296-8

Volume 6

J.F. Meyer, R.D. Schlichting (eds.)

Dependable Computing for Critical Applications 2

1992. 114 figs. XIV, 439 pages.
Cloth DM 172,–, öS 1204,–
ISBN 3-211-82330-1

Volume 7

H. Kopetz Y. Kakuda (eds.)

Responsive Computer Systems

1993. 96 figs. XI, 377 pages.
Cloth DM 150,–, öS 1050,–
ISBN 3-211-82458-8

Prices are subject to change without notice

Springer-Verlag Wien New York